Contents

AF234610

Identifying the challenges: participatory research and socio-economic studies

Challenging environmental issues

Gender challenges

Farmer knowledge, extension and training

Animal-drawn equipment and harnessing

Animal-based transport

Animal issues: donkey use, cow traction and feeding

Social and economic challenges in West Africa

Meeting the Challenges of Animal Traction

A resource book of the
Animal Traction Network for Eastern and Southern Africa (ATNESA)

Edited by
Paul Starkey and Pascal Kaumbutho

An ATNESA publication
made possible with the assistance of
The British Department for International Development (DFID)

Practical
ACTION
PUBLISHING

Intermediate Technology Publications Ltd
trading as
Practical Action Publishing Ltd
25 Albert Street, Rugby,
Warwickshire, CV21 2SD, UK
www.practicalactionpublishing.com

in association with

The Animal Traction Network for Eastern and Southern Africa (ATNESA)
PO Box BW540, Borrowdale, Harare, Zimbabwe

First published in 1999
Transferred to digital printing in 2008

A catalogue record for this book is available from the British Library & Library of Congress

ISBN 978-1-85339-483-6 Paperback
ISBN 978-1-78044-545-8 Digital book

Citation: Starkey, P. (1999) *Meeting the Challenges of Animal Traction: A resource book of the Animal Traction Network for Eastern and Southern Africa (ATNESA)*, Rugby, UK: Practical Action Publishing https://doi.org/10.3362/9781780445458

Since 1974, Practical Action Publishing has published and disseminated books and information in support of international development work throughout the world. All print editions are produced and distributed via ethical and sustainable print on demand global facilities.

Practical Action Publishing is a trading name of Practical Action Publishing Ltd (Company Reg. No. 01159018 | VAT 880 9924 76). All profits are covenanted back to its parent group, Practical Action (Charity Reg. No. 247257).

The views and opinions in this publication are those of the author and do not represent those of Practical Action Publishing Ltd or its parent charity Practical Action. Reasonable efforts have been made to publish reliable data and information, but the author and publisher cannot assume responsibility for the validity of all materials or for the consequences of their use.

Publication sponsored by The British Department for International Development (DFID) Eastern Africa, PO Box 30465, Nairobi, Kenya

DTP to camera-ready copy by Malcolm Starkey and Paul Starkey, Animal Traction Development, Oxgate, 64 Northcourt Avenue, Reading RG2 7HQ, UK

Cover photograph: women weeding drought-affected maize with a cow and an ox in Zimbabwe. © Paul Starkey

The manufacturer's authorised representative in the EU for product safety is Lightning Source France, 1 Av. Johannes Gutenberg, 78310 Maurepas, France. compliance@lightningsource.fr

National challenges and perspectives

Addressing the challenges: project experiences

Preface and acknowledgements

The Animal Traction Network for Eastern and Southern Africa—ATNESA—was formed in 1990 and aims to improve information exchange and regional cooperation relating to animal draft power. ATNESA is an open, multidisciplinary network coordinated by a regional steering committee that works with national networks. Among other activities, ATNESA stimulates collaborative actions, convenes international workshops and produces resource publications. Full details are available from the ATNESA Secretariat and much network information can be found in the book *'Improving animal traction technology'* published by ATNESA in cooperation with the Technical Centre for Agricultural and Rural Cooperation (CTA).

This publication has been developed following a series of ATNESA workshops, notably a workshop on *'Meeting the challenges of animal traction'* held from 4-8 December 1995 at Karen, Kenya. The Kenya Network for Draught Animal Technology (KENDAT) hosted this workshop. The core costs were met by the Eastern Africa Division of the British Department for International Development (DFID) which was then known as Overseas Development Administration (ODA). One hundred and thirty people from 27 countries participated in this workshop. Among the organisations sponsoring several participants were DFID, CTA, the Food and Agriculture Organisation of the United Nations (FAO), the Commonwealth Foundation and the regional Agrotec/Farmesa project based in Zimbabwe. ATNESA would like to thank these and many other organisations that sponsored participants and the preparation of thematic and technical papers.

A 55-page illustrated report of the ATNESA Kenya workshop, giving details of the workshop presentations, discussions and conclusions was published by ATNESA and is available from ATNESA and KENDAT. Also available from KENDAT are the proceedings of a national workshop on the same theme.

ATNESA decided to publish a series of resource books that would include edited papers from several sources. These would complement the detailed workshop reports. It was felt that such

resource books would have longer-lasting value than conventional proceedings publications. Some of the resource books have specific themes: *'Animal power for weed control'*, *'Donkeys, people and development'* and *'Conservation tillage with animal traction'*. This particular resource book has the broader theme of *'Meeting the challenges of animal traction'*. Papers address a number of important challenges to animal traction that relate to participation, environment, gender, extension, transport, equipment and animal husbandry. In addition, several papers describe national-level challenges and project attempts to address these. Together the various edited papers combined in this volume provide a wealth of experience that ATNESA believes will be valuable to all concerned with the development of animal power.

The editing of this publication has been the responsibility of Professor Paul Starkey (ATNESA Technical Adviser), assisted by Malcolm Starkey, Freda Miller and John Stares. Dr Pascal Kaumbutho (KENDAT Chair and ATNESA Treasurer) was the Chair of the Kenyan workshop organising committee. The delay in getting the publication finalised was caused by a range of editorial and publication problems for which the Chief Editor takes overall responsibility. He apologises to ATNESA members and commends to them this valuable resource book full of information that will be of interest and value to ATNESA for many years to come. ATNESA and the editors would like to thank all who were involved in the preparation of papers and the editing and publication of this resource book.

This publication has been made possible due to the financial support of DFID-Eastern Africa. ATNESA wishes to acknowledge and appreciate this assistance. Particular thanks are due to Jim Harvey and Martin Leach, Senior Natural Resources Advisers.

One of ATNESA's aims is better information exchange between all those involved in improving animal traction in the region. ATNESA looks forward to further close collaboration with individual members, national animal traction networks, supporting organisations and other networks. People using this book are encouraged to contact directly the authors and the organisations that have provided resource information, and to exchange information with them and with the national networks.

Contact addresses for ATNESA and its associated national networks are provided inside the back cover.

Professor Paul Starkey
Dr Pascal Kaumbutho

Abbreviations and units

ATNESA	Animal Traction Network for Eastern and Southern Africa
CGIAR	Consultative Group on International Agricultural Research, Washington DC, USA
CIRAD-SAR	Centre de Coopération Internationale en Recherche Agronomique pour le Développement, Départment des Systèmes Agro-alimentaires et Ruraux, Montpellier, France
COOPIBO	Coopération au Développement Ibo, Belgium (an NGO)
FAO	Food and Agriculture Organisation of the United Nations, Rome, Italy
FIT	Farm Implements and Tools, Kenya
d	Days
DAREP	Dryland Applied Research and Extension Project, Kenya
DFID	British Department for International Development (formerly ODA)
DTU	Development Technology Unit, University of Warwick, UK
FSRU	Farming Systems Research Unit, Zimbabwe
ha	Hectares
HP	Horsepower
IFAD	International Fund for Agricultural Development
ITDG	Intermediate Technology Development Group, UK
KARI	Kenya Agriculture Research Institute, Nairobi, Kenya
kcal	Kilocalories
kg, kgf	Kilograms, Kilograms force
km	Kilometres
kW	Kilowatts
l	Litres
m	Metres
N	Newtons
NAMA	Network for Agricultural Mechanization in Africa
NNRDP	North Namibia Rural Development Project
NRI	Natural Resources Institute, London, UK
ODA	Overseas Development Administration, London, UK (now DFID)
ODI	Overseas Development Institute, London, UK
OXETS	Oxenisation Extension and Training Services, Tanzania
PADPDP	Palabana Animal Draught Power Development Project, Zambia
PRA	Participatory Rural Appraisal
RIDEP	Rural Integrated Development Project, Mbeya, Tanzania
RRA	Rapid Rural Appraisal
s	Seconds
SAARI	Serere Agricultural and Animal Research Institute, Uganda
t	Tonnes
TPRI	Tropical Pest Research Institute, Arusha, Tanzania
UK	United Kingdom of Great Britain and Northern Ireland
USA	United States of America
W	Watts
WP-ADPP	Western Province Animal Draft Power Programme, Zambia

Photo (opposite): Women farmers discussing animal traction constraints and challenges in Kenya

Meeting the challenges of animal traction

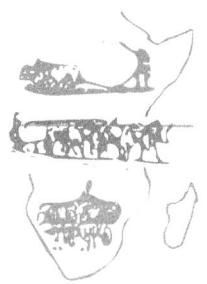

Identifying the challenges: participatory research and socio-economic studies

A farming systems approach to improving draft animal power in sub-Saharan Africa

Forbes Muvirimi [1] and Jim Ellis-Jones [2]

[1] Department of Agricultural, Technical and Extension Services, PO Box 1927, Bulawayo, Zimbabwe
[2] Silsoe Research Institute, Wrest Park, Silsoe, Bedford MK45 4HS, UK

Abstract

In Zimbabwe the use of draft animals is widespread and long-established outside tsetse-infected areas. Most farmers prefer to use oxen for plowing, especially on heavier soils, as they are faster and stronger than donkeys. However, lighter operations, especially weeding and transport, are increasingly being carried out by donkeys.

The 1991/92 drought reduced the cattle herd from 4 million to less than 3 million animals and the donkey herd from 400,000 to less than 300,000. Peak demand for animal power is for plowing at the end of the dry season when animals are in worst condition and feed resources at their lowest. As a result, availabilty of draft power is a limiting factor in many areas.

Productivity could be improved either through increasing the supply of draft animals or reducing the demand for draft animals by increasing their effectiveness. Increased use of donkeys and increasing the carrying capacity of communal land could increase the supply of draft animals. Conservation tillage systems and improved implements could reduce the demand for draft animals.

Farming systems in Zimbabwe are complex and vary geographically. If research is to be relevant to farmers it is essential that existing farming systems, rather than current extension recommendations, form the basis for research programmes.

Introduction

This paper is based on a research project financed by the Overseas Development Administration's (ODA) Livestock Production Programme entitled 'Increasing the productivity of draft animals in sub-Saharan Africa'. The project comprises three interrelated themes:

- Socio-economic studies to identify current draft animal power practices, to characterise specific target groups of farmers using draft animals and to determine priorities for draft animal power research.

- Nutrition, health and management aspects of draft animals of limited capability.

- An evaluation of draft animal power equipment and, if necessary, development of new implements for draft animals of limited capability.

This paper draws from the socio-economic component of the research project.

Animal traction in sub-Saharan Africa

Worldwide, it is estimated that 400 million draft animals (bovines and equines) are being used in agricultural operations. Starkey (1988) estimated that of these some 18.6 million animals are employed in sub-Saharan African agriculture. These are predominantly work oxen (in excess of 350 kg) but include donkeys, mules, horses and cows.

A review by Mrema and Mrema (1993) of the utilisation of animal power in sub-Saharan Africa showed that of the 11.3 million draft oxen in use, nearly 80% are found in five countries–Ethiopia (53%), Zimbabwe (7%), Kenya (6%) and Tanzania and Uganda each with 5%.

Little information is available on the use of smaller/weaker oxen, cows or equines as the preferred draft animal is the large ox. Even those countries with large populations of equines, particularly donkeys, have little information on current use, management and performance.

Animal traction in Zimbabwe

The role of draft animals within the farming system

In Zimbabwe the use of draft animals is widespread and long-established outside tsetse infected areas. There are some 900,000 households in the communal sector for whom mixed farming is the main activity. Arable plots are commonly 2–3 ha in size.

Livestock, particularly cattle, play a vital role in the farming system. Eckert and Mombeshora (1987) reviewed the functions of livestock in the communal farming areas of Zimbabwe. They identified socio-cultural and socio-economic functions. The socio-cultural functions include use for *lobola* (bride price), ancestor worship and funerals. In areas of higher rainfall, draft, manure, sales, milk and meat, in that order of importance, were found to be the major socio-economic functions. Sales are mainly to raise cash for buying food, paying school fees and other emergency expenditure. In low-rainfall areas cattle provide a substantial amount of draft, but increasingly donkeys are replacing cattle, especially for transport. However, they have limited value other than the provision of draft. A comparison of the relative economic values of outputs from cattle and donkeys is shown in Table 1.

Prior to the drought in 1991/92 the communal cattle herd exceeded 4 million animals (Central Statistics Office, 1990) with stocking rates exceeding sustainable carrying capacities in many areas. The number of donkeys was estimated to be 400,000. The drought reduced the cattle herd to less than 3 million and the number of donkeys to less than 300,000.

Mrema and Mrema (1993) identify the main benefits generally associated with using animal power as:

- increasing the productivity of labour
- expanding the area under cultivation
- increasing the intensity of land use
- improving the quality and timeliness of key farming operations
- reduction in the drudgery associated with hand tool agriculture which is used on 80% of the cultivated land in sub-Saharan Africa

Problems associated with adoption of draft animal power include (Geza and Reid, 1983; Mrema and Mrema, 1993):

- the lack of animals for traction, related to poor herd composition, low calving rates, late weaning and poor management. As a result there are insufficient animals for draft (especially plowing) and for manure

Table 1: Estimates of the relative economic values of outputs from cattle and donkeys in the communal areas of Zimbabwe

Output	% of total value	
	Cattle[1]	Donkeys[2]
Draft	64	95
Milk	14	–
Manure	4	2
Meat	8	–
Herd growth	10	3–5
Social value	?	–
Total	100	100

Sources: 1) Barrett, 1992; 2) Estimates from dicussions with farmers

- competing demands for livestock products
- disease problems, particularly trypanosomiasis in tsetse areas
- lack of available feed and environmental concerns of over utilisation of grazing areas (too many cattle for the amount of fodder available)
- lack of suitable implements
- increasing the work burden for manual operations, especially that of women
- a poor image of animal power among opinion formers and (urban) elites in sub-Saharan Africa, often resulting in a preference for tractors even when they are not cost-effective.

Draft animal power has the potential to play a major role in increasing agricultural production in sub-Saharan Africa providing the benefits can be realised and the problems avoided or minimised.

Project study areas

Three areas were selected that were broadly representative of conditions found in the semi-arid parts of Zimbabwe. A rapid rural appraisal, formal survey and monitoring of typical farmers have been undertaken. The project has used participatory methods involving farmers and manufacturers in both problem identification and seeking solutions.

In each area farmers' objectives tend to focus on food security, giving priority to production of food crops under risk minimisation strategies, generating cash from food surpluses in good years, sometimes growing a cash crop when conditions allow (cotton in one area), selling livestock when no crop is available for sale or seeking non-farm income when farming cannot provide enough. Cattle provide an opportunity for capital accumulation so cash surpluses are often invested in cattle. There is a constant demand for more cattle as few farmers are satisfied with their present herd size. Data from the Ministry of Lands, Agriculture and Rural Resettlement (1993) confirms that non-farm incomes are important in most areas and contribute nearly 50% to total income outside the cotton-growing areas.

The farming systems are broadly similar in the three areas, with similar cropping systems, management practices, livestock and implement ownership patterns even though natural resource conditions vary considerably. Important differences in the drier areas are:

- larger areas are cropped
- donkeys are more common
- small grains assume greater importance even though maize is still grown.

In all areas the farming systems are geared towards crop production, with the role of cattle and donkeys being to support this activity. Traditionally, hand hoe cultivation was practised widely, but as available labour decreased due to increased schooling, animal power assumed the importance now attached to it. Peak demand for animal power is for plowing from the end of September to early December when animals are in worst condition and feed resources at their lowest.

Components of the farming systems have a high degree of interdependence. Crop enterprises are dependent on livestock for land preparation and provision of manure. Livestock are dependent on crop residues for survival during the dry months. Both crops and cattle provide outputs for domestic consumption and cash generation. Donkeys provide input for crop production and are playing an increasingly important role in transport.

Most farmers prefer to use oxen for plowing, especially on heavier soils, as they are faster and stronger than donkeys. Where donkeys are used for plowing, the furrow depth is often inadequate and moisture conservation poor. However, lighter operations, especially weeding and transport, are increasingly being carried out by donkeys. In households where cattle are available they are used for plowing and donkeys are used for other operations.

Communal grazing constitutes the main feed source for animals, although some farmers supplement their animals with stover, pumpkins or melons as grazing becomes scarce.

The Farming Systems Research Unit (1994) compared extension recommendations and farmer practices for livestock. Table 2 demonstrates the extreme difference between recommendations and practices. If research is to be relevant to farmers it is essential that existing farming systems, rather than current extension recommendations, form the basis for research programmes.

Elements of the farming system affecting utilisation of draft animals

Factors affecting draft power utilisation can be categorised within three major sub-systems.

Intra-household sub-system

The components of the intra-household sub-system include the animal, the implement and the operator. These have been described by Ellis-Jones and Panin (1992). Like other production processes farming operates in a system where resources are allocated so that expenditure does not exceed income. This is done within the general framework of farmer's objectives of provision of subsistence needs, risk minimisation, profit maximisation and drudgery reduction. Resources are always limited. Farmers therefore have to prioritise on expenditure with affordability being an important factor for new technology adoption. Cost should not be viewed in absolute terms but within the context of the household system.

Inter-household and community sub-system

Key to the inter-household sub-system are the production relationships between farmers, which enable access to draft by those not owning animals. Such arrangements involve cash, reciprocal labour, lending, payment in kind and other agreements between farmers. A major question is for how long and whether there is an obligation for those owning animals, to share them

Table 2: Livestock extension recommendations compared with farmer practice

Activity	Extension recommendation	Farmer practice
Cattle breeds	Exotic 'improved' breeds	Indigenous breeds
Objectives	Beef production	Multi-purpose use, especially draft power
Stocking rates	10 ha per livestock unit	2 ha per livestock unit
Grazing management	Rotational grazing in fenced paddocks -grazing schemes	Key resource grazing, use of high potential sites, eg drainage lines
Fodder management	Legume reinforced pastures in grazing areas	Browse management
	Supplementary feeding, salt licks etc	None, except in extreme drought
	Stover collection and preparation with urea	Stover collection and storage only
	Agroforestry, including planting of *Leucaena*, etc	Some agroforestry planting; browse management
Use of draft animals	Primarily oxen	Mixed spans often used
	Winter plowing	Most plowing done after first rains
Drought management	Destocking - early sales; movement discouraged and highly regulated	Movement to other areas; supplementary feeding, distress sales
Disease control	Weekly (wet season) and fortnightly (dry season) dipping; dosing; antibiotics; movement controls	Dipping regime followed except in drought; traditional herbs used to treat disease
Donkeys	Few recommendations with respect to management, disease control etc	Indigenous practices

Source: FSRU, 1994

with those without animals. Policies to encourage wider and more efficient use of existing draft are likely to be beneficial.

The physical and economic environment sub-system

This includes soil types, rainfall, the equipment industry, extension, credit availability and general policy environment. It determines the overall productivity of the system and has an important influence on the production modalities within the household system and the relationships of different households.

These sub-systems combine to form the components of the farming system and are essential determinants of household decision making. In planning interventions it is vital to incorporate them in the analysis and characterisation of farmers.

Farmer recommendation domains

The complexity of the farming systems means that animal traction research needs a new thrust to ensure the needs of all farmers are translated into appropriate technologies, national policies and support mechanisms. Researchers need to know and consult with their intended clients. Key questions include:

- what needs can be addressed by research?
- whose needs are addressed (Would anyone take-up the technology)?
- how would the entire farm system react to the technology?

Table 3: Extension recommendation domains of farmers

Rural households

No animals owned		Animal owners					
No access to draft animals	Some access to draft animals	Inadequte draft power			Adequate draft power		
		Donkeys only	Donkeys and cattle	Cattle only	Donkeys only	Donkeys and cattle	Cattle only

Identification of current practices and the characterisation of specific target groups of farmers (recommendation domains) are essential for addressing these questions. This will ensure:

- precise definition of farmer groups relative to their extension and research needs.
- better focusing of extension and research activities
- more efficient allocation of resources so they can be targeted at farmers who are able to respond because of the relevance to their circumstances
- setting achievable targets for each target group
- government policies relevant to particular regions or a group of farmers can be established
- relevant information for each group of farmers can be collated.

Table 3 shows how farmers have been classified into recommendation domains. This has been undertaken on the basis of research and extension needs.

Major draft animal power issues

Profitability

Animal traction remains the most economic form of draft for many farmers. However, there are costs which relate to the acquisition and maintenance of animals and equipment, such as repair bills and costs of veterinary products. Farmers aim to minimise the costs and maximise the benefits to maximise profits. In the long term, benefits will exceed costs, but in the short term low productivity and animal loss due to disease or drought may mean that costs are unaffordable.

Farmers' perceptions of the relative performance of cattle and donkeys have been described by Hagmann and Prasad (1994). Donkeys are more tolerant of disease and drought, require less water and feed supplementation and are in better condition at the end of the winter season. They require less training, are easier to handle and often preferred by women. However, donkeys are unable to deep plow, but can be more suitable for cultivation and transport. They are able to work for longer periods and have a longer working life.

A comparison has been made using indicative costs and benefits of owning and using alternative animals. Costs have taken into account differences in the purchase and resale values of the animals, length of working lives, depreciation of equipment, labour and some feed supplement costs. This shows that the cost of donkeys can be substantially less than that of cattle.

However, when the value of manure, milk, herd growth and other social benefits are taken into account, there is little difference between cattle and donkeys. This demonstrates the importance of other benefits of cattle in reducing animal power costs.

Potential benefits from cows exceed those of both oxen and donkeys, provided that fertility and milk production do not suffer.

Availability of draft animals

Prior to the drought of 1992, draft animal availability was a concern in some areas, but with up to 75% of cattle perishing in some localities, shortage of draft animals became the major issue in all areas. As a result of deaths of mature and larger animals there was a decline in older cows and oxen relative to younger and smaller animals. A similar trend occurred with donkeys. Cattle losses were generally higher than for donkeys. As a result there has been a general increase in the use of donkeys for animal power in many areas.

Table 4: Ownership of draft animals in semi-arid areas of Zimbabwe

	% of farmers (n=248)
No animals	37
Inadequate draft: donkeys only	10
cattle and donkeys	2
cattle only	4
Adequate draft: donkeys only	4
donkeys and cattle	12
cattle only	31

Source: Muvirimi, 1995

Table 4 shows the present animal ownership patterns in relation to the farmer recommendation domains shown in Table 3.

Plowing is regarded as the most critical animal-powered operation. Where there are sufficient oxen, draft is supplied by oxen, but as numbers decrease, the burden of draft is shared between oxen, cows and donkeys. A large number of farmers (53%) do not have adequate animals and therefore have to rely on alternative sources.

Government plowing services do provide limited tractor plowing, but they are largely regarded as unreliable, non-viable and not sustainable. Draft animal power contracting services are not common and are only provided when the contractor has completed his own plowing. Various arrangements between farmers have been developed to gain access to animal power. These include:

- Barter, for instance herding cattle in exchange for plowing.
- Lending land in return for plowing services.
- Persuading close relatives to assist with plowing.

Inevitably operations are carried out late with resulting low yields.

For those with no access to animal power, zero tillage (holing out with hoe at planting time) is practised, but this is universally unpopular because of the high labour demand and low yields achieved in comparison with tillage with animal power.

Consequently cattle and donkey theft has become a widespread problem.

Management

Livestock management practices are generally poor due largely to a lack of resources, particularly finance and knowledge. Due to their higher economic value, cattle are generally better managed than donkeys. The lack of information on donkeys, their low economic value and their ability to withstand poor treatment contributes to their receiving little or no management. However, where donkeys are the major source of draft power, their value is increasing and management practices are improving. Due to the shortage of draft power, when timeliness of cropping operations is crucial, draft animals are often prescribed tasks exceeding their capabilities. This is further compounded, particularly for donkeys with inappropriate implements, such as the heavier ox-drawn plow being used due to lack of lighter implements. In many areas, farmers resort to using mixed spans of cattle and donkeys.

No routine veterinary practices are adopted even though donkeys suffer from intestinal worms, blackleg, or mouth and harness sores. Serious wounds may be treated. Males are usually castrated to stop them wandering, although no information on how and when donkeys should be castrated is available. There are no recommended breeding programmes, with the result that breeding is random and selection of males is based purely on phenotypic characteristics. Mistreatment of donkeys to make them work harder or faster is common.

Only in those areas with a long history of donkey use is there widespread use of breast band harnesses for donkeys and the traditional yoke for cattle. In other areas yokes are extensively used for donkeys. This is attributed to both the unavailability of harnesses, their high price and the belief by farmers that yokes are more efficient.

Nutrition

Nutrition is a widespread problem affecting optimum utilisation of draft animals in communal areas. This is mainly due to lack of grazing or supplementary feed and inadequate management of those resources.

In theory, animals have access to grazing throughout the day, but in many cases the grazing areas are distant or the animals are working so the feeding time is restricted, leading to low productivity. Donkeys are usually in better body condition than cattle at the start of the plowing season. This reflects the donkeys' ability to thrive in conditions of scarce nutrient supply. Extensive and strategic use of home-grown supplements, mainly stover and melons, would contribute to maintaining animals' body condition. Crop residues are usually removed from the lands and stored in racks above the kraals for easy access. Improvement in storage methods could improve quality. There is very limited use of bought-in supplements because of cost. However, the use of other supplements such as fruit pods and multipurpose tree species could alleviate seasonal nutritional deficiencies.

Health

Most farmers do not have a standard health management system. Treatment is ad hoc, and only carried out when the animal is in real danger of being lost. However health problems not necessarily leading to animal loss do affect animal performance. The losses are often blamed on nutrition rather than health, and the extent of these losses has not been quantified.

Injuries caused by use of cattle yokes and poor harnesses in donkeys are common. Although cattle are dipped regularly, this is not the case with donkeys. Problems of tick-borne diseases and internal parasites are not apparent, though the effect of these on the overall productivity of draft animals, particularly donkeys, needs further investigation.

Ownership of equipment

Ownership and use of agricultural equipment in the three areas is similar. Despite the fact that 53% of farmers do not have adequate animal power, 91% own a plow. Ownership of other implements (cultivators, harrows and planters) is limited. Table 5 shows current estimates of ownership of equipment.

Most equipment comes from Bulawayo Steel Products and Zimplow, the large-scale animal-drawn equipment manufacturers although village blacksmiths have assisted in maintenance and are capable of providing spare parts and assisting in the design of modifications.

Table 5: Percentage of farmers owning animal-drawn equipment in semi-arid Zimbabwe in 1995

| Implement | % of farmers owning/not owning equipment (n=248) | |
	None	At least one
Plow	19	91
Cart	60	40
Cultivator	69	31
Ridger	98	2
Harrow	71	29

Source: Muvirimi, 1995

Most equipment is purchased using non-farm income and is regarded as an asset to be passed on from father to son. This investment is unlikely to be replaced by new equipment unless major low-cost improvements can be made that provide significant benefits over existing equipment. Factors like durability, lightness and low maintenance should be of primary concern to developers.

Gender and age considerations

Labour operations tend to be gender specific as demonstrated in Table 6.

Men undertake most work with animals with women undertaking manual operations. However, if no men are available women will handle the

Table 6: Some labour differentiation between men and women

Operations undertaken mainly by:

Men	Women
Plowing with animals	Planting behind the plow
Planting with a planter	Planting by hand
Weeding with a cultivator	Hand weeding
Transport by cart	Head transport
Marketing produce	Most domestic tasks

animals. Donkeys are favoured by women because of their easier handling.

There appears to be a direct correlation between cattle ownership and age of the farmer. Farmers with inadequate animal power are likely to comprise younger families. Farmers between 40 and 50 years old, appear to own greatest numbers of draft animals, with older farmers having smaller numbers, having given animals to their sons (for *lobola*, bride price) or slaughtering animals for funerals.

Environmental concerns

Early work (Cleghorn, 1966; Sandford, 1982) estimated communal grazing to be seriously overstocked, with livestock numbers exceeding the carrying capacity and much of the grassland bare or in poor condition. Such work has been criticised as being based primarily on range condition and assessment in commercial farming areas and has not taken into account the strategic use of high potential sites such as dambos or vleis, drainage lines and browse.

The concept of carrying capacity is unpopular because it alludes to overstocking and possible destocking. The concept that carrying capacity is greatly exceeded has been questioned from both economic and ecological standpoints. Nevertheless, pressure on land and deforestation due to expanding livestock and human populations is a matter for concern. At the same time shortage of animal power is recognised as being a major constraint to increased crop production.

Opportunities for increasing productivity

The means of improving productivity are either through increasing the supply of draft animals or reducing the demand for draft animals by increasing their effectiveness. Farmers in different recommendation domains will consider technologies in the light of their own circumstances and ability to balance resources with expenditure. Those without animals are likely to use more resources for acquiring or hiring animals. They will be less receptive to technology that demands substantial resources if this will only effect a marginal improvement. Profitability of investing in new technology should therefore be assessed for each recommendation domain.

Increasing the supply of draft animals
Using existing animals more effectively

The available draft is less than that required to provide every farmer with adequate draft. The strongest demand for increasing animals is from those not owning adequate animals. As incomes rise there will be more investment in animals, thus increasing grazing pressure.

Improved management practices including strategic feed supplementation and low cost disease prevention will substantially increase draft power availability. However, this will not necessarily assist those with inadequate draft or with no animals.

Greater use of donkeys

There is a tradition of using only oxen as draft animals. Only recently have donkeys and cows been considered seriously for draft. The increased interest in these animals necessitates greater attention to developing recommendations for improved donkey and cow management.

However, there is a controversy surrounding the role of donkeys in the driest areas. Whereas donkeys are basically kept for draft (often related to crop production), cattle are multipurpose. The driest areas in Zimbabwe have been classified suitable for extensive livestock production only. In this situation the value of donkeys is primarily for transport and encouragement for crop production may not be viable.

Encouraging draft animal power contractors

The fact that some contracting already occurs is an indication of an existing market. It could be expanded by encouraging higher market related prices. The provision of government-subsidised tractor hire schemes undercuts both private tractor and draft animal contractors, preventing the emergence of a market-orientated service by private contractors. Such a service would need to reflect market demand as it is unlikely that those with more than adequate draft power would regard social obligation as a sufficient incentive to provide plowing services.

Increasing carrying capacity

Despite the widely-held view that communal grazing is overutilised there is no consensus on this issue. Cousins (1987) states that there is no evidence that the communal herd has been declining due to over-exploitation of grazing resources. It is, however, important that total biomass production and use by animals is optimised as one strategy for increasing draft animal availability. Investigations are required to ensure that available fodder resources can be sustained. Cycles of increased livestock numbers in good rainfall years and deaths in drought years need to be avoided.

Reducing the demand for draft animals

Conservation tillage systems

Tillage operations undertaken by farmers in all areas are similar, with emphasis on plowing with the mouldboard plow. Unfortunately, this has a high draft requirement, increasing the risk of not completing plowing in time. Extending the plowing period through early (winter) plowing is therefore important and methods of promoting this need encouragement. Conservation tillage offers the greatest potential for increasing productivity. The use of no-till tied ridges and ripping into crop residues, tine planting or direct drilling of seed has been widely recommended (Elwell and Norton, 1988; Elwell, 1993). Although there has been some success on heavier soils with no-till tied ridges with the crop planted in the furrow (Nyamudeza et al, 1991), adoption by farmers on lighter soils remains very low (Sarapinda, 1989, 1990; Huchu 1990).

No precise figures of the numbers of farmers in Zimbabwe who have adopted conservation tillage are available. Probably less than 1% of communal area farmers and perhaps 5–10% of commercial farmers have adopted conservation tillage (Contil, 1990). The reasons for low adoption include: shortage of labour, lack of draft power, lack of suitable equipment and the fact that most soil and water conservation systems are not compatible with the wide range of technical and socio-economic problems faced by communal area farmers. These include a high labour requirement for construction and maintenance, difficulties in planting and weeding, poor crop establishment and increased weed problems. Vogel (1993) concluded that although erosion under no-till tied ridging is negligible, the system may generate micro-environments that result in delayed crop establishment and poor crop stands. These problems help to explain low adoption rates.

Some of these problems may have been overcome as a result of recent work by the Cotton Research Institute as reported by Mashivira et al, 1995 and Ellis-Jones et al, 1993. This involves crop establishment and fertilising on the flat and ridging-up by plow or cultivator at weeding when plants are large enough not be damaged.

Improved implements

The implements that are readily available are not the most suitable for smaller animals and other alternatives are presently being evaluated. This involves on-station and on-farm testing of:

- a light donkey plow
- tine and rippers that can be attached to the beam of the ox plow. Farmers with inadequate animals will appreciate attachments requiring less draft rather than a completely new plow
- light donkey weeders
- single animal drawn weeders and tie makers
- low-cost modification to existing carts to make them more suitable for donkeys.

Small firms are able to manufacture low-draft equipment such as cultivator and ripper tines and innovative implements such as a low-cost planter.

Conclusions

Draft animal power technology is appropriate and relatively cheap for most farmers. However, research and extension needs a new approach to ensure that current farmer needs are met. Problems should be approached in an multidisciplinary manner considering the socio-economic circumstances of farmers, the existing use and management of draft animals and implements. To promote and ensure adoption of new technologies research should ensure:

- an enabling policy environment
- research, training and extension targeted at specific farmer recommendation domains with technology options for each
- involvement of farmers, extension workers and manufacturers (large- and small-scale) in problem identification, research project design and evaluation.

Acknowledgements

The authors would like to acknowledge the financial support of the Overseas Development Administration (ODA), UK and the Department of Agricultural, Technical and Extension Services, Zimbabwe in funding this research.

References

Barrett J C, 1992. The economic role of cattle in communal farming systems in communal farming systems in Zimbabwe. ODI Pastoral Development Network. *Network paper 32b*. Overseas Development Institute, London, UK.

Central Statistics Office, 1990. *Quarterly Digest of Statistics*. Ministry of Finance, Economic Planning and Development, Harare, Zimbabwe.

Cleghorn W B, 1966. Report on the condition of grazing in Tribal Trust lands. *Rhodesian Agricultural Journal* 63(3): 57–67.

Contil, 1990. *Conservation tillage for sustainable crop production systems*. IAE/GTZ Working document, Institute of Agricultural Engineering, Harare, Zimbabwe. 10p.

Cousins B, 1987. *A survey of current grazing schemes in the communal lands of Zimbabwe*. Centre for Applied Social sciences. University of Zimbabwe, Harare, Zimbabwe. 96p

Eckert M and Mombeshora B, 1989. Farmer objectives and livestock functions. pp213–252 in: *People, land and livestock*. Proceedings of a workshop held by the Centre for Applied Social Sciences, University of Zimbabwe, Harare, Zimbabwe.

Ellis-Jones J and Panin A, 1992. Profitability of animal draft power. pp 94–103 in: Starkey P, Mwenya E and Stares J (eds). *Improving animal traction technology*. Proceedings of first workshop of the Animal Traction Network for Eastern and Southern Africa held 18–23 January, 1992, Lusaka, Zambia. CTA, Wageningen, The Netherlands, 480p. ISBN 92-9081-127-7

Ellis-Jones J, Twomlow S, Willcocks T, Riches C, Dhliwayo H and Mudhara M, 1993. Conservation tillage/weed control systems for communal farming areas in semi-arid Zimbabwe. Paper presented to Brighton Crop Protection Conference. p 1161–1166.

Elwell H A, 1993. Development and adoption of conservation tillage practices in Zimbabwe. *FAO Soil Bulletin* 69: 129–190. Soil Resources, Management and Conservation Service, Land and Water Development Division, FAO, Rome, Italy.

Elwell H A and Norton A J 1988. *No-till tied ridging, a recommended sustained crop production system*. Institute of Agricultural Engineering, Harare, Zimbabwe. 40p.

Farming Systems Research Unit (FSRU), 1994. *Coping with risk and uncertainty in Zimbabwe's communal lands*. A summary report. Farming Systems Research Unit, Department of Research and Specialist Services, Ministry of Lands, Agriculture and Water Development, Harare, Zimbabwe. 40p.

Geza S and Reid M, 1983. Environmental conservation in communal lands with special reference to grazing lands. *Zimbabwe Science News* 17 (9/10): 148–151.

Hagmann J and Prasad V L, 1994. *The use of donkeys and their draft performance in smallholder farming in Zimbabwe*. Project Research Report II, Conservation Tillage for Sustainable Crop Production Systems. AGRITEX Insititute for Agricultural Engineering, Borrowdale, Harare, Zimbabwe. 8p.

Huchu P, 1990. *A comparison of crop performance in the 1989/90 season between practising Gutai farmers and non-registered farmers*. Report No. 8, Monitoring and Evaluation Unit, AGRITEX, Insitute of Agricultural Engineering, Borrowdale, Harare, Zimbabwe.

Mashivira T T, Hynes P, Twomlow S and Willcocks T, 1995. *Lessons learned from 12 years of conservation tillage research by the Cotton Research Institute under semi-arid smallholder conditions*. Paper presented at a conference on Soil and Water Conservation Tillage for Smallholder Farmers in Semi-arid Zimbabwe held 3–7 April 1995, Masvingo, Zimbabwe. 9p.

Ministry of Lands, Agriculture and Rural Resettlement (MLARR), 1993. *The second annual report of farm management data for communal farming areas farm units: 1989/90 farming season*. Ministry of Lands, Agriculture and Rural Resettlement, Harare, Zimbabwe. 64p.

Mrema G C and Mrema M J, 1993. Draft animal technology and agricultural mechanisation in Africa: its potential role and constraints. *NAMA newsletter* 1 (2): 12–33. Network for Agricultural Mechanisation in Africa.

Muvirimi F, 1995, in preparation. Survey results from Semukwe, Chikwanda and Sebungwe Communal Farming Areas.

Nyamudeza P, Mandiringana O T, Busangvanye T and Jones E, 1991. The development of sustainable management of vertisols in the lowveld area of south-eastern Zimbabwe. *International Board for Soil Research and Management (IBSRAM) Newsletter* 20: 6–9. International Board for Soil Research and Management.

Sandford S, 1982. *Livestock in the communal areas of Zimbabwe*. Report prepared for the Ministry of Lands, Resettlement and Rural Development, Harare, Zimbabwe by the Overseas Development Institute, London, UK.

Sarapinda C D, 1989. *Attitude of Gutai and non-Gutai members and adoption of some recommended cultivation techniques by Gutai farmers*. Report No. 4, Monitoring and Evaluation Unit, AGRITEX, Institute of Agricultural Engineering, Borrowdale, Harare, Zimbabwe. 9p.

Sarapinda C D, 1990. *Adoption of dry season CARD production recommendations*. Report No. 7, Monitoring and Evaluation Unit, AGRITEX, Insititute of Agricultural Engineering, Borrowdale, Harare, Zimbabwe. 11p.

Starkey P, 1988. *Animal traction directory: Africa*. Vieweg for German Appropriate Technology Exchange, GTZ, Eschborn, Germany. 151p.

Vogel, 1993. Effect of tillage on topsoil temperature and strength in coarse sands with special reference to tied ridging systems. pp 17–29 in *Conservation for sustainable crop production systems*. Report No. 6, Monitoring and Evaluation Unit, AGRITEX, Institute of Agricultural Engineering, Harare, Zimbabwe.

Participatory technology development for animal traction: experiences from a semi-arid area of Kenya

by

David Mellis [1], Harriet Matsaert [1] and Boniface Mwaniki [2]

[1]*Dryland Applied Research and Extension Project (DAREP), Kenya Agriculture Research Institute (KARI), RRC-Embu, Box 27, Embu, Kenya*
[2]*Rural Technology Development Unit (RTDU), Box 82, Siakago, Kenya.*

Abstract

As part of the general trend towards involving the target group more in development and research activities, participatory approaches are being used more frequently. Participatory Technology Development has been advocated recently as the only research and development process applicable to resource-poor, marginal and complex farming systems such as those found in semi-arid areas of sub-Saharan Africa. Empowerment of the participants, increased confidence of farmers and artisans in their own knowledge, improved capacity of clients to innovate and experiment, and an enhanced ability to cope with change are often claimed to be more achievable using participatory methods than through traditional technology transfer methods. This paper draws on experiences of participatory development in a pilot programme with metal workers and farmers in the semi-arid areas of Lower Embu and Tharaka, Kenya. Institutional issues are explored through the involvement of an NGO (Farm Implements and Tools, FIT), a government applied research and extension project (The Dryland Applied Research and Extension Project, DAREP), government extension staff (Rural Technology Development Unit, RTDU) and local commercial traders. The opportunities and obstacles of participatory technology development are explored in the context of animal traction and options for meeting the challenges are suggested.

Challenges remain in:

- *improving the quality control, standardisation and raw material supply of locally made tools*
- *continuing the feedback from farmer to toolmaker beyond the prototype stage and into the market relationship*
- *coping with the low demand pull exerted by farmers in semi-arid areas*
- *including engineers into the process, even though they may belong to non-local*

institutions such as universities or large-scale manufacturing concerns

- *improving communication over the large distances in semi-arid areas*
- *increasing empowerment of farmers and artisans to demand appropriate services.*

Options for meeting these challenges may be found by:

- *working with farmer groups who have higher incomes, are nearer to the manufacturers and can bear higher risks*
- *exploring ways of working with blacksmiths who make cheaper tools and are nearer to farmers in semi-arid areas*
- *involvement of local stockists or traders to provide credit, raw materials, transport and marketing outlets*
- *the use of fact sheets with input from engineers, artisans, and farmers and paid for by advertising from traders to improve quality while consolidating and disseminating the process of product development*
- *improving the participatory facilitation skills of participants.*

Introduction

As part of the general trend towards involving the target group more in development and research activities, participatory approaches are increasingly being used. Participatory Technology Development has been advocated recently for the types of farming systems found in semi-arid areas. Three general types of agricultural system have been identified by recent literature on agricultural development: industrial or commercial, green revolution and a third type characterised by resource-poor, complex and risk-prone farming systems (Scoones and Thompson, 1994). In the first two types agricultural research has traditionally been top-down, with the assumption that technology can be transferred from research institutions to farmers. However, in the third type of agriculture, which is the most common in sub-Saharan Africa, this has not worked, since researchers have not been able to replicate the complex and marginal physical and socio-economic environment of these farmers. Therefore, an alternative approach has been sought for these areas. It has been variously labeled 'Participatory Technology Development', 'Farmer Back to Farmer', 'Farmer First', and 'Farmer Participatory Development' (Hudson and Cheatle, 1993), but in general the aim is to increase the involvement of the beneficiaries in the research and development process. The semi-arid agriculture found in the Dryland Applied Research and Extension Project (DAREP) mandate area falls within the third type of farming system and attempts have been made by DAREP to develop a participatory methodology for agricultural research, together with farmers, artisans, traders, extension workers, researchers and NGOs.

Furthermore, empowerment of the participants, increased confidence of farmers and artisans in their own knowledge, improved capacity of clients to innovate and experiment, and an enhanced ability to cope with change are often claimed to be more possible under participatory development than through traditional technology transfer methods. This paper draws on our experience of participatory technology development in a pilot programme with *jua kali* (informal sector) metal workers and farmers in semi-arid Lower Embu and Tharaka. Institutional issues are explored through the involvement of non-governmental organisations (Farm Implements and Tools, FIT), a government applied research and extension project (the Dryland Applied Research and Extension Project, DAREP), government extension staff (Rural Technology Development Unit, RTDU) and local commercial traders. We discuss the opportunities and obstacles within participatory technology development specific to animal traction.

Background

Context of technology change

Semi-arid areas such as Lower Embu and Tharaka present special opportunities and challenges to animal traction. Although agriculture is labour limited, rather than land limited, specific hindrances prevent the resource of animal power from being utilised fully. These include low population density (less than 100 persons per square kilometre), poor infrastructure, uncertain returns from farming and harsh technical working conditions for animals and tools (Mellis and Mwaniki, 1995).

Furthermore, farmers in this area are experiencing a period of change which is having a dramatic impact on their traditional farming system of shifting cultivation and pastoralism. Population growth and the consequent land demarcation are causing weed control and soil conservation to become important management issues. There is also a high rate of migration as young men go for off-farm income and older children are away at school. This means that labour is a severe constraint. Farmers are therefore very interested in new tools (including animal power options) which could allow them to adapt to their changing circumstances.

Traditionally, farmers in the area have used tools made by local blacksmiths, such as axes, and the *miro* (a digging stick with a metal blade at the end). Recent changes include the importation of tools and materials to the area, so whereas, for example, blacksmiths used to smelt their own iron to manufacture tools, they now use scrap metal and do repairs only.

The main source of technology change has been inward migration and external travel by the local population. For example in Tharaka, *Victory* plows came into the area from their Akamba neighbours to the south in the 1940s, and more recently from locals working off-farm in towns such as Embu, Chuka and Meru to the north. Illustrating this fact, an ICRA study in 1984 found 23% of farmers owning plows, whereas in 1993 DAREP found the figure had increased to 44% of farmers. Mobile traders at local markets have also brought in new technologies such as *jembes* (dutch hoes) and forked *jembes*. Handtools are sold in many weekly markets, but larger items like plows and wheelbarrows are only available from large towns, necessitating expensive travel and transportation by farmers.

Institutional approaches

It would seem that agencies promoting improved technologies have had little impact in producing technology change in the area. From 1982 to 1988 the Embu, Meru and Isiolo (EMI) project gave free tools to groups of farmers to carry out soil conservation. From 1989 the Dryland Applied Research Project (DARP) offered some shop-bought and locally-made tools such as wheelbarrows and weeders, at cost price to farmers, displaying them at nine sites through the region. These tools were tested extensively with farmers, but there was no adoption. This may have been due to the fact that DARP was not able to respond to farmers comments, nor to deal with associated issues of credit and supply, and, on their part, the farmers had become used to being lent or given free tools.

As project designers became aware of the need to develop technologies *with* farmers rather than *for* them, the Dryland Agriculture Research and Extension Project, was conceived to work using a more participative methodology. The main objective was to 'develop and evaluate sustainable agricultural technologies and participatory research methodologies'. The project operates within a decentralised research infrastructure and is a collaborative project between Kenya Agriculture Research Institute (KARI), Kenya Forestry Research Institute, Ministry of Agriculture, Livestock Development and Marketing and the Natural Resources Institute (NRI). Research scientists work within an interdisciplinary team and technology components include cropping systems, soil and water management, livestock and agroforestry.

DAREP's work on tools has been carried out within the soil and water management programme which looks at the constraints facing farmers in tillage generally and this has led to research on tillage tools, land husbandry and water harvesting (Skinner and Mwaniki, 1994).

Paticipatory technology development

Getting started: participatory rural appraisal

The entry activity for the soil and water management programme was a focused participatory rural appraisal which included stratified samples and group meetings (Skinner and Micheni, 1993).

During the participatory appraisal, farmers expressed that, in terms of tools, their agricultural production was constrained by:
- shortage of plows at the optimum time for land preparation
- lack of transport to carry manure to the field
- lack of labour for weeding
- lack of tools to build soil conservation structures.

This problem of access to tools was caused by:
- their high cost
- the lack of credit
- poor returns from farming
- high transport costs.

Possible remedies were identified during the participatory appraisal including strengthening supply networks, credit support and training of local blacksmiths. In the discussions between the researchers, farmers and extension workers, the participants concluded that research should be carried out on low cost tools which could be made or repaired locally.

Finding what to try: farmer groups

Following the work in the field with farmers, the researcher carried out an initial literature and project review of the potential technologies and methodologies to alleviate the tool supply and tillage constraints.

To keep farmers involved in the process, the results of the participatory appraisal were publicised at the nine DAREP field stations during open days and farmers were invited to participate. At this point the researcher decided to work from only two field stations in order to enable a more in-depth involvement with farmers while keeping within their resources. Farmers were selected on the basis of their interest, wealth rank, and gender. This enabled two enthusiastic groups of about ten farmers to be formed which represented the resources and tillage systems of the local community.

DAREP was initially concerned about the group composition. It was thought that perhaps the differences in wealth within the focus groups would make their interests too divergent to allow them to work together in technology development. However, it was found later that hand cultivators may occasionally hire a plow or aspire to plow ownership and plow owners often cultivate some of their land by hand. Furthermore, although representation by women was low at first (considering women do most of the farm work in this area), over subsequent seasons, more women have replaced their husbands and groups are now balanced. However, it was recognised that efforts need to be made to ensure that meetings are scheduled at convenient times for women to attend (eg when older children are back from school and can manage the farm). The tools project tried working with the womens' groups used for some initial tool evaluations, but found them more interested in gaining access to inputs rather than in research and technology development.

The group meetings were extremely useful. Women and men farmers were able to debate and reach consensus on the selection of technologies and evaluation of trials, and the groups also took responsibility for presenting the results of trials to the wider community at project open days. Focus groups chose to test tools for land preparation (*Bukura* plow and *Mutomo* plow), planting (rotary injection planter, jab planter) and weeding (chiefs cultivator, emivator and pye hoe). An ard plow was rejected as looking too weak for their hard soils (it broke during demonstration). It should be noted that tools were only one element in a range of technology options for soil conservation and weed control which are being tested by farmers. They are also investigating zai pits, contour furrows, tied ridges, water harvesting and manure placement. The broad range of technologies available has been important in sustaining the participatory process, as will be seen below.

Trying out: farmer testing

Trials were designed and implemented by farmers in their own fields. Farmers selected criteria for monitoring their trials and both quantitative (labour inputs, accuracy of seed placement, weed counts etc) and qualitative (durability, ease of use etc) data were collected.

Focus group meetings were held to search for and screen the available options, to initiate the research, to monitor the progress and results with farmer-to-farmer visits and to evaluate the groups' experiences at the end of the season. Farmers in the groups particularly liked the farmer-to-farmer monitoring visits and as a result farmers continue

to try new technologies which had previously not appealed to them. After experimenting with new technologies in the first season, farmers had suggestions for developments to the technologies. For example one farmer suggested developing the zai pits that were initially presented, into furrows which would be easier to make with a plow (or by hand). Farmers were also concerned about issues of supply and maintenance of the tools that they had been testing.

Improving and innovating: *jua kali* group

To respond to farmer evaluations and concerns with new agricultural tools, it was necessary to work closely with tool designers and manufacturers. During the first season of trials DAREP attempted to involve local blacksmiths, *jua kali* artisans, large-scale tool manufacturers, Ministry of Agriculture extensionists, and researchers from the University of Nairobi. It was found that *jua kali* tool manufacturers form larger towns such as Kivaa and Embu were able to respond rapidly and enthusiastically to farmers' evaluations with a small amount of external input. Although engineers from the local government Regional Technology Development Unit were involved in the process, and had their doubts about the durability and efficacy of some of the implements, it was difficult for them to respond practically to the farmers needs due to the lack of tools and equipment at their workshops and limitations on transport. The University was also involved, but was unable to respond rapidly to farmers requests or *jua kali* needs within their existing academic programme.

Having no experience or expertise of working with small manufacturing businesses, DAREP collaborated with a Nairobi-based NGO called Farm Implements and Tools (FIT) and a local Peace Corps worker to link the farmer focus groups to a number of *jua kali* workers in Embu district. FIT conducted a survey of the metal workers to assess their capabilities and skills and then invited them to join farmer focus groups at one of the field stations. During this meeting, it was found that farmers were able to communicate their agricultural implement needs very effectively, and the *jua kali* workers were invited to respond. One farmer (a local chief) presented his innovation of a drag hoe on which rocks could be placed to adjust the penetration and work load.

At this point the *jua kalis* decided to form an Agricultural Tools Group. To achieve this, the provincial officer for Applied Technical Training was invaluable in liaising with the umbrella Embu Jua kali Association and was able to smooth over the suspicions of the Jua kali Association officials. In response to the farmers' demands, the sectoral group was able to share raw materials, equipment and ideas. One member procured most of the scrap material, enabling buying in bulk, and another member did most of the forging work due to his skill and equipment in that area.

With a small grant for working capital (about US$10–20, provided by FIT), and some ideas for tool design in the form of books (Intermediate Technology's *Tools for agriculture*; ITDG, 1992) and drawings (from a local artist), *jua kalis* quickly made some new and adapted tools based on farmers' recommendations. These included an improved version of the *Mutomo* wooden-beamed plow, a light one-handled plow, two adapted versions of the chief's drag hoe, two spray pumps and an improved jab planter. The artisans were thus able to incorporate farmers' existing knowledge, external knowledge in the form of a book on tools, and their own expertise in metal-working processes. The tools were then presented back to farmers by the *jua kalis* before the next season at a tool show. A panel of farmers (four women and four men) judged the tools and awarded a prize to the best according to the criteria below (Tanburn and van Bussel, 1995):

- type of materials and quality of work
- function (flexibility and efficiency)
- durability
- number of operators required
- applicability to different soil types
- portability.

FIT suggested the addition of two criteria:

- originality/innovation
- suitability for use by women in particular.

The prize was given to a wooden plow since the panel said:

- *'It was light to use, can be used by women and older people.'*
- *'The penetration is very good.'*
- *'It can be used for wet or dry plowing, and for weeding.'*

- *'It looks easy to repair.'* (Tanburn and van Bussel, 1995)

Sustainability: marketing and quality control

During the meeting, farmers agreed to invite *jua kalis* to their DAREP open days where they could display and sell their tools. Since the relationship between *jua kalis* and farmers seemed established and the groups had arranged to communicate, no more funds were given to the process in order to see how sustainable it might be.

In Embu, the artisans sold some of the prototypes (including the wooden plow) and obtained orders from farmers in the neighbouring high potential areas. The following season the artisans received invitations to the open days at Mutuobare and Kajiampau, but did not sell any tools or get any orders from farmers in the drylands. These are an example of the comments made by farmers (Mwaniki, 1995):

A woman farmer who often rents a plow and weeds with a panga: *'Some implements were expensive, others seemed poorly made, others were complicated and we did not understand them and therefore could not buy.'*

A male plow owner who weeds with oxen: *'The items are a bit expensive; others we gave our contribution on how they [the prototypes] can be modified, but the changes were not as per our comments; others were viewed as unsatisfactory and thus purchasing was not possible.'*

Thus, although the prototypes had been enthusiastically received, the production models were less good, showing up the difficulty of copying what was now the third generation tool. However, agreements were made that farmers would order tools either directly from the artisans or by writing letters through DAREP. By this method artisans were also invited by the farming community to display their tools at three other DAREP stations. Some drag hoes were bought, but only by extension or project staff. Since then, some of the Mutuobare farmers have visited the *jua kali* workshops in Embu, and a plow, wheeled hoe and drag hoe have been bought by farmers in semi-arid areas. Tools continue to be sold to farmers in the higher potential area around Embu.

The apparent lack of demand from dryland farmer groups has not dampened their interest in research, however. They either want to save enough money to buy the tools they like, or to see improvements to *jua kali* tools. The fact that they feel identity as a group has kept them seeking solutions to their problems from their own resources. One group has formed a savings group and meets every month to donate money to one member. Members have used the money for buying *jembes* and *pangas* or paying school fees. Both groups continue to research methods of improving water and soil conservation with both animal traction and hand tillage methods. These include post-plant ridging with a plow, plow weeding, zai pitting, tied ridging and water harvesting, Having contact with the DAREP multidisciplinary team has meant that the same farmers are trying out other technologies such as new crops and crop varieties, water harvesting for trees and crops, and cultural pest control methods. The diversity of technological options has kept the farmers' interest and the momentum of the participatory process going even though one option (access to tools) was being delayed by the other group (the artisans).

A follow-up evaluation of the progress so far was sponsored by FIT and carried out by engineers from the Rural Technology Development Unit (Mwaniki 1995). They found that some tools made by *jua kalis* such as manure forks and water sprinklers were performing well, and appreciated by their buyers, while a wheeled hoe and drag hoe were found not to be functional due to poor manufacture. Furthermore, the Rural Technology Development Unit continues to observe problems of design quality in other tools circulating around Embu. Plows bought by local NGOs and stockists have wrong share positions, weak beams, poor adjustment mechanisms and wrong mouldboard shapes. It appears that even if producers know how to make good tools, the customers are not able to demand the quality they need.

However, both farmers and artisans claim that they have benefited from the process. Typical comments were (quotes from Mwaniki, 1995):

- *'It was good, near to the farmers.'* Jennifer Kiura (Farmer Research Group)

- *'Farmers were able to tell producers their problems. I gained in knowledge and would*

like to participate in future.' Andrew Gatiti (Chairman, Farmer Research Group)

- *'I have new customers, new marketing ideas and have gained knowledge in manufacturing the jab planter. In the future I would like to have jigs and fixtures put in place.'* David Kamau (Artisan)

- *'I have learnt new marketing skills and have more confidence.'* Gerald Ngugi, (Blacksmith)

Farmers have been exposed to a number of new tools and new sources of tool supply. Both male and female farmers have learnt to evaluate tools on their own farms and at open days. They have learnt that they can communicate their needs effectively to tool makers and now know where these artisans are to be found. *Jua kalis* have gained confidence in adapting and inventing tools, increasing their skills and product range. They have become more aware of the farmer market and of methods to gain market information and advertise their wares. They also state that they are now more aware of the importance of standardised production and of the use of jigs and fixtures.

Present and future: repeating the process and developing linkages

Organisations involved in supporting the artisans met with them in Embu and made progress in identifying the constraints and opportunities that face the participatory technology development process. The first meeting highlighted the constraints in credit, marketing, group development and quality control. The meetings suggested several ways of overcoming obstacles, trying to identify solutions which would require little external input, and would build on the achievements so far. Examples include:

- working more closely with stockists or middlemen to obtain raw materials on credit and improve the marketing channels

- exchange visits and training would help directly in improving design and production skills while also improving group identity

- design and quality of tools could be improved through more field testing involving farmers, artisans and engineers.

In summary, the artisans felt that the methods used so far should be repeated, with a few extra linkages, in order to develop the products further.

Table 1: Some challenges and options for animal traction development in Embu

Challenges	Options
There is a need to improve the quality and standardisation of tools produced by local manufacturers.	Introduce branding of tools as a marketing strategy and to improve accountability for poor standards. Use of fact sheets to educate consumers to demand quality in products. Training courses for artisans.
Plows may be a difficult technology for an entry activity, as the cost is high and quality hard for small-scale manufacturers to achieve.	The project could concentrate initially on simple, cheap tools such as hand tools. As capability and confidence develops the programme could begin to tackle more complex technologies.
Large distances and poor infrastructure between artisans and farmers makes the constant feedback needed for paticipatory development difficult.	Strengthen farmer and artisan organisations.
The low population and buying power of farmers in the area means they exert little demand pull in product development.	Develop tools which are also appropriate to small farmers in the high potential areas. These farmers can more easily bear the initial risk during product development. Development of farmers' organisations in the semi-arid areas may exert more demand pressure on manufacturers. External input may be needed to support the market in area receiving famine relief food.
Manufacturers, engineers, and universities need to develop a more flexible approach which will allow them to be involved in participatory technology development and contribute to its success.	Encourage their closer involvement with small target groups and formalise the linkages. Raise awareness of the benefits of participatory technology development for these organisations.
Farmers and artisans still lack confidence in their ability to demand external input and consequently still play a relatively passive role in the participatory technology development process.	Place more emphasis on group management skills and empowerment. NGOs or extension workers can help with training/support. Field workers need good community development and facilitation skills and training in participatory approaches.
Different agendas of supporting organisations can confuse issues. For example DAREP's interest in 'researchable options' limits it to technologies unproved in the locality and FIT's desire to develop 'innovative methodologies with 100% recovery' limit the approaches they are prepared to facilitate.	Farmer and artisan agendas should remain central to the process. Strong group organisation should allow farmers to make vertical and horizontal linkages to fulfill needs.

Following these suggestions, FIT organised and sponsored a meeting between the *jua kalis* and local stockists using a small business advisor as facilitator. A good relationship was started between these two groups, and interest was shown by the stockists in *jua kali* products.

Discussions continue to be held by the stakeholders as to how tool quality can be improved in a sustainable manner. Recently FIT introduced an idea pioneered by Voluntary Service Overseas in Mombasa. Product information useful to artisans was put on a single A4 sheet and the printing costs were paid for by advertising on the reverse side. In the Embu tool development process, plow design information could be printed on such fact sheets. This may enable artisans to be more aware of the critical parameters in making a plow and customers would know what characteristics to look for in good quality tools.

Lessons learnt

Animal traction technology development, especially in semi-arid areas, is a complex activity, involving linkages between farmers, artisans, engineers, business advisers etc. This complexity makes animal traction particularly suited to a participatory technology development approach.

The main external inputs required have been in group selection, identifying ideas to try, small grants of working capital (less than US$200 in total; Tanburn and van Bussel, 1995), and facilitating communication between the users, sellers and producers of technology. The participatory approach has encouraged researchers, engineers and farmers to interact, since the technical staff have actually gone to the field, talked with farmers and responded to them. This has been important to enable the synthesis of local and external knowledge.

The development of organisational structures such as farmer and artisan groups is an important part of the participatory process. These structures increase the self confidence of the participants and allows them to address associated issues of supply, credit and also to share resources.

Tools are best developed as part of an integrated and 'problem-orientated' approach, rather than as aim in itself. This is especially important with different, interdependent groups. Thus if one solution or group fails or stalls, the (other) target group can still continue developing solutions without losing momentum.

Participatory technology development improves the effectiveness of technology selection and screening as farmers and artisans expert local knowledge is involved from the early stages.

When farmers are involved in designing their own research activities, the quality of trial management, monitoring and evaluation is high. Farmer groups were able to make very specific

recommendations for evaluated technologies. For example, farm type, soil conditions and detailed instructions for use. Farmer-to-farmer transfer of technologies is correspondingly powerful.

Participatory technology development is an iterative activity, involving constant learning and replanning. Flexibility, creativity, as well as regular monitoring and evaluation (by all stakeholders) is an important part of the process.

In Embu there are many challenges still to be met (see Table 1) and the authors welcome input from others to overcome them.

Conclusions

A pilot programme has identified methods of developing prototype tools using a participatory approach. These and other methods will be needed to take these tools into production and out to the market place. Facilitator and market approaches should be used where possible, but these may need to be supplemented by external inputs to strengthen the market in semi-arid areas.

References

Hudson N and Cheatle R J (eds), 1993. *Working with farmers for better land husbandry*. Intermediate Technology Publications, London, UK.

ITDG, 1992. *Tools for agriculture*. Intermediate Technology Publications, London, UK.

Tanburn J and van Bussel P, 1995. The potential for development of improved agricultural equipment by jua kali metal-workers: a case study in Embu, Kenya. pp 131–137 in: Kaumbutho P et al (eds) *Meeting the challenges of draft animal technologies in Kenya*. Proceedings of the second workshop of the Kenya Network for Draught Animal Technology (KENDAT) held 27–31 March 1995, Nairobi, Kenya. KENDAT, c/o Department of Agricultural Engineering, University of Nairobi, PO Box 30197, Nairobi, Kenya. 170p.

Mellis D and Mwaniki B M, 1995. Challenges to Draught Animal Technology in Semi-arid Areas: Experiences from Lower Embu and Tharaka.

Mwaniki B, 1995. *Evaluation of Jua kali Agricultural Tools Programme*. FIT Programme Report, FIT/ILO, 4 Route des Morillons, CH-1211 Geneva 22, Switzerland.

Scoones I and Thompson J, 1994. *Beyond farmer first*. Intermediate Technology Publications, London, UK

Skinner H and Micheni A N, 1993. *Tools and tillage survey of Lower Embu and Tharaka-Nithi*. Dryland Agricultural research and Extension Project, Box 27, Embu, Kenya.

Skinner H and Mwaniki B M, 1994. *DAREP/RTDU Tools and tillage on-farm research: Sept. 1993-August 1994*. Dryland Agricultural research and Extension Project, Box 27, Embu, Kenya.

Socio-economic aspects of animal power: a diagnostic study in Zimbabwe

by

R Tsimba [1], S Chawatama [1], L R Ndlovu [1], J Mutimba [2], P Ndlovu [1], K Dzama [1], J H Topps [1], D Hikwa [3] and M Mudhara [3]

[1] *Department of Animal Science,* [2] *Department of Agricultural Economics*
University of Zimbabwe, Box MP 167, Mount Pleasant, Harare
[3] *Agronomy Institute, Department of Research and Specialist Services*
Private Bag CY594 Causeway, Harare.

Abstract

A diagnostic participatory and questionnaire survey of 451 farmers was conducted in Tsholotsho, Chinyika and Mutoko Communal Areas in Zimbabwe to understand farmers' crop production constraints in relation to draft power. Nearly half (49%) of all the farmers had no draft animals but 55% of these had access to neighbours' or relatives' animals. The rest hired or worked in others' fields in exchange for draft use. Animal diseases and shortage of grazing land were noted as the major problems in animal production. Both farmer education and experience as well as household herd were found to vary between districts (p<0.001). Tsholotsho had the largest number of farmers (53%) with more than 10 years' experience and Mutoko had the largest proportion (43%) of 'master farmers'. Farmer education and experience were found to be independent of gender. The average land holdings per farmer for Tsholotsho, Chinyika and Mutoko were 6.2, 4.2 and 1.6 ha, respectively. Tsholotsho is a dry area whilst Chinyika is a resettlement area. Mutoko is a typical communal area with a high population pressure. Maize is the main crop in Chinyika and Mutoko whilst millet is common in Tsholotsho.

Introduction

Animal traction is a critical component of smallholder farming systems. Animals provide the energy source for plowing, ridging, transport and cultivation for many small-scale farmers. In addition, they provide manure, which is the main source of fertiliser for crops for some farmers. Farmers with cattle were found to have larger arable holdings, and did more winter plowing and manure application (Shumba, 1984).

Scientists have often been accused of top-down approaches to problem-solving. In many cases the prescriptions they have given have ended up either aggravating the problem, underestimating it, or farmers refuse to adopt it due to problems of affordability, appropriateness and sustainability. In either case, these approaches have been expensive failures. Farmer participation is, therefore, important in research efforts in communal areas since it can identify their problems and then develop appropriate interventions.

It is in light of these facts that a diagnostic survey was conducted to study the draft power situation in three districts of the communal areas of Zimbabwe. The objectives of the study were to establish:

- the role of animal traction in small holder farming systems
- linkages of draft power supply and demand with other sub-systems
- possible areas of intervention.

It is further intended that this work will provide reliable data on draft power that will aid in the development of an animal traction model for smallholder farming systems in Zimbabwe that use oxen, cows or donkeys for crop production.

Materials and Methods

Data collection

A participatory diagnosis was conducted using a questionnaire survey of farmers in three districts in the communal areas of Zimbabwe. The survey was carried out during the 1994–95 growing season in Mutoko, Tsholotsho and Chinyika (Chinyudze in particular). Chinyika and Mutoko are in agro-ecological zones II and III, respectively, and Tsholotsho is in zone IV. These areas are fairly representative of the smallholder sectors of

Zimbabwe which consist of communal, resettlement and small-scale farming areas.

The questionnaire consisted of questions on farmer, crop production, and livestock production characteristics. To complement the survey, an informal diagnostic discussion was held with the farmers. The farmers cited their production problems and suggested possible solutions. This was done to establish the farmers' perceptions of their problems so that research priorities would be formulated in light of their needs.

Statistical analysis

A total of 452 records were used in this analysis. The data were analysed using the SAS computer programme's General Linear Models Procedure. Frequencies, means and Chi-square tests for association were calculated for the respective characteristics.

Results and discussion

Farmer characteristics

A statistically significant ($p < 0.001$) association was found between farmers' levels of education and their district. This may be due to differences in extension efforts and arrangements in the three districts. The level of education of the farmers has a bearing on the agronomic practices and the ease with which new technologies can be adopted.

In Tsholotsho 12% of the respondents ($n=126$) were master farmers, 11% were trainee farmers and 78% were ordinary farmers. Chinyika ($n=173$) had 4% master farmers, 35% trainees, and 61% ordinary farmers. On the other hand, Mtoko ($n=153$) had the largest proportion of master farmers (43%) with 3% trainee farmers and the remainder ordinary farmers.

Farmer experience was also associated with district. Tsholotsho had the most farmers with greater than ten years experience (53%) compared to 4% for Chinyika and 43% for Mtoko. The low level of experience in Chinyika may be attributed to the fact that it is a newly-commissioned resettlement area. There were no significant associations between farmer education or farmer experience and gender implying that farmers tend to receive the same education opportunities regardless of gender.

There was also an association ($p < 0.001$) between household heads and district. In Tsholotsho 62% of

the households were headed by males compared with 84% in Chinyika and 80% in Mtoko. The results are shown in Table 1. The household head is usually the principal decision maker and this has direct implications on the production system such as the type of crop grown. Crops such as groundnuts and beans are usually regarded as "women's crops" (Truscott, 1991). Female-headed households have been found to be poorer in terms of arable land ownership, draft animal ownership, implement ownership and access to agricultural loans (Shoko and Sithole, 1995).

Production problems and coping strategies
Ownership

Most of the farmers had a major shortage of draft power. Out of 451 respondents, 49% did not have any draft animals. In Tsholotsho, Chinyika and Mutoko, 33%, 51% and 36%, respectively, had at least five animals. The majority (55%) of those who did not have draft animals were allowed access to relatives' or neighbours' draft animals.

On the other hand, 36% of the respondents with fewer than two draft animals teamed up with relatives. The remainder hired draft power or worked in other people's fields in return for use of draft animals. The high proportion of cows in Tsholotsho (Table 2) seems to indicate that cows are utilised more for draft. However, cows have been reported to be 20–30 % less efficient than oxen (Howard, 1980). This will then lead to decreased crop productivity. A span of two oxen requires about three and a half days to plow a hectare of land on wet soil (Francis, 1993).

According to the farmers, the implications of having no or inadequate draft were delays in planting and failure to carry out winter plowing. This had an adverse effect on crop yields. They said that crops grown in winter-plowed fields had better chances of survival in seasons of drought. However, most farmers did not winter-plow their fields since crop residues form an important livestock feed source, particularly in the dry season.

Disease

The majority of farmers cited disease as the main problem in cattle production. Of those who cited disease, 79% ranked it as the number one problem (Table 1). The diseases that were mentioned as problematic are red water, heart water and eye

Table 1: Farmers' ranking of problems affecting cattle ownership in all three villages (n=451). The figures show the percentage of farmers mentioning each constraint who ranked it 1, 2 or 3 in severity.

Rank (%)	Grazing	Disease	Ticks	Dipping	Labour	Drugs	Theft
1	61	79	78	41	40	50	19
2	32	16	13	45	40	40	65
3	7	5	9	14	20	10	16
n mentioning	79	152	23	56	20	58	43

infections in summer. Six farmers complained that they had problems in contacting veterinary officials or that their response to emergencies was too slow. Other problems mentioned are shortage of drugs, dipping, cattle rustling (a recent problem arising from hardships due to the economic structural adjustment programme), and drought.

Grazing land

Grazing shortage was the second important problem. Of those who mentioned it as a problem, 61% ranked it as the number one problem (Table 1). Most of the farmers asserted that grazing was particularly a problem in the dry season. They claimed that the communal grazing areas were apportioned the most infertile and desolate areas.

However, in Chinyika each farmer is allocated some land and there is no competition between grazing land and arable land. In a follow-up discussion with 11 farmers in Mutoko all the farmers indicated grazing land shortage as one of the main problems in the area. This is likely to be a result of shortage of grazing land. Most respondents suggested the use of winter supplements and paddocking of grazing areas as possible solutions.

Others

One farmer mentioned the cost of hiring draft as prohibitive, and two farmers cited each of restocking, labour shortages and low calving rates as problems. Fertiliser was also cited as one of the production problems due to its prohibitive cost. Those who have cattle use manure as an alternative.

Relationships between draft animal use and tillage practices.

Most farmers in Mtoko and Tsholotsho plow their fields only once whilst the majority of farmers (76%) in Chinyika plow twice (Table 3). This is most likely to be a result of the high adoption of winter plowing in this area. Even though ridging has been found to be very useful in low rainfall areas (in terms of moisture conservation), quite a substantial number of farmers in Tsholotsho (93%) were found not to have adopted this practice at all. This may be due to the low crop yields obtained in this region relative to the effort required to construct the ridges, as well as the shortage of draft power.

In Chinyika, maize cultivation was noted to be very common. This is likely to be because of the availability of cultivators for most families in Chinyika, unlike Mutoko where only one in every 12 farmers owned a cultivator. The absence of the

Table 2: Mean draft animal ownership in the three districts in 1994 (number of animals per household ± standard error)

	District		
	Tsholotsho	Chinyika	Mtoko
Steers	1.4±0.2	2.4±0.15	0.8±0.1
Bulls	0.1±0.03	0.4±0.07	≈0
Cows	1.9±0.3	0.3±0.22	0.8±0.01
Donkeys	0.9±0.2	≈0	≈0

Table 3: Frequency of plowing, ridging, and weeding with animal power in Tsholotsho, Chinyika and Mtoko (% of farmers in each sample)

Frequency (per year)	Tsholotsho (n=126)			Chinyika (n=172)			Mtoko (n=153)		
	Plowing	Ridging	Weeding	Plowing	Ridging	Weeding	Plowing	Ridging	Weeding
0	28	93	48	11	19	12	5	28	77
1	71	6	39	13	67	37	55	69	22
2	1	1	14	76	14	51	38	2	1
3	0	0	-	0	0	-	3	1	-

ox-drawn cultivators tends to constrain the timely implementation of critical operations. The large proportion of plows relative to cultivators implies that weeding is still largely done by hand. However, some farmers said that they used ox-drawn plows in place of cultivators.

Crop yield and area cultivated

Although maize is the staple crop in Zimbabwe, millet is the most popular in Tsholotsho (Tables 4 and 5) mainly due to the low rainfall in the area. Other crops are considered minor. In Chinyika, maize/bean intercropping is very common. The highest maize yields were obtained in Chinyika (Table 4). This is mostly because of the better rainfall and soil in Chinyika compared to the other two areas.

Farmers from Mutoko were found to have the fewest arable land holdings (Table 5). The average land holding was found to be 6.2 ha, 4.2 ha and 1.6 ha for Tsholotsho, Chinyika and Mutoko, respectively. The low area in Mutoko is likely to be due to the high population pressures, unlike Tsholotsho (a dry area) where the population is much less and Chinyika which is a resettlement area. The low draft animal ownership per family in

Mutoko (average of 2 draft animals/family) may also be due to the high population pressure.

For maize production, the majority of the farmers in Tsholotsho (69%) used a team of four animals, unlike in Chinyika and Mutoko where use of a pair was most common. The use of pairs in Mutoko is most likely to be a result of the low draft animal numbers. The main span size in the communal areas is therefore two animals. There was a significant (p<0.001) association between span size and total area cultivated in 1994. This indicated that farmers with larger holdings tended to use larger teams of animals.

For millet, 68% of the farmers in Tsholotsho used a team of four animals whilst 100% and 90% of farmers in Chinyika and Mutoko, respectively, used pairs of animals.

Conclusion

The majority of smallholder farmers do not have adequate draft power. Oxen and dry cows are predominantly used for draft work like plowing, planting, cultivation, ridging and transportation. Donkeys are relatively uncommon with the exception of Tsholotsho.

Table 4: Mean crop yields for the 1993/94 growing season (kg/ha)

Crop	Tsholotsho	Chinyika	Mutoko
Maize	218	4500	336
Millet	383	0	0
Sorghum	49	0	0
Sunflower	16	367	49
Groundnuts	88	119	24
Beans	4	138	27
Tobacco	0	114	8

Table 5: Mean area cultivated per household for each crop (ha)

Crop	Tsholotsho	Chinyika	Mutoko
Maize	1.8	3.1	1.0
Millet	3.2	≈0	0.2
Sorghum	0.6	0	0.1
Sunflower	0.1	0.5	0.2
Groundnuts	0.3	0.2	0.2
Beans	0.1	0.4	0
Total area	6.2	4.2	1.6

Farmers with adequate draft power can carry out timely work and are able to do winter plowing. Availability of draft animal power also reduced demand for labour for operations like planting, weeding and transport. Draft animals are an important source of fertiliser (manure) for crops.

The animal power problem has two components - unavailability (absolute numbers) and inadequacy (animal quality). Those with inadequate draft power usually planted late.

The main problems of animal production were disease (mainly tick-borne), shortage of drinking water and grazing. Animals were therefore weak for draft work. Suggested solutions were boreholes and supplementary feeding (crop residues). Other problems identified were cost of veterinary drugs, cattle rustling and a high cost of hiring draft animals or restocking after drought.

This diagnostic study indicates that there may be a vicious circle in the smallholder system which links low crop yields, low crop residues (nutrition) and low draft power. Crop yields can be improved by efficient use of feed resources or by improving draft animal management

References

Francis J, 1993. *Effects of strategic supplementation on work perfomance and physiological parameters of mashona cattle.* MPhil Thesis, University of Zimbabwe, Harare, Zimbabwe.

Howard C R, 1980. The draft ox: management and uses. *Zimbabwe Rhodesia Journal* 77(1): 19–34.

Shoko T and Sithole P N, 1995. *Rural women in agricultural development.* Proceedings of a workshop held 10–12 January 1995 at Mandel Training Centre, Harare, Zimbabwe. Institute of Agricultural Engineering, Borrowdale, Harare, Zimbabwe. 18p.

Shumba E M, 1984. Yields of maize in the semi-arid regions of Zimbabwe. *Zimbabwe Agricultural Journal* 81: 91–94.

Truscott K, 1991. *Women and tillage, strategic issues posed by farmer groups.* pp 31–43 in: Proceedings of a workshop held 14–5 November 1989 at the Institute of Agricultural Engineering, Borrowdale, Harare, Zimbabwe.

Animal traction and market conditions: a case study from south-western Tanzania and northern Zambia

by

Torben Birch-Thomsen

Institute of Geography, University of Copenhagen, Oster Voldgade 1, 1350 Copenhagen, Denmark

Abstract

Historically the adoption pattern of animal traction in Rukwa, Tanzania, and Northern Province, Zambia, has been closely linked to the history of maize cultivation. In particular the introduction of improved maize varieties along with government subsidised fertilisers and pesticides in the mid-1970s increased the comparative advantage of animal traction, especially for plowing. Permanent cultivation became possible without decreasing yields. The removal of input subsidies and the national pricing system, as part of the structural adjustment programmes implemented in the late 1980s and early 1990s, affected the adoption pattern of animal traction. Studies were undertaken in two villages between 1990 and 1993, and found that the effect varied depending on the degree of animal traction adopted prior to the structural adjustment programme. Factors such as intensity of cultivation, population density and degree of market integration are shown to be of great importance.

Introduction

Like most of sub-Saharan Africa the agricultural systems represented in the south-west of Tanzania and northern Zambia have experienced major changes within the last five decades. There have been many different forces causing the changes and their effects have been many, and have varied in time and place. Typically, the process of change within the agricultural systems in response to changed production conditions includes an intensification, in terms of increased frequency of cultivation, and adoption of innovations like new cultivars, cultivation techniques and/or new implements (Boserup, 1965; Ruthenberg, 1980; Pingali, Bigot and Binswanger, 1987). In trying to explain the process of change, some writers have based their analysis on theories relating production to household needs and wants (usually population driven) under conditions of 'subsistence' (Chayanov, 1966; Boserup, 1965, 1981). Others

have based their analysis on theories that relate production to demands from the market (Hayami and Ruttan, 1985). The main impetus for change in both types of theory derives from the pressure of increasing scarcity, particularly land availability.

In accordance with recent works by other writers (Goldman 1993; Tiffen and Mortimore, 1994), the point of view in this paper is that intensification and adoption of innovations are not merely driven by scarcities and constraints, but equally by opportunities and advantages. The latter is often as significant as constraints where good market access exists, through a cash crop infrastructure to facilitate both supply of inputs and outlets of products, leading to an expansion of economic opportunities.

Although perhaps simplifying the diversity within sub-Saharan agricultural systems, a scenario of change is given in Figure 1, and the possible implications for resource management are listed in the boxes. One of the important factors restricting the generality of the scenario is the variation in agro-climate, especially in relation to the adoption af animal traction (McIntire, 1992). Furthermore, to reduce the complexity, no specific comments are made on the importance of animal transport, but, as stressed by Starkey (1994), it can be crucial for the profitability of animal traction.

An important distinction between forces causing the process of change and intensification, in relation to the adoption of animal traction in the present case study, relates to what Lele and Stone (1990) refer to as "autonomous intensification" and "policy-led intensification". The former refers to the Boserupian process of intensification and adoption of innovations: changes evolve over a realtively long period of time in response to increasing population pressure through experimentation, adaptation and adoption of new techniques within the indigenous traditional

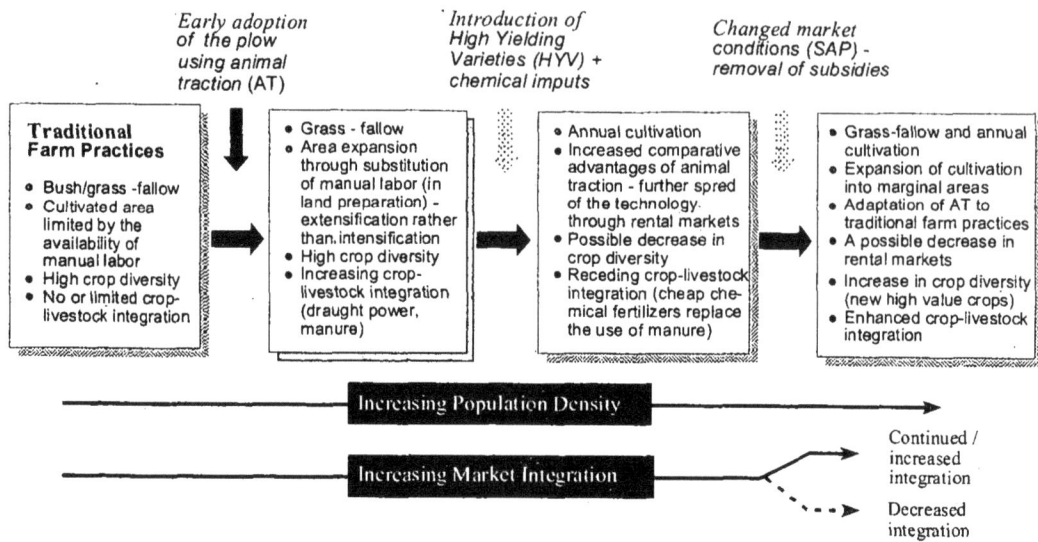

Figure 1: A possible scenario of changes within agricultural systems in sub-Saharan Africa

knowledge system (Swift, 1979). The latter refers to the effect of government policies which may either act by widening or restricting the farmer's sphere of activity: widened where opportunities are created, for example through subsidies and/or improved infrastructure; restricted due to factor scarcity, for example through changes in land laws or tenure. Changes as a result of alterations in national policies may occur more or less 'over-night'.

The objective of this paper is to illustrate how the adoption pattern of animal traction, especially related to field operations, may vary according to the significance of the different forces driving the process of change. Furthermore, I aim to portray the impact of recent alterations in government policies, in this case the effect of changing marketing conditions as a consequence of the Structural Adjustment Programme on the use and dissemination of animal traction.

Field data and study area

The paper is based on three periods of fieldwork between 1990–93 in the south-western part of Tanzania, Rukwa Region, and north-eastern Zambia, Northern Province. Firstly, a household survey was carried out in Ulinji Village, Sumbawanga District, Tanzania, and David Chikoti Village, Mbala District, Zambia in 1990. It involved 31 and 30 households, respectively.

Secondly, both villages were revisited in 1991 (Birch-Thomsen, 1993). Thirdly, minor fieldwork was carried out in Ulinji Village in 1993.

Ulinji Village is located on the Fipa Plateau in Sumbawanga District, on the main road approximately 15 km south-east of the district and regional centre, Sumbawanga. The mean annual rainfall is between 803–960 mm and the natural vegetation is dominated by grassland with scattered trees and bushes. David Chikoti Village is located on the Mambwe-Mwenzo Dissected Plain directly south of Ulinji across the border to Zambia. During the dry season the district centre, Mbala, can be reached by car along the gravel road in approximately 1.5–2 hours. No public transportation service is available. The average annual rainfall is between 1000–1100 mm, and the vegetation varies from grass/bush fallow close to the settlement area to dense *miombo* woodland in remoter areas.

In addition to the difference in location, the two villages vary in several important respects (Table 1). First of all, bearing the difference in location in mind, it is not surprising that the highest population density is found in Ulinji, but the difference becomes even more pronounced when calculated on the basis of the actual land area possessed by farmers. Part of the explanation is that large parts (36%) of the village area in

Table 1: Population and cultivation differences between Ulinji and David Chikoti villages based on surveys of 31 and 30 households, respectively

| | Population density (people/km^2) | | | | | | |
Study site	Per total area	Per cultivable area	Intensity of cultivation (R-value)[1]	% area planted with maize	Maize sales as % of total production	% farmers using animal traction for plowing	% of cultivated area plowed with oxen
Ulinji	47	310	86	61	39	87	86
Chikoti	28	61	53	43	87	53	51

1) R-value = cultivated area/(cultivated + fallow area)*100

Ulinji are non-cultivable due to rock outcrops and steep slopes. Furthermore, supporting the assumptions made by Boserup, the intensity of cultivation (measured by the R-value; Ruthenberg, 1980) reflects the difference in population density, with permanent cultivation in Ulinji and a fallow system in David Chikoti. However, this broad calculation conceals variations within the villages in crops grown, techniques used in land preparation, and the location of the fields both on the topo-sequence and distance from homestead. This variation is by far most evident in David Chikoti where the majority of farmers follow a dual-farming strategy closely related to specific crops: permanent cultivation of maize close to the homesteads, and a fallow rotation system (a grass-mound-system; Stromgaard, 1988) of finger millet, beans and cassava on more distant fields. For a further discussion on the measures of density and intensification, see Kates et al (1993). In terms of area cultivated, maize was the dominant crop in both villages in 1990. But, whereas in Ulinji maize was both a cash and food crop, it was almost entirely a cash crop in David Chikoti. Finally, the general assumption about changes in cultivation techniques with changing intensity of rotation (Ruthenberg, 1980; Pingali, Bigot and Binswanger, 1987) seems to be supported by this study. However, an important point to emphasise is that the percentages of upland area plowed using oxen (86% in Ulinji and 51% in David Chikoti) conceal a variation in the pattern of use of animal traction in each village. This variation is to some extent linked to the discussion on intensity mentioned

above. While 85% of the farmers using animal traction for plowing in Ulinji plowed all their fields, only one farmer did so in David Chikoti. Moreover, half of the farmers in David Chikoti plowing with animals cultivated between 20–85% of their holdings using manual labour–typically in the traditional grass-mound-system. The dual strategy of farming was both crop and technology related.

Factors affecting the adoption pattern of animal traction in 1990

In order to understand the dissemination and adoption pattern of animal traction in the study area, it is necessary to look at the changes which have occurred in the past four decades. For that purpose, a brief description of the transformation, its major causes, and its effects on cultivation techniques and cropping pattern, will be given with reference to Figure 2.

In the 1950s and early 60s, agriculture in the two study villages was dominated by the traditional farming systems prevailing among the Fipa in southern Tanzania and the Aisa Mambwe in northern Zambia. The grass-mound-system was practised both in Ulinji and David Chikoti, but, in addition, farmers in David Chikoti practised the slash-and-burn system called *chitemene* (Pottier, 1988; Stromgaard, 1989). During this period the first introduction of animal traction took place in both villages, and in both cases it was introduced by a single farmer. The process of change during the early adoption of animal traction refers to the autonomous intensification described earlier in this paper. Given the relatively low population density,

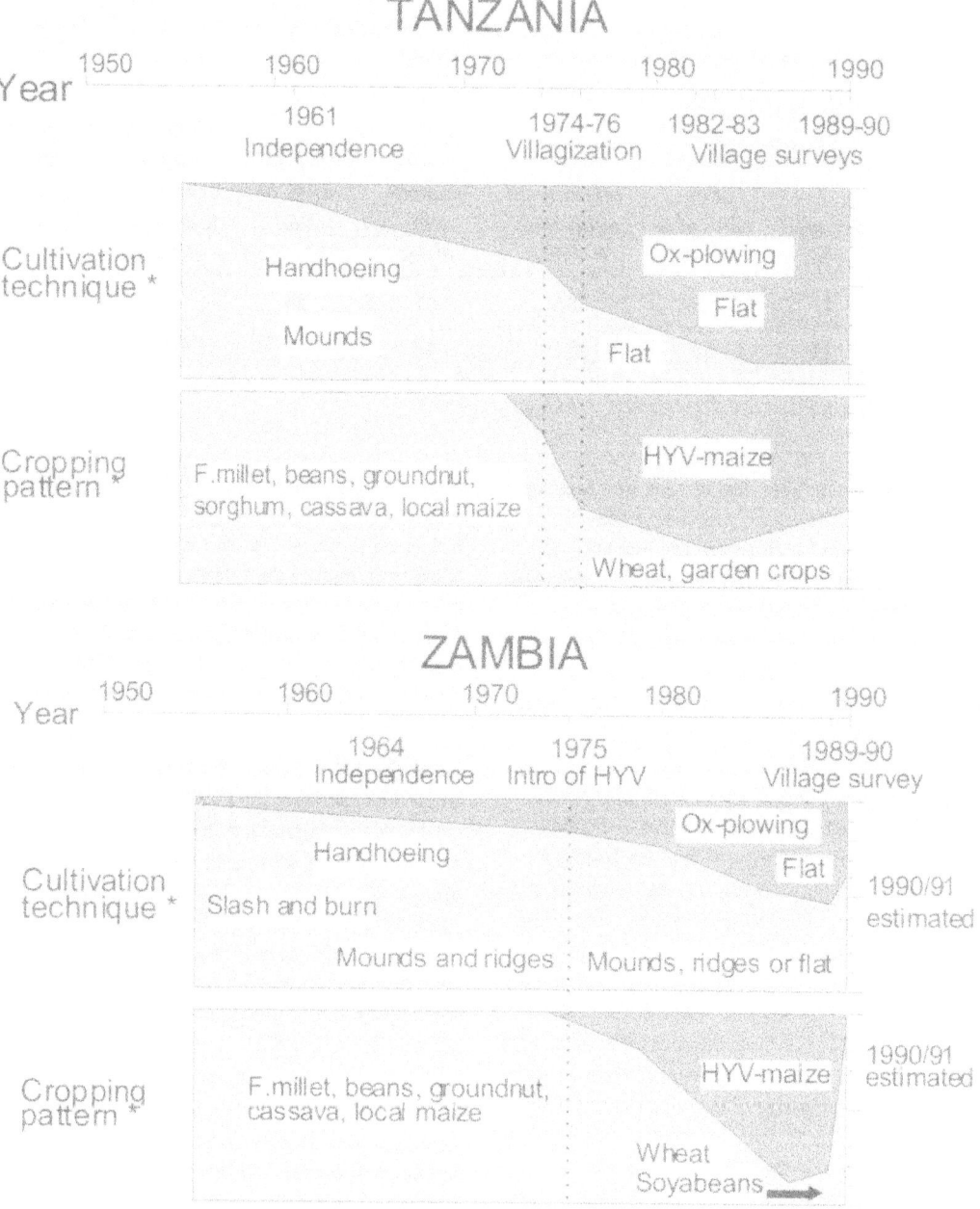

* The height of each box represents 100 % of farmers
  (sample size 31 and 30 households respectively)

Figure 2: Changes in cultivation techniques and cropping patterns in the two villages

the substitution of manual labour by animals enabled farmers adopting this new technology to expand their land under cultivation. The dissemination of ox-plowing in the two villages within this period reflects the general difference in population density and intensity of rotation. By the mid 1970s approximately 40% of the farmers in Ulinji and 20% in David Chikoti were using animal traction for plowing, with little effect on the general cropping pattern.

Between 1974 and 1976 a major change in land use and tenure occurred in the Tanzanian village as a consequence of the national re-settlement programme. People were moved into a nucleated settlement pattern, and the traditional tenure system was abolished. During this period the adoption of animal traction accelerated, and by 1976 approximately 60% of the farmers used the technology when plowing their fields. The technology continued to spread during the late 1970s and early 1980s, not just in terms of farmers owning plows and trained oxen but also induced by the emerging rental market (used by 26% of the farmers). As observed elsewhere in Tanzania (Kjærby, 1989) this increase occurred despite the government policy supporting tractorisation. Restrictive policies on land tenure and labour (reallocation of land, communal fields and work) increased the factor scarcity and added to the effect of the general population increase. This is believed to be part of the explanation for the increased adoption of animal traction. In addition, the change towards a more intensive land use pattern, caused by villagisation, was enhanced through the National Maize Project which started in 1974–75 and included the introduction of subsidised hybrid maize seeds, fertilisers and pesticides. With the use of these inputs, permanent cultivation with high outputs became possible. While reducing the importance of the previous traditional practice of fallow, it increased the comparative advantage of ox-plowing. Furnished with the heavily subsidised input prices and the national pricing system, the production of maize for the national market became economically feasible even in remote regions such as Rukwa (Raikes, 1988). Despite the problem of irregular deliveries of inputs, market integration increased rapidly during the 1980s.

The picture is somewhat different in David Chikoti. Despite the early introduction of animal traction in the mid-1950s, the dissemination evolved at a slower speed than in Ulinji. Two main explanations are evident. Firstly, the dominance of extensive farm practices, which left stumps of trees in fields, acted as a constraint to mechanisation. Secondly, the introduction of HYV-maize and the establishment of a marketing system did not become efficient until 1982 (Sano, 1989). As a result of the government subsidies related to maize production, and the deliveries of seeds and fertilisers, as well as collection of harvest through the Northern Cooperative Union, permanent cultivation became possible. Farmers started to clear large fields close to the village. Because of the labour-saving capability of ox-plowing in land preparation, the dissemination of the technology increased in close relation to the adoption of the permanent cultivation of maize. Maize and ox-plowing became a new way of cultivating fields in parallel to the traditional farming system.

In their conclusions Pingali, Bigot and Binswanger (1987) state that high yielding varieties and fertilisers are not a precondition for mechanisation, nor is mechanisation a precondition for the adoption of high yielding varieties and fertilisers. Whereas the latter has been found valid in the case from David Chikoti, these findings indicate a positive relation between the income-generating maize cultivation and ox-plowing in both villages. Bearing this in mind, the outcome of adopting animal traction in the 1980s was an intensification as well as an expansion of the area cultivated. In Ulinji, with an increasing scarcity of cultivable land, only a very moderate expansion took place. On the other hand, farmers adopting the new technology in David Chikoti were able both to expand and intensify their cultivation. On average, farmers using oxen cultivated twice as much land as hoe-cultivators - an average of 7 ha compared to 3.6 ha. The same trend was evident when comparing the size of the individual fields. Furthermore, although extensification rather than intensification is the norm when animal traction technology is new (Starkey, 1992), the close relation between maize cultivation, including the use of fertilisers, and animal traction in David Chikoti led to an intensification.

To sum up the changes prior to 1990, the transformation of the agricultural systems in Ulinji was brought about by an increasing factor scarcity. In addition, farmers responded to the new opportunities related to the market conditions. On the other hand, despite the lower population density in David Chikoti the intensification and adoption of animal traction took place in response to the 'artificially' improved access to the market, with farmers expanding the areas given to the marketed crop, maize. The high dependency on the government policies, in terms of marketing, increased the vulnerability of farming in David Chikoti.

Changed market conditions in the 1990s and the effect on animal traction

By the late 1980s and early 1990s farmers in both villages were facing major changes in market conditions. In Tanzania the period of structural adjustment was started by the agreement between the government and the World Bank on the Economic Recovery Programme of 1986–89. But it was not until the implementation of the Economic and Social Action Programme (1990–93) that the liberalisation within agriculture really took off. The removal of the national pricing system, a general abolition of subsidies, and legalisation of the negotiation of prices between private dealers and farmers are some of the main changes that affected farmers, especially in a remote region such as Rukwa. In the case of Northern Province, the failure of the Northern Cooperative Union to collect the harvest from the crop-season 1989/90 and to pay farmers and deliver seeds and fertilisers caused a breakdown in the marketing system. Large amounts of the uncollected maize had been burned by the time of the second field work in 1991. Furthermore, farmers experienced a declining maize economy because of the removal of subsidies.

As illustrated by the data from the second visits to Ulinji (1993) and David Chikoti (1991), farmers reacted rapidly to these policy-induced changes, causing an immense impact on land use, cropping patterns, and the use and dissemination of animal traction in both locations.

Ulinji

During the fieldwork in July 1993, major changes in land use were observed. Firstly, cultivation had expanded into an area not previously identified as part of the village area. In addition, a number of new fields had been opened for cultivation in an area previously reserved for grazing (Birch-Thomsen & Fog, 1996). The virgin grassland had been tilled at the end of the rainy season (April–May) preparing it for cultivation in November/December. The previously common technique of making mounds was observed, but most of the new fields were plowed using animal traction – an adoption of the 'new', and labour-saving, technology into the traditional practice of farming widely used prior to the villagisation process. During a group meeting it was explained that some farmers had shifted their cultivation to the old grazing area and left the fields close to their homesteads for fallow. These fallow fields were either used for grazing or planted with eucalyptus trees – the availability of firewood and building poles had decreased dramatically. Second, there had been a change in the cropping pattern within the village. Wheat had been introduced as a new cash crop at the expense of maize, though the latter continued to be an important staple food crop. In addition, farmers stated that the area cultivated with beans, sunflower and finger millet had increased. In the case of finger millet, this was especially true on the newly opened fields. Thirdly, garden cultivation along small streams and depressions had been expanded further.

David Chikoti

The dramatic drop in maize marketing observed during the fieldwork in 1991 forced farmers to increase the cultivation of more distant areas. Farmers who had been marginally involved in HYV-maize production returned to the cultivation of the traditional crops finger millet and beans. The regrowth of vegetation in these parts of the village area favoured the traditional grass-mound-system and even lopping of branches (*chitemene*). Farmers previously hiring other farmers to plow their permanent fields (44% of farmers using animal traction) stopped using this practice. On the other hand, some farmers owning one or more ox-teams were actually observed to use ox-plowing in the traditional system. After approximately 15 years of experience with ox-plowing, these farmers were reluctant to give up the labour-saving technology. They adopted animal traction into the traditional system by substituting the previous hoed mounds with winter plowing in between trees and bushes. In David Chikoti, the survey revealed a reversal in the development of the farming systems, contrary to general theories on the evolution of farming systems. Because the conditions affecting maize cultivation have been breaking down, with diminishing economic returns to farmers and insufficient supply systems, farmers are returning to more extensive cultivation practices.

Acknowledgments

The collection of field data has been made possible through the joint project 'Agricultural and ecological consequences of deforestation and afforestation in Tanzania and Zambia based on satellite remote sensing' undertaken by the Institute of Geography (University of Copenhagen), Institute of Resource Assessment (University of Dar es Salaam) and Department of Botany and Department of Geography (University of Zambia). The project has been funded by DANIDA under the Programme for Enhancement of Research Capacity in Developing Countries (ENRECA).

References

Birch-Thomsen T, 1993. Effects of land use intensification - introduction of animal traction into farming systems of different intensities. Part I-III, p352 + appendices. *Geografica Hafniensia* A2, Publications - Institute of Geography, University of Copenhagen, Denmark.

Birch-Thomsen T and Fog B, 1996. Changes within small-scale agriculture in Tanzania - implications for the future. *Danish Journal of Geography*, Institute of Geography, University of Copenhagen, Denmark (forthcoming).

Boserup E, 1965. *The conditions of agricultural growth: the economics of agrarian change under population pressure.* Aldine, London, UK.

Boserup E, 1981. *Population and technological change - a study of longterm trends.* The University of Chicago Press, USA.

Chayanov A V, 1966. 'Peasant Farm Organazation. In: Thorner D et al (eds), *A V Chayanov on the Theory of Peasant Economy.* R D Irwin, Homewood, Illinois, USA.

Goldman A, 1993. Agricultural innovation in three areas of Kenya: Neo-Boserupian theories and regional characterization. *Economic Geography* 69(1): 44–71.

Hayami Y and Ruttan V, 1985. *Agricultural development: an international perspective.* Johns Hopkins University Press, Baltimore, USA.

Kates R W et al, 1993. Theory, Evidence, Study Design. pp. 1–40 in: Turner II B L, Hyden G and Kates R W (eds), *Population growth and agricultural change in Africa.* University Press of Florida, Florida, USA. 461p.

Kjærby F, 1989. Villagization and the crisis - agricultural production in Hanang District, northern Tanzania. *CDR*

Project Paper 89.2, Centre for Development Research, Copenhagen, Denmark.

Lele U and Stone S W, 1990. Population pressure, the environment and agricultural intensification - variations on the Boserup hupothesis. *MADIA Discussion Paper 4.* The World Bank, Washington DC, USA.

McIntire J, 1992. Evolution of farming systems and what it means for animal traction research in Africa. pp 83–87 in: den Hertog G and van Huis J A (eds), *The role of draught animal technology in rural development,* Proceedings of the International Seminar held at Edinburgh, Scotland 2–12 April 1990. Pudoc Scientific Publishers, Wageningen, The Netherlands.

Pingali P, Bigot Y and Binswanger H P, 1987. *Agricultural mechanization and the evolution of farming systems in Sub-Saharan Africa.* World Bank, The Johns Hopkins University Press, London, UK.

Pottier J, 1988. *Migrants no more - settlement and survival in Mambwe villages, Zambia.* International African Institute, Manchester University Press, London, UK.

Raikes P, 1988. *Modernising Hunger.* Heinemann, Portsmouth, UK.

Ruthenberg H, 1980. *Farming systems in the tropics.* Clarendon Press, Oxford, UK.

Sano H O, 1989. From labour reserve to maize reserve: the maize boom in the Northern Province in Zambia. *CDR Working Paper 89.3.* Centre for Development Research, Copenhagen, Denmark.

Starkey P H, 1992. Changes in animal traction in Africa and Asia: implications for development. pp. 11–22 in: den Hertog G and van Huis J A (eds), *The role of draught animal technology in rural development.* Proceedings of the International Seminar held at Edingburgh, Scotland, 2–12 April 1990. Pudoc Scientific Publishers, Wageningen, The Netherlands.

Starkey P H, 1994. A world-wide view of animal traction highlighting some key issues in eastern and southern Africa. pp. 66–81 in: Starkey P, Mwenya E. and Stares J (eds), *Improving animal traction technology.* Proceedings of the first workshop of the Animal Traction Network for Eastern and Southern Africa (ATNESA) workshop held 18–23 January 1992, Lusaka, Zambia. Technical Centre for Agriculture and Rural Cooperation (CTA), Ede-Wageningen, The Netherlands. 490p.

Stromgaard P, 1988. The grassland mound-system of the Aisa Mambwe of Zambia. *Tools and Tillage* VI(1): 33-46. National Museum of Denmark, Copenhagen, Denmark.

Stromgaard P, 1989. Adaptive strategies in the breakdown of shifting cultivation: The Case of Mambwe, Lamba, and Lala of northern Zambia. *Human Ecology* 17(4): 427–444. Plenum Press, New York, USA.

Swift J, 1979. Notes on traditional knowledge, modern knowledge and rural development. *IDS Bulletin* 10(2): 41–43.

Tiffen M and Mortimore M, 1994. Malthus controverted: the role of capital and technology in growth and environment recovery in Kenya. *World Development* 22(7): 997–1010.

Draft animal power potential and utilisation in the Tonota District of Botswana

by

J A Karim-Sesay

Department of Agricultural Engineering and Land Planning
Botswana College of Agriculture, Private Bag 0027, Gaborone, Botswana

Abstract

This paper examines the status of draft power availability for crop production on traditional farms in the Tonota Agricultural District of Botswana with particular reference to draft animals. The results of a survey of 9 extension areas and secondary data from station records showed that 36% of the farmers owned donkeys while 17% and 2% owned cattle and tractors, respectively. Another 17% obtained draft power through hiring: donkeys (7%), cattle (3%), and tractors (7%). A substantial proportion (28%) of the farmers used other sources of power or a combination of various available resources.

Out of a total of 12,000 hectares plowed and planted in the 1990/91 cropping season 69% was plowed by animal traction, 25% by tractors and 6% by unspecified sources. However, in spite of the large potential power available on the farm only 37% of the available arable land was cultivated. Tractors constituted only 9% of the power resources but contributed 25% to the power needs in the district for crop production.

It was generally concluded that the available power sources are currently either unutilised or grossly under-utilised. Furthermore, the potential use of the fresh dung from draft animals as a source of renewable energy for fuel and manure as done in other parts of the world has not been exploited. The amount of dung available could produce substantial quantities of biogas or dung-cakes for fuel.

Introduction

Draft animals are the main sources of farm power in Botswana, especially for small-scale farmers. The animals are used mainly for field operations such as plowing, harrowing, planting and weed control. The major off-farm activity the animals are used for is transportation of water, farm produce, firewood and the farm family. For instance Baker (1988) reported that in the Shoshong and Makwate Districts of Botswana,

66% of animal traction hours were used for transportation during the 1983–84 season. In general, animals, especially donkeys, are used throughout the year for these off-field activities whereas tractors are used mainly for 3–4 months of the year, making them a grossly underutilised source of farm power.

Draft animals as potential sources of fuel and manure

Draft animals are also a valuable source of renewable energy as a potential source of biogas and manure from their wastes. Sahu (1986) reported that in the North-East Hill Region of India a total fresh dung output of 8134 tonnes yielded about 309×10^6 litres of biogas (927×10^6 kcal) or 2036 tonnes of dung-cakes (479.5×1^{06} kcal) per day. The quantity of manure available per day came to about 1952 tonnes. However, in Botswana, this potential source of fuel and manure is currently not being utilised. Development work being carried out by the Rural Industries Innovation Centre (RIIC) in Kanye on biogas technology may address this problem in future.

Current use of draft animals in Tonota District

Over 80% of the population in the Tonota Agricultural District are engaged in farming operations of one sort or another. The traditional system of farming, like in the rest of Botswana, is based on animal traction. Recent surveys (Chipo, 1992) have shown that over 80% of the farmers in Tonota still use animal traction for growing their crops. Animal traction contributesg 89% of the power needs for crop production in the district. Consequently the proportion of land area actually planted to available land is small. For instance Chipo (1992) reported that out of a total of 32,000 ha of de-stumped land available for crop production in the district during the 1990/91 period only 37% was actually plowed and planted.

Tractor mechanisation

Tonota Agricultural District lies in the north-eastern part of the country where the predominantly clay soils have good potential for a fully mechanised farming system. However, the high purchase and hiring cost of tractors and the low levels of returns on investment because of low yields have tended to retard the development of tractor mechanisation. This is true not only in this region but throughout the country. The major contributing factor to the low yield is the low, poorly distributed and generally unreliable rainfall.

Materials and methods

Data were collected from the records of the Agricultural Demonstrators and surveys conducted in the District during the 1991/92 and 1992/93 cropping seasons. Ten farmers were randomly selected from each of the 9 'extension areas' within the District. Information was sought on ownership and use of draft resources and their contribution to the area of land cultivated. The total available power in the district was estimated using values previously quoted by Kemp (1987).

Results

Ownership and usage of draft power sources

The survey on draft power ownership and usage, presented in Tables 1 and 2, shows that approximately 63% of the farmers in the district owned or hired animal draft power (cattle 20%, donkeys 43%) for cultivating their fields. About 17% of farmers owned cattle while the remaining 3% obtained cattle draft power through hiring. A high proportion of the farmers (36%) owned and used donkey draft while 7% of them hired donkey power from local farmers. Another 9% of the farmers used tractor draft even though only 2% of the farming population owned tractors. A substantial proportion (29%) of the farmers used other sources or a combination of various resources.

The Matshelagabedi extension area had the lowest number of farmers who owned tractors (0.4%) followed by Matsiloge (1%). More farmers used their own tractors in the Mabesekwa area

Table 1: Ownership and use of draft power resources in Tonota District

		Percentage of households using the following power sources						
		Cattle		Donkeys		Tractors		
Extension area	No. households	Own	Hire	Own	Hire	Own	Hire	Other
Matsiloge	400	8	0	80	8	1	3	2
Mabesekwa	250	30	13	17	0	5	11	24
Matshelagaberdi	260	15	0	31	8	<1	4	42
Tonota West	550	28	4	27	9	2	5	25
Tati Siding	380	21	5	26	11	2	7	28
Shashe	650	15	3	33	6	1	14	27
Thalogang	500	18	1	40	10	2	6	23
Tonota East	435	13	2	29	7	2	2	44
Shashe Mooke	377	3	1	38	3	1	9	45
Means	422	17	3	36	7	2	7	29
Overall means	422	20		42		9		29

Table 2: Available draft power in Tonota District

Extension area	Total power available (kW)	Cattle		Donkeys		Tractors		Power/unit of cropped area (kW/ha)
		Power (kW)	% of total	Power (kW)	% of total	Power (kW)	% of total	
Matsiloge	552	240	44	211	38	101	18	0.6
Mabesekwa	580	175	30	0	0	405	70	0.4
Matshelagaberdi	310	52	17	224	72	34	11	0.4
Tonota West	1207	385	32	484	40	338	28	0.6
Tati Siding	732	228	31	268	37	236	32	0.7
Shashe	954	260	27	458	48	236	25	0.8
Thalogang	834	300	36	264	32	270	32	0.3
Tonota East	862	218	25	306	36	338	39	1.0
Shashe Mooke	551	151	27	265	48	135	25	0.7
Total	6582		30		39		31	0.6

Notes: Calculated on the assumptions that cattle can develop 0.5 kW but only 10% are available for work, donkeys can develop 0.4 kW and 70% are available for work whereas tractors develop an average of 40 kW and 75% are in use.

(5%) than the other areas. Shashe, with a farming population of 650 households showed the highest number of people (14%) who hired tractor power for plowing purposes.

Current status of draft power

The estimated draft power resources in the district are shown in Table 2. A total of 6,582 kW of power are available for crop production with an average of 730±270 kW per extension area. Sixty-nine percent of the known sources of power are provided by draft animals (30% from cattle and 39% from donkeys). Tractors contribute 31% to the total available power in spite of the fact that only 2% of the farming population own tractors. A maximum of 89% of the power available in Matshelagabedi was obtained from animals while Mabesekwa obtained 70% from tractors.

Tonota West had the highest amount of available power followed by Shashe while the lowest was from Matshelagabedi. The average draft power available per cropped area was 0.6± 0.2 kW/ha.

Tonota East had the highest available power per cropped area (1.0 kW/ha) while Thalogang had the least (0.3 kW/ha).

Contribution of draft sources to cultivated areas

A total of 11,900 hectares, constituting 37% of the available arable land area, was cultivated during the period under review (Table 3). Draft animal power was used for cultivating 69% of the land area (23% by cattle and 46% by donkeys) while tractors cultivated 25% of the area. The remaining 6% was cultivated by other sources including hand tools and a combination of tractors and animals.

In the Matsiloge extension area 79% of the land was cultivated using donkey power with 8% and 11% being contributed by cattle and tractors, respectively. Tractor power contributed 51% to the land area cultivated in the Mabesekwa extension area whereas cattle contribution was highest (40%) in the Thalogang area.

Table 3: Contribution of draft power sources to area cultivated, by extension area

Extension area	Cultivated area (ha)	% of total area	Percentage of area cultivated with:			
			Cattle	Donkeys	Tractors	Others
Matsiloge	870	25	8	79	11	1
Mabesekwa	1560	43	32	13	51	5
Matshelagabedi	840	31	24	56	12	8
Tonota West	2150	47	12	61	2	4
Tati Siding	1000	27	30	50	316	4
Shashe	1250	36	20	44	29	7
Thalogang	2500	37	40	40	14	6
Tonota East	900	30	16	35	38	12
Shashe Mooke	820	26	7	49	34	10
Mean per extension area	1320	37	23	46	25	6

Potential of draft animals as sources of energy for fuel and manure

The estimates for fresh dung produced by draft animals and its potential energy values are presented in Table 4. A total of 59,400 tonnes of fresh dung were produced which yielded 230×10^6 litres of biogas with a heating value of 680×10^3 kcal per day. The dung cake that could be produced amounted to 14,800 tonnes with a heating value of 390×10^3 kcal per day. The quantity of manure to be obtained per day amounted to 14,300 tonnes. There were large variations in the total amount of fresh dung produced in extension areas; the mean production was 6,600 tonnes with a standard deviation of 2,800 tonnes per day.

Discussion

The ownership and use of tractors is at a low level in Tonota and throughout the country and will continue to be so for quite a while. This is largely due to the high purchase price and running cost which most rural farmers cannot afford. The tractor is also considered uneconomical because of the low yields obtained as a result of the low and unreliable rainfall patterns (MoA, 1991). However, due to the drought relief programme in which the government pays for plowing, planting and support services more farmers are now hiring tractor for cultivating their lands. For instance in Shashe,

14% of the farmers use tractors even though only 1% of them own tractors.

The use of cattle for draft purposes will continue to drop because of the animals' susceptibility to drought and the high prices the Botswana Meat Commission is now offering for high quality meat. Farmers are turning to donkeys and tractors for plowing their fields. The power available per unit area of cropped land is quite high (0.6 kW/ha) as compared to the 0.4 kW/ha suggested by Giles (1975) for optimum crop production. This may be misleading as only 37% of the land is cultivated. The actual power for cultivating the available land area averages about 0.2 ± 0.06 kW/ha per year, the highest (0.3 kW/ha) being observed in the Tonota East extension area.

The potential of the animals as sources of renewable energy and manure is enormous especially when the non-draft animals are included in the calculation. The large amount of fresh dung and its corresponding heat value serve as potential sources of domestic heat for cooking and other activities in the home. The dung-cake is a vital source of heat in India and other parts of the world where conventional fuel is scarce or where the continued use of wood fuel poses a threat to the environment. This could be an important alternative to the use of wood in the rural areas of Botswana where the vegetation is under constant

Table 4: Potential alternative energy sources from draft animals in Tonota District

	Fresh dung[1] (t/day)	Energy through biogas/day[2]		Energy through dungcake/day[2]		Manure (t/day)
		Biogas (x10³ l)	Useful heat (x10³ kcal)	Dung cake (t/day)	Useful heat (x10³ kcal)	
Matsiloge	7100	270	809	1800	417	1700
Mabesekwa	5200	197	590	1300	304	1200
Matshelagaberdi	1500	59	175	400	90	400
Tonota West	11400	433	1300	2800	669	2700
Tati Siding	6700	256	769	1700	396	1600
Shashe	7700	292	877	1900	449	1800
Tlialogang	8900	337	1000	2200	521	2100
Tonota East	6400	244	733	1600	378	1500
Shashe Mooke	4500	170	509	1100	262	1100
Total	59,400	2258	6762	14,800	3488	14,300
Average/extension area	6600	251	753	1600	388	1600

Notes:
1) Calculated on the basis that cattle with an average size of 200 kg produce 5400 kg of dung per year, while a donkey weighing 150 kg produces 2700 kg/year (Stout, 1979)
2) 1 tonne of fresh dung yields 38×10^3 litres of biogas and 250 kg of dung-cake. One litre of biogas yields 3 kcal of useful heat while 1kg of dung-cake gives 235 kcal of useful heat (Pandya, 1980)

threat of drought, overgrazing by domestic animals and damage by wildlife such as elephants.

Conclusions

Although a large amount of power exists in the Tonota District for draft purposes, only a small proportion of the total land area is cultivated because of the unfavourable climatic conditions which restrict the plowing activities to a very short period, and the poor economic status of the rural farmers which prevents them from purchasing tractors to address this shortfall. However, as tractors are considered inefficient because of the high operating costs due to limited annual use, an appropriate solution would be to improve the animal traction technology through the development of appropriate tools and the selected use of power machine systems. As the continued use of firewood for fuel is detrimental to the already fragile environment, it is necessary to develop technologies that fully utilise the vast potential of draft and non-draft animals as sources of renewable energy, not only for the farm but for domestic use as well. To meet the future power needs of the farmers in the district requires an efficient use of the presently available power from draft animals, improved farm implements and a well organised tractor hiring service.

References

MoA, 1991. The 1990 agricultural statistics. Central statistics office, PO Box 0024, Gabarone, Botswana.

Pandya A C, 1980. Energy for agriculture in India. Technical Bulletin, Central Institute of Agricultural Engineering, Bhopal, India.

Sahu S D, 1986. Draught animal potential in North-Eastern Hill Region of India. Agricultural Mechanisation in Asia and Latin America, 17 (4): 29–31.

Stout B A, 1979. Energy for world agriculture. Food and Agriculture Organisation of the United Nations (FAO), Rome, Italy. 286p.

Users in control: farmer participation in technology research and development

by

Simon Croxton*

*International Programme Manager, Food Production
Intermediate Technology Development Group, Myson House, Railway Terrace, Rugby CV21 3HT, UK*

Abstract

This paper discusses approaches to technology development. It starts from the premise that animal traction technologies have broad similarities with most other rural technologies, in terms of the constraints and possibilities for development. Fundamental to the argument is that more conventional approaches to technology development do not have a history of promoting widespread innovations. An alternative approach that attempts to minimise the control of professionals and other external agents, and maximise the control of manufacturers and end users, is critical for successful technology development.

Definitive methodologies are not themselves the key to developing more participative ways of working. A key factor is the attitude of professionals, and this needs to be explicitly recognised if more participative approaches to technology development are to be put into practice. The software aspects of technology (skills, knowledge and forms of social organisation) need to be given at least equal weight to the hardware (techniques and equipment). Formal education rarely provides professionals with the skills they require to work in this way.

The paper uses a case study from Sudan to illustrate the discussion. It concludes that resources should be directed more at ways of enabling participative approaches to develop than at research controlled by professionals.

Introduction

Research into, and development of, technologies suited to small farmers in rural areas of developing countries is no easy matter. History is not littered in success stories. If it were, a paper such as this would not be of interest. This does not

apply ony to animal traction technologies, but to all areas of rural technology.

This paper examines approaches to technology development. It starts from the above assertion, and also argues that animal traction technologies are not a special case but are just one of a range of technical areas of interest to many small-scale farmers. The issues facing professionals working in this particular field are similar to those associated with other technologies. Therefore, it is useful to look at the possibilities and constraints facing research and development of animal traction technologies from a broad perspective. What lessons can be learned at a general level and how might these be incorporated into practical work in a particular technology?

To ground the discussion in practical experience, the paper illustrates some of the major points with an example drawn from ITDG's work with farmers and blacksmiths in western Sudan. This project has been chosen, not just because it illustrates an alternative approach to technology development, but also because it has been judged successful by external evaluators (Abu Sin and Hadra, 1994) in terms of supporting a major technical innovation—the use of donkey plows in an area where they were never used before.

What do participatory approaches offer technology research and development?

There is a growing recognition that conventional approaches to developing rural technologies have not delivered the goods. This recognition has developed into a well argued critique of the technology transfer model, and considerable interest in, and research into, alternatives based on a participatory approach. These participatory ideas and the associated rhetoric form a major component of what has been described as a new development paradigm (Jamieson, 1987;

---
* *Subsequent address:* Simon P Croxton
Sustainable Agriculture Programme
International Institute for Environment and Development
(IIED), 3 Endsleigh Street, London WC1H ODD, UK

Chambers, 1994). There is now a body of research and information on practical field experience that provides considerable documentation on farmer participation in technology research and development (see, for example, Haverkort, van der Kamp and Waters-Bayer, 1991; Hiemstra, Reijntjes and van der Werf, 1992; Croxton and Appleton, 1994; Okali, Sumberg and Farrington, 1994; Scoones and Thompson, 1994). This paper uses these arguments as a starting point to look at how best to translate theory into practice relevant to the development of animal traction technology.

The basic tenet of these arguments is that users need to be involved at all stages of technology development. Conventional approaches have relied too strongly on researchers and technical specialists identifying constraints and possible solutions, and then attempting to transfer them to rural settings. Local skills and knowledge are frequently not recognised and certainly rarely included, in this process which is managed and controlled by outsiders. In contrast, a participative approach uses existing local skills and knowledge as a starting point, and is built around a process that enables users (eg farmers) to control and direct research and development of technologies that meet needs prioritised by farmers themselves.

Unfortunately, the rhetoric of participation all too often fails to translate theory into practical suggestions. Although almost all donors and development agencies are now increasingly embracing the concepts of participation, this is not always reflected by participative practices in the field. In fact, there is a dearth of information on methodologies. This does not mean that there are inherent difficulties in doing this, but rather reflects the isolation of professionals from experience that will enable them to become involved in and facilitate such a process. Many of the skills involved are merely adapting ideas and methods of participatory rural appraisal (PRA) and community empowerment, such as the ideas of Freire (1968) and Hope, Timmel and Hodzi (1984), to enable farmers to identify and seek solutions to problems they face.

Associated with this is an implicit requirement that professionals recognise that they are equal partners on a voyage of discovery. All those involved have something to contribute. So, whereas there are methodological models that can be used as a guide through this process, the attitude of outsiders is vitally important. This is perhaps a more difficult issue to address. Professionals are inevitably the product of formal education systems. Formal education systems, in turn, are invariably based on assumptions concerning the strengths and validity of Western scientific thought, and do not necessarily validate other knowledge systems. Closely related to this is another assumption that predominates in formal eduction systems: that individuals who have had access to formal education will be better equipped to develop suitable technologies. Unfortunately for many professionals, formal eduction does not provide the opportunity to develop practical interaction skills that facilitate a participative process.

In addition, professional training tends to focus on methodologies: systematic ways of doing a certain activity or range of activities. Seeking the certainty of a formally systematised methodology also poses problems. Experience from the field suggests that there is not so much a definitive methodology, as a range of methods that can be used to develop and sustain a participative process. There is a real danger that if a 'blueprint' methodology is used, then, almost instantly, the dynamic interactions which characterise human behaviour become subsumed by a straitjacket that reinforces control by whoever best understands the practices that are predefined by that particular methodology.

The key concept is that outsiders (ie people from development agencies and technical specialists) are participating in a process that is managed by farmers' rather than farmers participating in something the outsiders are controlling and managing. But for this to happen satisfactorily, conventional views of the relative power and status of various stakeholders need to be challenged. This can be uncomfortable for rural people and professionals alike. Both have frequently been used to ascribing each other with a particular status. This, in turn, defines a certain way of interacting with each other. It can be as difficult for a farmer to relate as an equal to a professional as *vice versa*. Frequently the key skills which teams of professionals lack are those very ones that can enable such new, less paternalistic relationships to develop.

Experiences in Kebkabiya, Sudan

The work on developing donkey plows with farmers in Kebkabiya was part of a larger agricultural project. This project was based on the concept of community empowerment and around management guided by village level institutions (village committees) which were based on traditional decision-making bodies. Project staff (the outsiders) acted as a catalyst for, and facilitator of, discussion, rather than directing it. This framework provided an environment which enabled farmers to identify and prioritise their needs, and resulted in identifying draft as a major constraint which they wished to tackle.

This groundwork was laid in the mid-1980s and methods of working developed over time. Project staff were aware of approaches and methods being tried elsewhere (but mainly by reading rather than from hands-on experience). Ideas were borrowed that seemed appropriate to their goal of community empowerment. Support was provided to strengthen project staff's confidence and skill in working in this way. They in turn developed and strengthened their own skills so that they were able to support and strengthen farmers' and blacksmiths' own experimental efforts. Approaches to any activity were adopted or rejected on the basis of their potential contribution to increasing the level of control of project activities by farmers, blacksmiths and village committees.

Many of the more refined participative techniques (such as PRA) were not used at the time, but the specific techniques used are not the key issue. What was important, and what this work allowed to happen, was to set in motion a process that would permit farmers themselves to set agendas for action. The idea was to ensure that users would regard themselves as being in control of a technology development process that was addressing an important concern in their lives. This resulted in high levels of interest in issues surrounding draft power, and an atmosphere developed where farmers were keen to experiment.

In addition, there was recognition that manufacturers would be key stakeholders if any technology was to sustain itself on the market, without the support of an external agency. In Kebkabiya it soon became clear that local village blacksmiths were the best placed to manufacture plows. So these blacksmiths were encouraged to join in the experimental process. Project staff only provided initial ideas on plow designs. Blacksmiths' skills and knowledge were explicitly recognised, and they were allowed to lead the process of adapting and modifying designs. This raised the blacksmiths' status, and was key to generating a creative enthusiasm that was a vital ingredient in developing affordable plow designs that worked in farmers' fields.

The informal methods used in early survey work served as a means of building relationships with farmers as much as enabling project staff to learn more about the communities with whom they worked. Both of these factors were regarded as absolutely vital prerequisites of successful technology development. Relationships had to be built where traditional roles and status were re-evaluated. This process took time, but had a major payoff as it was the basis for real research in which all stakeholders were participating.

In truth, this process was not as participatory as it might have been. Survey methods drew on methods that were a reflection of the Rapid Rural Appraisal (RRA) which later developed into PRA techniques. But despite using survey techniques that might be regarded as extractive rather than participatory, there was still sufficient participation in practice to break down barriers between farmers and project staff, and include farmers and their representatives in all key decisions. Similarly, some approaches to training remained fairly formal, especially in the early years. The most important factor, and one that was always a clear goal, was to increase the capabilities and capacity of local people to determine the direction and pace of technology development. The result was that a constructive dialogue developed between farmers, project staff and local blacksmiths which addressed the various issues surrounding plowing.

The issues that needed to be addressed were, of course, complex. In an area where plows had never been in widespread use before, there were few farmers with experience of plowing, and no plows available on the market. In addition, issues of cost and access to draft animals predominated. How was any plow to find its way on to the local market, and who would be able to obtain it?

The implications of this technical change on women (traditionally responsible for the majority of field operations, but also often in an inferior economic position) needed to be addressed. What were the implications for the poorer families, which the project claimed to be concerned about most?

In the light of these factors, the finer technical details of plow or harness design were of less immediate importance than enabling people to start trying out plows. This would allow them to see if this 'new' technology really was likely to provide an answer to their cultivation problems. The project's objective was more to stimulate farmers to try plows in their fields and blacksmiths to adapt some basic designs to suit their resources. At the same time, the practical experience permitted all involved to gain a better understanding of how the 'soft' aspects of the technology (skills, knowledge and forms of social organisation) needed to be adapted or strengthened to enable the majority of farmers to make use of the 'hardware' (the plow). Prioritising farmers' access to plows to use in their own fields, rather than perfecting plow design, also meant that farmers were able to assess the value of the product (the benefits of plowing in terms of yields, productivity, reduced drudgery, etc).

Developing an atmosphere of research managed by farmers and blacksmiths was not a rapid process, but did develop over time. Project staff had key roles as motivators and in providing ideas. However, these relationships changed slowly, so that key decisions were made more and more by local farmers (and blacksmiths) and the role of the project staff became more and more that of facilitators.

Actual plow designs borrowed existing designs. Two main designs were looked at, one based on a wooden ard (nothing new here), the other on a steel mouldboard plow, a scaled-down version of a standard ox plow, suitable for donkeys. This latter design had been around in the region for a few years, originating in two large development projects based a few hundred kilometres away. However, because the approach to developing and disseminating plows had been different, with the focus on getting the design right, rather than on dissemination and how the market works, this plow had not previously been widely adopted. Technology development had been managed and controlled by researchers, and the process of dissemination had been given less consideration.

The more conventional approach did not seek to enable farmers to have any control over technology development and dissemination. In Kebkabiya this approach was turned on its head, with the focus of activity more on getting plows to farmers and letting them do the real experimentation. Manufacturers (in this case local blacksmiths) were able to fine tune basic designs in line with their own skills and resources and also take account of farmers' feedback on performance.

The obvious difficulty some senior project staff members had in working with, and relating to, farmers in a rather different way, hindered the process for some time. The change in status implied by giving equal appreciation of everyone's individual skills and knowledge, challenged conventional concepts of roles and responsibilities. Technical specialists, some with postgraduate degrees, were being required to concede that farmers' and blacksmiths' skills and knowledge were as valuable as their own, and sometimes more so. The tensions this created proved to be a major constraint for several months. However, the prevailing ethos of the majority of project staff was one of sharing experiences and learning together with the community. In this working environment there was no room for self-importance to flourish. Issues concerning the relative status of those with a high level of formal education and those without faded in importance. However, the short-term hiatus that developed was finally only dissipated by the resignation of one staff member.

Currently, project management is in the hands of representatives of local committees, and there is little external support any more (apart from a minimum level of funding to the local management board). However, plows continue to be used by more farmers each season. Farmers are continuing to experiment, and differing designs are being used to cultivate different soil types. Farmers are convinced of benefits. Women are increasingly using and owning plows. Blacksmiths from neighbouring areas are coming to local blacksmiths in Kebkabiya to learn basic designs. Plows are appearing on local and neighbouring markets. Even if local management of an institution that resembles a project fails to sustain itself in the longer term, plowing technology has established itself.

Lessons and implications

What lessons can be drawn from this rather specific experience? Probably very few concerning the specifics of particular plow designs, or even of the suitability of draft power to farmers elsewhere in similar agro-ecological regions.

What is more important is to note that the widely-recognised 'success' of the animal traction work in Kebkabiya (Abu Sin and Hadra, 1994) owes a great deal to an approach which, over time, enabled farmers and manufacturers to control and manage the technology development process. The technical issues turned out to be less important. It is obvious that a plow (or any other technology) has to be sufficiently well designed to work, but this on its own is not critical for widespread adoption. What seems to be far more important is to look at ways in which outsiders can help to develop an environment where experimentation by farmers and manufacturers is valued—where key decisions are made by farmers (or at least by their representatives on committees). The specifics of methodology are less important than a 'mind set' that seeks ways of ensuring that 'users' are in control of the technology development process. In Kebkabiya, the methods used developed over time, as an iterative process with all stakeholders continually learning from experience.

This has major implications for resource allocation. It suggests that resources are better invested in the 'software' side of technology development (ie, in the skills, knowledge, and forms of social organisation needed to use a particular bit of equipment or technique).

This also has great implications for engineers and technicians. It suggests that their role is less important than those of community development specialists. Engineers and technicians need to review their own attitude to their role and should also seek to acquire skills as facilitators who enable farmers and manufacturers to experiment and adapt technologies. Basic techniques and equipment already exist. Experience shows that these are rarely directly transferable from one situation to another, but they should be seen as easily accessible starting points. Less attention should be paid to perfecting these in research station environments, and more attention paid to ways of working with farmers.

It is interesting to note that in a recent newsletter on agricultural mechanisation in Africa (NAMA, 1994) the vast majority of contributions described work which was fixated on technical aspects. Apart from the isolated observation that simply transferring techniques is unlikely to be effective, not one article looks in depth at approaches to technology development. Where social aspects are discussed, the same old phrases appear. Talking about "providing advice to farmers" or discussing "cultural constraints" does little to move away from a paternalistic approach to an alternative, enabling one. A newsletter such as this is a fairly accurate reflection of the thrust of on-going work in rural mechanisation. The rhetoric of participation may become more commonplace in development literature, but there is clearly still a long way to go before alternative ways of working appear in practice. Yet while constraints are still identified in terms of insufficient resources for research (an argument unlikely to attract attention in these days of continual erosion of funds for agricultural support services generally), there is still a dearth of discussion, let alone practice, which focuses on alternative ways of deploying existing resources.

A common statement is that there is a lack of clear methodology to guide working in this way. Yet although PRA, adult learning and associated concepts can provide guidelines, the example of Kebkabiya shows that it is not so much a refined set of particular techniques that were used, but rather a mind-set on the part of outsiders that provided the guidance to determine approach and probably more important of all provided the guidance of 'what to do on Monday morning'. Looking for a blueprint is counterproductive, for it is only the philosophy and aspects of process that can be transferred from one context to another (Pretty et al, 1995). Our discussions need to focus on these issues far more than they do at present. There may be a role for formal research, but we need to be clear about its role in a wider technology development process.

It is highly unlikely that a major new technical breakthrough in animal traction technology will provide an answer to farmers' problems. The basic technical options already exist. We need to be looking more closely at, and giving more weight to, issues surrounding approaches to developing technologies that allow these basic, widely known technologies and techniques to be adapted and used by farmers in their own fields. We need to be identifying and looking at ways to best support and strengthen the skills that are required to work in this way. As development professionals we need to continue to look critically at our own values and approaches and recognise that these are as influential in technology development as our other skills as engineers, agriculturalists, economists, and so on.

References

Abu Sin A M and Hadra T O, 1994. *ITDG's support to Kebkabiya Smallholders Project: an evaluation.* ITDG (Intermediate Technology Development Group), Khartoum, Sudan, and Rugby, UK.

Chambers R, 1994. Participatory rural appraisal (*PRA*): challenges, potentials and paradigms. *World Development* 22(10)

Croxton S and Appleton H, 1994. The role of participative approaches in increasing the technical capacity and technology choice of rural communities. In: *Proceedings of a workshop on technology and rural livelihoods: current issues for engineers and social scientists, Natural Resources Institute, September 1994.* Silsoe Research Institute (Overseas Division), Silsoe, UK.

Freire P, 1968. *Pedagogy of the oppressed*, Penguin, London, UK.

Havercort B, van der Kamp J and Waters-Bayer A (eds), 1991. *Joining farmers experiments.* Intermediate Technology Publications, London, UK.

Hiemstra W, Reijntjes C and van der Werf E (eds), 1992. *Let farmers judge.* Intermediate Technology Publications, London, UK.

Hope A, Timmel and Hodzi C, 1984. *Training for transformation. A handbook for community workers.* Vols 1–3. Mambo Press, Gweru, Zimbabwe.

Jamieson N, 1987. The paradigmatic shift of rapid rural appraisal. In: *Proceedings of the 1985 international conference on rapid rural appraisal, Khon Kaen, Thailand.* Faculty of Agriculture, Khon Kaen University, Khon Kaen 40002, Thailand.

NAMA, 1994. *NAMA Newsletter.* NAMA (Network for Agricultural Mechanization in Africa), Agricultural Development Unit, Commonwealth Secretariat, London, UK. Vol 2, Nos 1 and 2, October 1994.

Okali C, Sumberg J and Farrington J, 1994. *Farmer participatory research: rhetoric and reality,* Intermediate Technology Publications and Overseas Development Institute, London UK.

Pretty J, Guijt I, Thompson J and Scoones I, 1995. *Participatory learning and action: a trainer's guide.* International Institute for Environment & Development (IIED), London, UK.

Scoones I and Thompson J (eds), 1994. *Beyond farmer first: rural people's knowledge, agricultural research and extension practice,* Intermediate Technology Publications, London, UK.

Photo (opposite): Men plowing with oxen in Machakos, Kenya

Meeting the challenges of animal traction

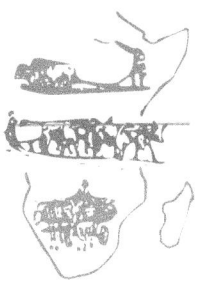

Challenging environmental issues

Animal traction in Africa: analysing its environmental impact

by

Roger Blench

Research Fellow, Overseas Development Institute, Portland House, Stag Place, London SW1E 5DP, UK

Abstract

The relationship between the adoption of animal traction and the environment in sub-Saharan Africa has so far been little explored in the literature. The paper argues that analyses should not be divorced from the broader socio-economic matrix of change; that where agricultural projects are involved, there is also a change in both crops and inputs. Similarly, where yields have declined in hand-hoe systems, traction may be a way of increasing surface area cultivated, in order to compensate. Animal traction should also be placed in a comparative frame with hand-hoeing and tractors; often all types of cultivation option have environmental costs.

Introduction

The relationship between the use of animal power, the environment and the sustainability of production systems remains an under-researched topic. There are two main reasons for this:

- chronological: research on all types of animal power is relatively recent
- technological bias: research has tended to emphasise technical aspects over socio-economic and environmental impact.

As a result there is remarkably little concrete information about the environmental impact of animal traction and most of that is anecdotal. There are also strong vested interests against in-depth exploration of these issues. On one side is a substantial NGO interest, promoting various types of 'alternative' technology; animal traction is almost ideal as a 'smokeless' and sustainable intervention. On the other, are large projects intended to increase the incidence of smallholder cash-cropping through animal power, sometimes at the expense of both the environment and food security.

This paper looks at the relationship between environment and farmers' animal power strategies and also the factors that might determine animal power's present-day distribution in Africa. A variety of charges have been laid against animal traction in terms of its impact on the environment; the validity of these are also examined.

Animal power in the context of projects

Throughout most of sub-Saharan Africa, where animal power is not traditional, its spread is strongly associated with colonial agricultural departments and in the post-Independence era, with agricultural development projects (Starkey, 1998). Plows have very commonly been made available at subsidised rates while training and loans are provided to make oxen more accessible to smallholders. Such activities are of course, not entirely disinterested; they reflect a desire to transform agriculture from subsistence into cash-crop production. This was quite explicit in the colonial era; the first plow introduced into Nigeria was the EMCOT ('Empire cotton') plow intended to expand smallholder cotton production to feed the mills of the north of England. This tradition very much continues with the addition of occasional mechanisation (see documentation in Tersiguel (1995) for the 1980s and 1990s in Francophone West Africa). The introduction of animal traction is thus usually bound up with major changes in the cropping system, often switching to high-input cash crops such as maize, cotton and groundnuts with a corresponding tendency to mine rather than manage the soil. In exploring environmental change it is therefore crucial to simultaneously narrate the social and economic pressures on farmers.

Evaluating alternatives

Animal power is never an isolated alternative in the present; it exists within a matrix of costs which include hiring additional hand labour or using tractors. As an investment it also reflects relative security of tenure; if a farmer is unsure about long-term access to land then investing in animal power is often unattractive compared with strategies that turn over cash within a single agricultural year. In other words, the impact of animal power use must be evaluated within the matrix of alternatives available to the farmer (see discussion in Munzinger 1982; Guibert 1984; Roupsard 1984; Bigot & Raymond 1991; Tersiguel 1995). For example, plowing on hillsides causes soil loss and increased runoff, but hand cultivation would cause much the same problems. The same is true with mechanical erosion around wells, or along transport routes.

Distribution of animal traction: historical versus environmental constraints

A striking aspect of the distribution of animal traction in Africa is the contrast between three major regions of Africa:

- 1) North Africa and Ethiopia
- 2) West-Central Africa
- 3) Eastern and Southern Africa

The approximate distribution of these regions is shown in Map 1. Animal traction is ancient in Region 1 and thus may be presumed to have spread to the limits of its ecological acceptability. In other words, farmers who do not use it are not limited by either unfamiliarity or chronology. It is striking, however, that much the same appears to apply in Region 2, West-Central Africa, where the further geographical spread of animal traction appears to be limited by ecology. Havard (1993)

Map 1: The distribution of animal-powered plowing in Africa

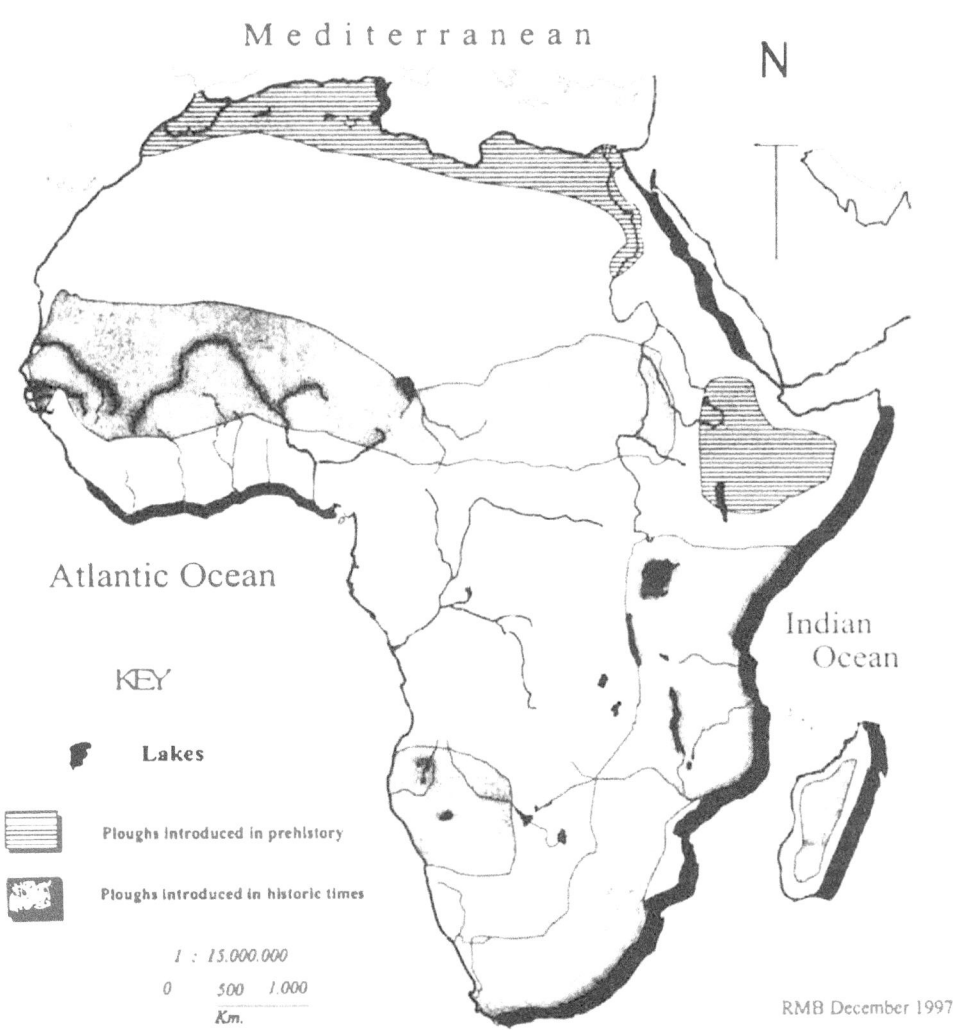

reports that saturation point has been reached in parts of Francophone West Africa. In other words, wherever animal traction is feasible, it is likely that users will exist.

Between regions 2 and 3 lie the Republics of Chad and Sudan, both of which have been severely affected by civil war in the last decades. Information about animal traction is sketchy in both cases, although it appears that there was no plowing in pre-colonial northern Sudan despite the presence of cattle-powered *shaqiyas* and camel-powered oil-mills (Croxton, personal communication). Plows were introduced in both Chad and western Sudan as part of cotton schemes in the colonial era and it is presumed that these persist.

Region 3, Eastern and Southern Africa, however shows a quite different pattern: animal traction is spreading out year by year from the centres of original introduction. Even though animal traction is not essentially of different vintage from West Africa in this region it has spread more slowly and more patchily. Somalia and Angola are excluded, because no current information is presently available.

This state of affairs is largely connected with the patterns of diffusion in the different regions. The spread of animal traction in Region 1 took place in prehistory and is thus largely irrecoverable. However, in Region 2, studies such as Roupsard (1984), Faure & Djagni (1989), Bigot & Raymond (1991), Guegen (1993) and Blench (1997a), have documented this process in some detail. A distinctive feature of the West African Region is that draft technologies are the same over large areas; in Nigeria, for example, the same ridger is used in all parts of the country. Much the same is true of the Niger Republic. Donkey-cart axles manufactured in Abidjan, Côte d'Ivoire are used in much of Sahelian West Africa. By contrast, in Southern Africa, a wide variety of animal power systems are found, often retaining a very local distribution (Starkey, 1995). Even technologies that appear to have a much wider regional application remain anchored in one area.

Socio-economic context of environmental change

The adoption or evolution of animal traction takes places within distinctive socio-economic matrices. Broader social and economic trends within a given region are often not clearly distinguished from those specific to animal traction. This section looks at some of the main issues relating to the context of traction.

Sustainability

Many development strategies of recent years have been based around the concept of sustainability; classically, for example, tractor programmes are 'not sustainable', whereas animal traction is. However, sustainability is a problematic concept. Usually intensification occurs when a system has become unsustainable; in other words it is often a catalyst for technological change. Classically, for example, low fertility due to shortening bush-fallows means that a household cannot support itself through hand cultivation (Pingali, Bigot & Binswanger 1987). This becomes an incentive to adopt animal traction and cultivate a larger area. However, when the boundaries of farms cultivated by animal traction begin to press on one another, traction may be dropped again in favour of more intensive systems of recycling nutrients, either for example, by adopting pigs, as in some parts of highland Kenya, or the Communal Areas of Zimbabwe or on the escarpments of the Jos Plateau in Nigeria. Technological change is therefore consequent on non-sustainable systems; farming moves to a different phase.

Deforestation

Deforestation and land clearance usually take place whether cultivation is manual, animal-powered or mechanical. However, since one of the significant advantages of traction is that a larger surface area can be cultivated, farmers tend to clear more land. Indeed animals can assist in stumping, weeding or in other ways accelerate this process. This type of clearance is driven by higher population densities, by the consequent fall in soil fertility or by land consolidation following wealth stratification, ie wealthier farmers accumulating larger plots of better land. When one group of farmers succeeds, they are able to buy up or otherwise acquire the better land, usually level and accessible lowlands. Poorer farmers are left with the choice of either moving up hills to cultivate more marginal slopes or moving further out to drier areas or regions of uncultivated bush. In both cases, farmers can be stimulated to adopt animal traction, either to produce more surplus for sale or simply to keep subsistence production to former levels. Once the best plains land has been consolidated, animal traction may well be used to open up hillsides. This can be a potential source of erosion, largely because farmers are not immediately familiar with the soil conservation techniques necessary to exploit such slopes effectively. However, without the intervention of effective extension services, a period of learning intervenes before new soil management practices are evolved.

Persistent poor rainfall also motivates an increase in the size of mean holdings as more plants must be sown to maintain the overall yield. This process was already reported in Zambia by Lancaster (1981) for a dry period in the 1960s. Even without wealth stratification, drought can motivate the exploitation of marginal and fragile ecosystems.

Plows and trees

Tractorisation always implies complete land-clearing. In particular, all the stumps must be cleared from a piece of land for a tractor to operate. In traditional savannah farming systems this is often problematic. In West-Central Africa, trees such as the locust, *Parkia biglobosa*, the shea, *Vitellaria paradoxa,* the baobab, *Adansonia digitata* and the oil-palm, *Elaeis guineensis,* play a crucial role in household economic strategies. They assist fertility regeneration by reducing soil erosion as well as improving soil structure (Kessler, 1992). Most species also have marketable fruits and the sale of these provides income, usually to women (see for example Bigot, 1983; Peltre-Wurzt, 1984).

Hand-hoes have always been able to work round such economic trees and to a limited extent animal plows have the same capacity. Farmers who wish to plow their land usually get rid of saplings and bushes and retain the larger trees. The result is a land use pattern often known in the literature as 'farmed parkland', levelled land with evenly dispersed mature economic trees. Animal power therefore usually creates an intermediate state of tree conservation, reducing biodiversity but retaining a pool of selected species which help maintain soil fertility.

Animal power and changes in cropping systems

Apart from changes in area and location, animal power is often associated with changes in the cropping system. A new cash crop, such as cotton or groundnuts, may be introduced and farmers come under administrative pressure to increase the area cultivated. Such systems have never been designed with sustainable agriculture in mind and are often associated with inappropriate application of inputs and soil erosion. Charrière (1984) reports on this situation in Chad and concludes that widespread adoption of animal traction has resulted in erosion, leaching and eventual desertification. Tersiguel (1995) recounts some of the strategies adopted to try and restore fertility in systems of intensive cotton production in Burkina Faso, but concludes that in most cases, the cultivation methods cannot return as much as they take out. In such cases, the use of animal power is essentially secondary to the overall environmental impact, which is primarily a result of the socio-economic changes induced by new crops.

Traction technologies

Animal power strategies are often determined by the availability of specific technologies, or the agendas of different types of development agency. Thus, if a local manufacturer is producing a specific type of heavy share designed for oxen, it is difficult for a farmer to acquire lighter shares more appropriate for donkeys. Nigeria is a good example of this, where a single type of ridger dominates almost all agricultural work in almost all environments (Haynes, 1965). The farmer may therefore use a tool which will increase erosion in some agro-ecological zones.

Some types of animal traction cause erosion, especially through transport of goods and people. Carts and sledges cause tracks to become wide and muddy and occasionally cause soil loss on slopes. Carts have varying levels of technological sophistication; pneumatic tyres can coexist with heavy iron wheels, as in Mali. Where heavy wheels are very numerous, they can be responsible for highly visible erosion. Sledges in Eastern and Southern Africa, and animal tracks near deep wells in the Sahel are similarly held responsible for environmental damage. However, it could be argued that *any* route of communication which is sufficiently popular will become eroded and that the formation of this type of track is probably preliminary to the construction of a surfaced road. It is difficult to see what mechanisms could be invoked, aside from aeroplanes, that would not cause at least comparable damage to the environment. Focusing on traction without considering the alternatives is to avoid putting animal traction solutions in a cost-benefit frame.

Different technologies are appropriate for different situations; socio-economic and environmental factors must be weighed up before making policy choices between tractors, animal power and hand tools. Such policy decisions must also be informed by an appreciation of household economics; there is little point in making technological choices for households too poor to implement them.

Changing species in response to environmental change

Broad climatic trends in sub-Saharan Africa are hard to determine although the decades since 1970 appear to have been a story of ever-decreasing precipitation, especially in Eastern and Southern Africa. Similarly, pasture degradation through overgrazing, once thought to be a certain environmental trend has been increasingly called into question as the elasticity of response of Sahelian pastures becomes better-known. Nonetheless, the vegetation of sub-Saharan Africa has never come under such extreme anthropic pressure as at present. More land has been cultivated and larger herds of cattle than ever before are grazing the rangelands. The consequence has been a low horizon of visibility for many pasture species, notably the grasses and sedges most suitable for bovine nutrition.

This has had two consequences; as the capacity of the semi-arid regions to support cattle declines, new savannahs are simultaneously opened up through the cutting down of the forest in the subhumid zone (Blench, 1994). This situation is less marked in Eastern and Southern Africa, where the potential for creating anthropogenic savannahs from rainforest appears to be less. Moreover, beyond the southern limit of the Maasai, there are no pastoral peoples with very large herds to place the same type of pressure on the environment familiar from West-Central Africa.

The effect in West Africa has been a general shift southwards of all cattle populations and a consequent shift southwards of the 'traction line', the southern limit of widespread animal traction (Blench, 1997a). As the plows move gradually southwards, maintaining cattle for traction at the northern limit becomes increasingly difficult, due to the disappearance of pasture grasses. One response can be to switch to cattle breeds more specialised in digesting browse. In West Africa, this has led to the widespread adoption of the Sokoto Gudali breed in preference to others (Blench, 1997b).

Another response, however, is to switch to species specialised in browse, notably the camel and the donkey. These are generally hardier than cattle and can largely be relied upon to find their own food. The disadvantage is that they do not reproduce through most of the region where they are required for traction and so must be sold for meat at the end of their working life and a new animal bought. Professional camel and donkey breeders usually live in the arid zone proper and there is a permanent flow of males or castrates from this region further south.

In Eastern and Southern Africa the situation is somewhat different. Drought has certainly affected most of the region since approximately 1980; in many regions it has been so severe that farmers' herds of cattle have all died. Farmers frequently have no capital to rebuild their herd, nor the willingness to risk continuing aridity. The response has therefore been to switch to donkeys. Donkeys have been in use throughout most of South Africa proper since before the drought, but they are presently spreading northwards from Zimbabwe and southwards from Tanzania as a traction option.

Does the spread of donkeys affect the environment?

The belief that donkeys are damaging to the environment appears to be somewhat local in Africa and thus probably more a reflection of the culture of those who assert it than a well-considered empirical observation. Starkey (1995) reports that this belief is extremely widespread in South Africa and has had the somewhat unfortunate consequence that administrators have initiated campaigns to shoot donkeys, to the dismay of their owners. It can safely be said that such a belief would be regarded as absurd in most parts of West-Central Africa. Donkeys can digest a wide variety of browse

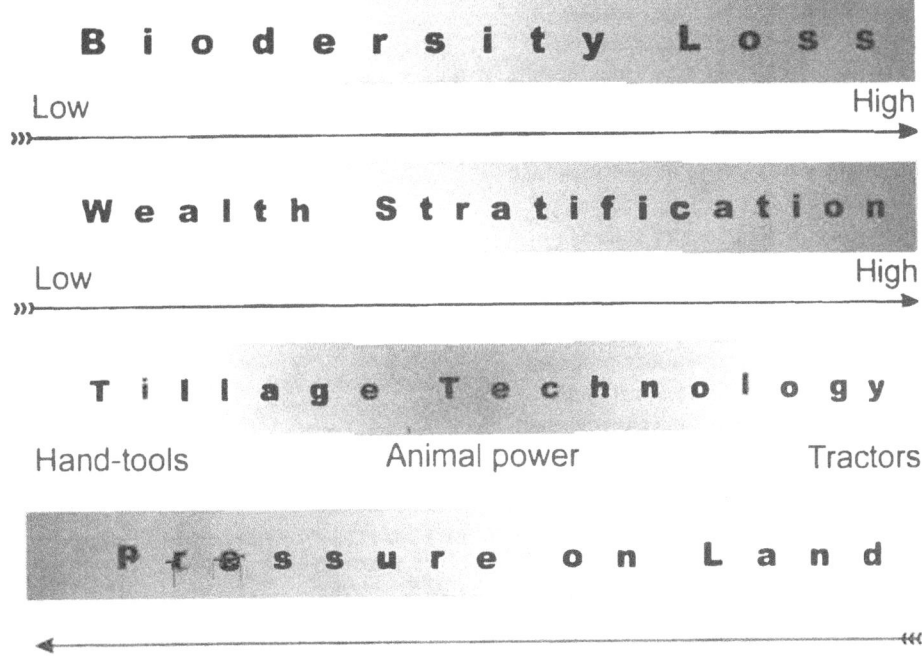

Figure 1: Schematic parameters for analysis of animal-power interventions

including extremely thorny plants, thereby making use of vegetation that few other species eat. Moreover, unlike goats, donkeys do not have the tendency to uproot the bushes they browse and may thus be an environmentally safe option.

Traction animals and humid environments

Animal traction in Africa is largely confined to highlands, semi-arid and northern subhumid regions (Map 1). This reflects the dominance of large cattle as traction animals, as the other traction species, camels and donkeys are even more constrained by ecozone. However, cattle of some type are found in all ecological zones, even in south-east Nigeria, which, with a rainfall of over 4000 mm annually, is one of the wettest regions of the world. Experiments in Sierra Leone have shown that traction with trypanotolerant Ndama cattle is perfectly feasible, and in Guinea-Bissau even very small Muturu are used for traction (Starkey, personal communication).

Given these examples, it is unclear what has constrained the development of traction in these regions. The answer may well lie in vegetation density rather than mean aggregate rainfall. Clearing secondary forest in very wet areas is an extremely time-consuming and energetic task. Traditional vegecultural farming systems in these regions, without cereals and based around yams and tree-crops, did not require open plots of cleared land and thus the motivation to make use of traction was traditionally limited. If the land is clear in a high humidity zone for whatever reason then farmers will experiment. Farmers in Benin Republic, where the open savannah comes nearly down to the coast of West Africa, are using crossbred Muturu x Zebu for traction far south of the limit in Nigeria (Blench, 1997a).

Conclusions and recommendations
Cost/benefit analysis for policy research

All types of intensification must cause some environmental change and usually some damage. However, blaming any specific technology for its effects without prior evaluation of the alternatives cannot be rational planning. All proposed interventions should be subject to cost/benefit analysis in comparison to the alternatives. Figure 1 shows schematically some of the relationships such analysis would have to take into account. In the real world, of course, boundaries are never so neat; almost all systems consist of mixture of these variables.

Information flow

Animal traction is a technology in transition and as such, many householders are in a situation of incomplete information. In a country such as Ethiopia, where the use of the ard is an ancient technology, farmers can be presumed to have tested a wide variety of strategies for improving productivity. The fact that interventions proposed in this century have had very little impact on Ethiopian farmers suggests that they have reached a sort of equilibrium of knowledge, balancing their socio-economic situation with the technical constraints of agronomy.

However, in the rest of sub-Saharan Africa, both available technologies and equipment are driven by contingent economic circumstances and knowledge of long-term effects on the soil and the environment remain limited. There is evidence that inappropriate tools and species are used in many parts of Africa. A very simple programme of testing and extension could help substitute more appropriate technologies, if these can be disentangled from the vested interests of NGOs and the goals of agricultural projects. This in turn leads to political decisions which are far outside the technological realm: the choice between rural food security and cash-crop production.

The relation between animal traction and the environment remains an under-researched topic. If this paper has any conclusion it is that animal traction cannot be taken as a 'smokeless' intervention simply because it does not use motorised power. As development projects increasingly include environmental assessment in their design it will become more and more important to evaluate the impact of particular animal power technologies. One method of moving the subject forward would be calling a small workshop to focus on this topic with the specific brief of gathering together existing research and making concrete recommendations for new research and monitoring practice.

References

Bernardet P, 1984 *Association Agriculture-Elevage en Afrique*. Harmattan, Paris, France.

Bigot Y, 1983. La culture attelée et ses limites dans l'évolution des systèmes de production en zones de savanes de Côte d'Ivoire. *Machinisme Agricole Tropicale* 84:44-52.

Bigot Y and Raymond G, 1991. *Traction animale et motorisation en zone cotonnière: Burkina Faso, Côte d'Ivoire, Mali*. CIRAD, Montpellier, France. 95p.

Blench R M, 1994. The Expansion and Adaptation of Ful°e Pastoralism to Subhumid and Humid Conditions in Nigeria. *Cahiers d'études Africaines*, 133-135:197-212.

Blench R M, 1997a. *Animal traction in West Africa: categories, distribution and constraints on its adoption and further spread: a Nigerian case study*. ODI Working Paper 106. Overseas Development Institute, London, UK.

Blench, R.M. 1997b. *Traditional livestock breeds: distribution dynamics in relation to the ecology of West Africa*. ODI Working Paper 108. Overseas Development Institute, London, UK.

Charrière G, 1984. La culture attelée: un progrès dangereux. *Cahiers ORSTOM, Séries Science Humaine*. XX (3-4):647-656.

De Wilde J C, 1967. *Experiences de developpement agricole en Afrique Tropicale*. [3 vols.] Maisonneuve & Larose, Paris, France.

Faulkner O T and Mackie J R, 1936. The introduction of mixed farming in Northern Nigeria. *Empire Journal of Experimental Agriculture* 4(1): 89-96.

Faure G and Djagni K, 1989. L'expansion de la culture attelée dans la région des savanes au Togo: facteur de progrès social sans progrès économique marqué. *Les Cahiers de la Recherche Développment*, 21:60-72.

Guegen R, 1993. *La traction animale en zone Mali-Sud*. Compagnie Malienne de Développment de Textiles, Mali.

Guibert H, 1984. *Dynamique du passage à la culture attelée des exploitations sénoufo dans l'ouest du Burkina Faso*. CIRAD, Montpellier, France. 30p.

Havard M, 1993. *La traction animale au Siné-Saloum, Sénégal*. CIRAD, Montpellier, France. 30p.

Haynes D W M, 1965. The development of agricultural implements in Northern Nigeria. *Proceedings of the Science Association of Nigeria* 6:101-107.

Kessler J J, 1992. The influence of karité (*Vitellaria paradoxa*) and néré (*Parkia globosa*) trees on sorghum production in Burkina Faso. *Agroforestry Systems*, 17:97-118.

Lancaster C S, 1981. *The Goba of the Zambezi*. Norman, University of Oklahoma Press, USA. 350p.

Munzinger P, 1982. *Animal Traction in Africa*. GTZ, Eschborn, Germany.

Peltre-Wurzt J, 1984. La charrue, le travail et l'arbre. *Cahiers ORSTOM, séries Science Humaine*. XX (3-4):633-646.

Peltre-Wurzt J and Steck B, 1979. *Influence d'une Societé de Developpement sur le Mileu Paysan. Coton et culture attelée dans la Region de la Bagoue (Nord Côte d'Ivoire)*. ORSTOM-CIDT, Abidjan, Ghana.

Pingali P, Bigot Y and Binswanger H P, 1987. *Agricultural mechanization and the evolution of farming systems in Sub-Saharan Africa*. Johns Hopkins University Press, Baltimore, USA.

RIM, 1992. *Nigerian National Livestock Resource Survey.* (VI vols). Report by Resource Inventory and Management Limited (RIM) to FDL&PCS, Abuja, Nigeria.

Roupsard M. 1984. Le point sur la culture attelée et la motorisation au Nord-Cameroun. *Cahiers ORSTOM, séries Science Humaine.* **XX** (3-4):613-631.

Starkey P, 1995 (ed). *Animal power in South Africa: empowering rural communities.* Development Bank of Southern Africa. Gauteng, South Africa. 160p. ISBN 1-874878-67-6.

Starkey P, 1996. *La traction animale en Mauritanie: situation et perspectives.* Project SPFP/MAU/4051. Food and Agriculture Organization of the United Nations (FAO), Rome, Italy.

Starkey P, 1998. The history of working animals in Africa. In Blench R M and MacDonald K (eds), *The origins and development of African livestock: archaeology, genetics, linguistics and ethnography.* University College Press, London, UK. (in press).

Starkey P and Faye A (eds), 1990. *Animal traction for agricultural development.* Proceedings of the Third Regional Workshop of the West Africa Animal Traction Network, held 7-12 July 1988, Saly, Senegal. Technical Centre for Agricultural and Rural Cooperation (CTA), Ede-Wageningen, Netherlands. 475p. ISBN 92-9081-046-7

Starkey P and Ndiamé J F (eds), 1988. *Animal power in farming systems. Proceedings of workshop held 17-26 Sept 1986, Freetown, Sierra Leone.* Vieweg for German Appropriate Technology Exchange, GTZ, Eschborn, Germany. 363p. ISBN 3-528-02047-4

Tersiguel P. 1995. *Le pari du tracteur: la modernisation de l'agriculture contonnière au Burkina Faso.* ORSTOM, Paris, France. 280p.

Environmental influences on the adoption of animal traction

by

Michel Havard and Gérard Le Thiec

Département des systèmes agro-alimentaires et ruraux (SAR)
Centre de coopération internationale en recherche agronomique pour le développement (CIRAD)
73 rue Jean-François Breton, BP 5035, 34032 Montpellier, France

Abstract

In sub-Saharan Africa, variations in the natural environment have a strong influence on the spread of animal traction. There is a long tradition of animal traction use especially in the high-altitude zones of Ethiopia. It has spread to the semi-arid and sub-humid zones but is still marginal in the arid and wet zones. In the semi-arid and sub-humid zones, animal traction is used for extensive farming and not for agricultural intensification; this is not as was planned by development programmes. Irrational agricultural practices, including improper use of animal traction, have often aggravated environmental degradation. Animal traction has generally led to an expansion of cropped area and the development of cash crops, but it has rarely raised crop yields per unit area. It has often been used for other, more profitable, purposes including transport, livestock production, land ownership and contract labour.

Introduction

In sub-Saharan Africa, the gap between population and agricultural growth rates has widened over the past three decades. In 1989, the annual growth rate was 3% for population, compared with 1.8% for agricultural production (World Bank, 1989).

In many agricultural situations, the traditional long-duration fallows, which ensured sustainability, cannot be continued because of land pressure due to population growth. In certain areas, the situation has led to irrational land use (excessive land clearing, overcultivation), which is an important cause of erosion, desertification, loss of soil fertility, and ultimately the degradation of natural resources.

Strategies that can check this downward trend are complex. They should be based on an analysis of local farming systems and their changes. Animal traction can serve as a basic component of these strategies as it has a positive effect on labour productivity, production factors, and soil fertility. It has become an imperative for raising the currently low agricultural performance in developing countries, where more than 70% of the farmers have only hand tools.

In 1990 sub-Saharan Africa had 12 million of the 400 million draft and pack animals in the world (FAO, 1990). Animal traction is suited only to certain physical conditions and markets; it is not a panacea. Animal traction has rarely been used by farmers for intensive farming, as originally planned by research and development programmes. Despite this it has a significant, though varied, impact.

During the past 20 years, in certain francophone countries of western Africa, animal power technology has grown four-fold and has almost reached saturation point—it is used on more than 80% of the farms in southern Mali (Gueguen, 1993) and in the groundnut basin of Senegal (Havard, 1993). These figures, however, obscure the wide disparities between different regions, where the natural environment is a decisive factor in the spread of animal traction.

Environmental influences on the development of animal traction

Sub-Saharan Africa can be divided into four general ecological zones based on average annual rainfall. These are arid (<400 mm), semi-arid (400–800 mm), sub-humid (800–1200 mm) and humid (>1200 mm). In addition, there are the high-altitude zones, where the average day temperature is lower than 20 °C (Higgins et al, 1978).

Constraints linked to the physical environment

Dense vegetation, stumps and roots are physical obstacles to the use of draft animals and cultivation implements. However, unlike tractor

Table 1: Distribution of draft animals in Senegal, 1980

	Semi-arid zone[1] (%)	Sub-humid zone[1] (%)	Total ('000)[2]
Horses (total)	96	4	222
Donkeys (total)	93	7	206
Draft cattle	74	26	90

1) Percentage based on regional distribution of draft animals
2) Estimated proportion of female animals: 45% for horses, 35% for draft cattle in the semi-arid zone
Source: Livestock department, Senegal

cultivation, animal traction can be used even if there are some trees. Limited shrub cover and possibilities of livestock production are therefore two factors that favour animal traction in the high-altitude, semi-arid, and sub-humid zones. In the high-altitude zones, animal traction requires less land preparation and road work than tractor cultivation.

Heavy soils can be cultivated more easily with animal traction than manually. Very heavy soils, however, need more powerful mechanised equipment.

Despite its dense vegetation and sparse population, the humid zone is not suitable for most large domesticated animals because of tsetse fly infestation; also there is often inadequate forage. The possibilities for animal traction improve near the edge of the zone and outside it.

Constraints linked to disease risk

Fodder supply, cleared land availability, and health conditions are satisfactory at the edge of the most humid zone; in this zone the only livestock are trypanotolerant oxen. Other animals (zebu-ox cross breeds, donkeys, humped oxen, horses, camels) can be used in less wet areas. Livestock distribution between the semi-arid and sub-humid zones in Senegal is linked to the disease situation (Table 1).

Large numbers of horses are found in the semi-arid zone, whereas donkeys predominate in the sub-humid zone. Both animals can be used as draft animals, as in Botswana where more than 75% of the horse and donkey population is used for traction (FAO, 1992). Trypanotolerant cattle predominate in the sub-humid zone, and the humped ox in the semi-arid zone. The potential for developing animal traction is high as the

proportion of trained animals in many countries is very low, often about 5% (Table 2).

Impact of climate on cultivation technology

The choice of cultivation techniques depends on climatic conditions and, in Senegal, their distribution is evidenced from the type of equipment in use, see Table 3 (Havard, 1993). Efficient control of weed growth (particularly difficult in the sub-humid zone) makes it necessary to perform plowing or ridging before planting. More plows and ridgers are found in this zone. In the semi-arid zone, the rainy season being shorter, farmers are obliged to plant rapidly without tillage; the number of seeders is therefore much higher compared with that of plows.

Development of animal traction and population growth

Improvement of public health conditions has led to high population growth in Africa. Large populations and the high rate of increase make population growth one of the most important factors of change in sub-Saharan Africa. It also influences the spread of animal traction. The first consequence of population growth is more intensive hand clearing of land. Conditions are thus improved for livestock production as fodder supply increases and disease risk is reduced. The shift to animal traction is facilitated when stump removal has developed.

As population density increases even further, land and labour management also change. Animal traction does not always develop in areas with high population density and low wages. Different combinations of hand-tool, animal traction and mechanical-power technology and a wide variety of equipment can be observed in these areas. The

Table 2: Estimated numbers of potential traction animals and the proportion used for draft in sub-Saharan Africa in 1990. All population estimates in thousands.

	Cattle		Donkeys		Horses	
	Total	% traction	Total	% traction	Total	% traction
Animal traction in the semi-arid zone						
Botswana	2616	14	152	92	33	75
Senegal	2740	5	310	50	400	50
Mali	5000	5	550	27	62	48
Zambia	2861	9	2			
Burkina Faso	2900	3	450	12	70	7
Chad	4173	3	240		200	
Cameroon	4697	1	40	12	26	
Gambia	400	5	nd	nd	nd	nd
Niger	3609	<1	512	2	302	
Total	28996	4	2256	23	1093	25
Animal traction in the sub-humid zone						
Ethiopia	30,000	20	5000		2650	
Kenya	13,793	5	nd	nd	2	
Zimbabwe	6711	10	103		24	
Tanzania	13,047	4	174			
Uganda	4200	14	nd	nd		
Angola	3100	10	5	100	nd	nd
Madagascar	10,254	3	nd	nd	nd	nd
Lesotho	530	34	130		122	
Nigeria	12,000	1	700		250	
Guinea	1800	5	1		2	
Mozambique	1370	7	20		nd	nd
Côte d'Ivoire	1046	6	1		1	
Malawi	1100	5	2		nd	nd
Benin	951	4	1		6	
Ghana	1250	2	10		2	
Togo	250	4	3		2	
Central African Republic	2595	<1	nd	nd	nd	nd
Guinea Bissau	340	1	3		1	
Sierra Leone	330	<1	0	0	0	0
Zaire	1550	<1	nd	nd	nd	nd
Total	106,217	9	6153		3062	
Grand total	**135,213**	**8**	**8409**	**7**	**4135**	**6**

Sources: FAO, 1992; Goe 1990   *Notes: nd - no data, Sudan not included*

Table 3: Regional differences in the types of implements used in Senegal, 1958–1980

Equipment	Semi-arid[1] (%)	Sub-humid zone[2] (%)
Toolbars (Houes)	45	30
Seeders	42	22
Plows	2	37
Lifting implements	11	5
Ridger bottoms		6
Total number	650,000	150,000

1) Estimates based on equipment distributed
2) Based on annual reports from the Department of Agriculture, Senegal

rural market for trade, labour, and even land ownership is also better structured in these areas.

Adoption of animal traction and population pressure do not necessarily lead to intensification of the cropping systems. The objective is often to reduce the work load, in which case animal traction may not be a suitable response to the challenges of agricultural development in high population density areas.

Impact of animal traction at farm level

Development programmes often recommend animal traction for the sole purpose of improving cultivation practices (tillage, planting, crop maintenance), whereas it could add value in many other ways on the farm (livestock production, available labour, transport) or outside it (contract labour). This study of the impact at farm level was based in western Africa and Madagascar.

Impact on cropping systems and land ownership

Farms using animal traction are usually larger and grow more cash crops, but yields per hectare are not higher than those on farms using hand-tool technology (Table 4). In many cases, expansion of cropped area is a prerequisite for adopting animal traction. Such capital accumulation is necessary to meet the high investment in animals and equipment.

Large farms also have abundant labour so that animal traction and hand-tool technology can be combined easily. They often have large herds of animals as well. Animal traction allows better management over a larger area (Table 5).

These changes are typical in the less populated cotton-growing areas of southern Mali, western Burkina Faso, and northern Côte d'Ivoire, particularly among indigenous farmers (Dugué, 1993; Gueguen, 1993). It is less common among immigrant farmers, who usually have smaller households and limited agriculture rights. But in densely populated areas, confronted by land shortage, expansion of cropped area does not have a significant effect and farmers' strategies are aimed at reducing the work load and at providing services, as in northern Togo (Fauré and Djagani, 1989). This strategy is observed in cases where access to land ownership is difficult, even if population density is low, as in northern Cameroon, where contract labour is frequent. Farm households that possess equipment but lack labour

Table 4: Comparison of farms using animal traction and farms using manual cultivation in sub-Saharan Africa

Parameter	Total number of farms	Positive effect of animal traction	No difference	Positive effect of manual labour
Yield/ha	14	4	8	2
Farm area	17	17	0	0
Area/labourer	19	19	0	0
Cash crops	19	12	7	0

Source: Pingali, Bigot and Binswanger, 1987

Table 5: Characteristics of farms using hand-tool technology and animal traction in the cotton-growing areas of Burkina Faso, Mali and Togo

	Burkina Faso[1] (1984)		Mali[1] (1986)		Togo[2] (1985)	
	Hand	*Animal*	*Hand*	*Animal*	*Hand*	*Animal*
Area cultivated (ha)	3.4	8	3.3	12.1	5.7	7.5
Manpower	8	12	6	19	10	15
Area/labourer (ha)	0.4	0.65	0.5	0.65	0.57	0.5
Cotton area (%)	16	23	31	38	12	17

Sources:

1) Bigot and Raymond, 1991 (data of Guibert, 1985 and Persoons, 1987); 2) Fauré and Djagni, 1989

may also resort to contract labour, even if there is no ceiling on acquiring land for new crops.

The decision to acquire equipment may also be part of a land ownership strategy. Such a purchase is probably necessary to support an application for clearing land or to reinforce a right to cultivate land by plowing, or it may be prerequisite for acquiring or purchasing a plot (Lassaux and Garin, 1994). Farms may therefore be over-equipped in terms of technical ratios.

The small size of the fields, their irregular shape and scattering are often cited as constraints to mechanisation. However, farmers have often shown considerable adaptability in reorganising their land when they see profit in animal traction. In cases of extreme land scarcity, for example due to population growth, or hilly terrain, adoption of animal traction is unfeasible.

Animal traction can serve to promote contract labour on other farmers' land. Sharecropping and similar solutions are growing more common, particularly in northern Cameroon and Madagascar (Lassaux and Garin, 1994).

Expansion of cropped area is often observed for cash crops, particularly when farmers shift to animal traction (Table 3). Such is the case in western Africa for groundnut (Havard, 1993) and cotton (Bigot and Raymond, 1991). These crops had a significant influence on the number of machines introduced. The spread of animal traction was also linked to the economic and financial organisation of the cotton and groundnut

subsectors. In eastern Africa, the same occurred with maize, and in Madagascar with rice.

The farm operations that are a priority tend to be mechanised first and so determine the distribution of animal traction practices and associated equipment. Planting is most common in the dry zone, weeding in the wetter zone, and tillage in the very humid zone, with a diversity of situations lying outside considerations on crops, equipment, and soil types.

Contrary to popular belief, animal traction can be used for soil tillage in intercrops and catch crops. Subsequent operations are the same for both hand-tool and animal traction technologies.

On-station trials have often demonstrated that, all conditions being equal, plowing has generally positive (15–70% surplus yield), although variable, effects for different crops (Le Moigne and Nicou, 1990). However, there is less evidence of this effect in farmers' fields (Table 3). Farmers do not usually follow recommendations on equipment, weed control, and fertiliser application. There is no improvement in per hectare yields although cropped area per labourer and overall output may increase following the introduction of animal traction.

Observations reveal that there is no difference in the planting period on farms using hand tools and those using animal traction to extend the cropped area. Planting dates and yields do not differ for the two types of farms (Faure, 1994). Animal traction does not necessarily improve work quality; manual soil tillage is sometimes done more carefully. In practice, farmers are less concerned with the quality of the work, their main objective is to cover a large area, as in the lake Alaotra region (Lassaux and Garin, 1994).

Impact on livestock production

The presence of draft animals can modify existing livestock systems and even contribute to changes in the land ownership system. Farmers possessing animal teams and fields managed for animal traction do not have the same interests as those with traditional forms of livestock and land management. Crop by-products and fallows serve as a source of fodder. Such farmers have a more individualised approach. In certain cases, farms using animal-draft technology form a relatively interdependent group with respect to others.

Fattening of draft animals at the end of their lifetime performance can modify the livestock market. However fattening possibilities do not necessarily combine with animal traction. Herd composition therefore changes around the areas where farmers rely on animal traction.

The growing fodder market is gradually integrated in the livestock or general distribution channels. Forage crops and the use of crop residues have also developed. Demand for feed concentrates has fallen although they are preferred to forage crops. Preference for concentrates depends on product availability and adequate farmer income.

The use of draft animals helps constitute a livestock production basis within the farm. In such a case, females can also be trained for traction, which is technically easy. Positive examples of this can be found in central Senegal and northern Cameroon.

Impact on labour

The introduction of animal traction in agriculture has an impact on labour productivity and distribution (between individuals and over time). The total amount of labour rarely decreases because not all the operations can be mechanised. As farms grow larger, additional manual labour is required for planting, weeding and harvesting. Although the per hectare duration decreases, the overall workload increases (Table 6). Animal traction makes certain tasks easier. This aspect is greatly appreciated by farmers although it is difficult to evaluate the advantage.

Peak periods and the distribution of tasks are often modified. Sometimes, planting on the expanded farms involves less labour than weeding and harvesting, which are done by women. The workload is heavier during the dry season when jobs include watering and feeding of draft animals, stump removal, land clearing of large areas, equipment maintenance and transport. Children have more work as they have to guide the draft animals and watch over them. One adult (usually a man) and two children are needed per animal team. Much needs to be done to reduce labour requirements to just one person per team.

Labour productivity is higher for mechanised operations (plowing, planting). For manual operations, the quantity of work increases with the area. Farms made up of several households are better able to manage the additional work involved in adopting animal traction. They can easily reorganize themselves and redistribute the tasks (including herd management) among the household members.

The dependency relationships created for hand-tool technology are reinforced within and between farms that supply or receive labour. The

Table 6: Average labour requirements and productivity in the cotton-growing area of northern Côte d'Ivoire

	Hand cultivation	Animal traction
Total person-days per year	185	225
Cultivated area per labourer (ha)	0.85	1.20
Person days per hectare	220	190

Source: Bigot and Raymond, 1991

introduction of animal traction tends to emphasise socioeconomic differentiation. Farmers using animal traction no longer have the same interests as others for exchanging work. Work exchanges take place within the same group of farmers (ie, either between farmers using animal traction, or between farmers relying on hand tools), and any exchange between groups is increasingly organised on the basis of monetary payment.

Impact on transport

The introduction of animal traction has boosted transport activities. Acquisition of carts has become a priority even if they are expensive and are not covered by bank credit. Transport of goods has increased because of the growing scarcity of natural resources (wood and water), remoteness of fields from the homestead (harvest transport), and more frequent exchange between towns and villages (migration, food aid).

Carts are also needed for certain technical innovations, for example construction of stone barriers to control erosion, sedentary livestock production based on forage crops, production of farmyard manure and compost from straw, crop residue processing, water transport. The use of carts motivates farmers to introduce and maintain draft animals even in areas where agricultural output is low. However, in areas where animal traction increases agricultural output, transport equipment is given lower priority than farm equipment.

Conclusions

Although sufficiently large numbers of livestock that can be used for animal traction are found in the arid zone, they cannot be used as agriculture is not well-developed. They are used for drawing water and for transport. In the humid zone, disease incidence is a major constraint. Animal traction is most developed in the high-altitude regions, where it has the longest tradition in sub-Saharan Africa.

Developed farming systems and livestock availability in certain parts of the semi-arid and sub-humid zones are favourable to the spread of animal traction. In western Africa, it was first introduced in the semi-arid zone and continues to develop. The sub-humid regions, despite the high potential, were slower to adopt animal traction, but now record the highest development rates. Horses

and donkeys are more suitable for the arid zone, and trypanotolerant cattle for wetter zones.

The introduction of animal traction has not led to crop intensification as planned by most development programmes. Instead, it has responded to farmers' objectives of extensive farming, reduction of the work load at certain periods, and control over the land. The total volume of work is rarely lower on farms using animal traction because the operations are not mechanised entirely. Additional manual labour is needed for planting, weeding, and harvesting when farmers expand their cropped area.

The share of cash crops increases when the area is extended; yields, however, remain the same as on farms using hand tools. In densely populated areas with a shortage of available land, introduction of animal traction does not necessarily increase the area cultivated per labourer. Contract work can then be a source of substantial income.

In traditional livestock regions, conditions are clearly more favourable for animal traction and various solutions exist for distributing livestock to other areas. Meat production and animal traction do not necessarily go hand in hand. Special breeding programmes for draft animals are not needed in traditional livestock regions.

Animal traction has aggravated environmental degradation, although it is also observed on farms using hand tools. The reasons are: shorter fallow, disappearance of woody plants, increased soil erosion, and reduced soil fertility.

References

Bigot Y and Raymond G, 1991. *Traction animale et motorisation en zone cotonnière: Burkina Faso, Côte-d'Ivoire, Mali.* CIRAD, Département Systèmes Agraires, Montpellier, France. 95p.

CIDT, 1970–1990. *Rapports annuels 1970 to 1990.* Compagnie Ivoirienne pour le Développement des Textiles, Bouaké, Côte d'Ivoire.

Dugué P, 1993. *Traction asine ou bovine. Quelles alternatives techniques pour une relance de la culture attelée en zone semi-aride? Le cas du Yatenga au Burkina Faso.* CIRAD-SAR, Montpellier, France. 35p.

FAO, 1979–1992. *FAO Production Yearbook 1979, 1990 and 1992.* Food and Agriculture Organization of the United Nations, Rome, Italy.

Faure G, 1994. Mécanisation et pratiques paysannes en région cotonnière au Burkina Faso. *Agriculture et Développement* 2: 3–13.

Faure G and Djagni K, 1989. L'expansion de la culture attelée dans la région des savanes au Togo : facteur de progrès social sans progrès économique marqué. *Les Cahiers de la Recherche Développement* **21**: 60–72.

Goe M R, 1990. Overcoming constraints to animal traction through a collaborative research network. pp136–143 in: Starkey P and Faye A (eds), *Animal traction for agricultural development*. Proceedings of the third workshop of the West Africa animal traction network, 7–12 July 1988, Saly, Sénégal. Technical Center for Agricultural and Rural Cooperation (CTA), Ede Wageningen, The Netherlands. 479 p.

Gueguen R, 1993. *La traction animale en zone Mali-Sud.* Compagnie Malienne de Développement des Textiles, Bamako, Mali. 40p.

Havard M, 1993. *La traction animale au Sine-Saloum, Sénégal.* CIRAD, Montpellier, France. 30p.

Higgins et al, 1978. Report on the agro-ecological zones project. Vol.1, Methodology and results for Africa. *World Soil Resources Report* **48**. Food and Agriculture Organisation of the United Nations (FAO), Rome, Italy.

Jahnke H E, 1982. *Livestock Production Systems and Development in Tropical Africa.* Kieler Wissenschafts Verlag Vauk, Kiel, Germany. 253 p. French version: Jahnke H E, 1984. *Systèmes de production animale et développement de l'élevage en Afrique Tropicale.* Kieler Wissenschafts Verlag Vauk, Kiel, Germany. 279p.

Lassaux J C and Garin P, 1994. Mécanisation sur les grands périmètres irrigués à Madagascar. *Les Cahiers de la Recherche Développement* **37**: 47-62.

Le Moigne M and Nicou R, 1990. *Efficacité agronomique de la mécanisation des opérations culturales en zones de savanes au sud du Sahara.* CIRAD, Montpellier, France. 43p.

Pingali P, Bigot Y and Biswanger H P, 1987. *Agricultural mechanization and the evolution of farming systems in sub Saharan Africa.* Johns Hopkins University Press, Baltimore, USA. 216 p.

World Bank, 1989. *World Development Report 1989.* World Bank, Washington DC, USA.

Environmental impact of animal traction in Rukwa Region, Tanzania

by

A M Kilemwa

Rukwa Region Agriculture Department, PO Box 8, Sumbawanga, Tanzania

Abstract

There is a relatively high level of adoption of animal traction in Rukwa Region, western Tanzania. This has brought benefits in terms of reduced drudgery, increased labour productivity and expansion of area cultivated. However, the technology has some negative effects, especially increased soil erosion due to farmers plowing along slopes and the widespread use of sledges.

It is important to see these negative impacts in relation to the positive benefits of the technology. Animal traction is an appropriate technology for the region and its potential contribution to agricultural development cannot be overemphasised. Lack of credit for purchase of oxen and implements is a major constraint to further expansion of the technology.

Introduction

Rukwa Region lies in the extreme west of Tanzania, bordering Zambia and Zaire. It occupies a total area of 70,000 km². The region has a population of about 900,000 people with an annual growth rate of 4.3% (figures based on 1988 census). More than 90% of the people in Rukwa Region live in rural areas with agriculture, livestock keeping and fisheries being their main occupations. About 70% of the total area is arable land, of which 3% is utilised for agriculture. This means that the enormous agricultural potential of the region is yet to be exploited. Nonetheless, the region is famous for its maize production and the use of oxen for plowing crop fields.

Mechanisation in Rukwa

Whether in human, animal or mechanical form, power is an essential component of all production processes in agriculture. Mechanisation plays a large role in ensuring land and labour productivity. Mechanisation increases yield by expanding the area under cultivation and, if treated in isolation from other production techniques such as fertilising, could well increase yield per unit area by improving the quality of tillage.

In Tanzania in general, and Rukwa Region in particular, humans still provide most of the power needed for farming activites. The introduction and use of draft animal power occured about 50 years ago (BRALUP, 1977). However, it has never replaced human power per se. In contrast, it has contributed a substantial workload to most farmers in carrying out post-tillage operations as the oxen are often used only for primary tillage (Kilemwa, 1993). It is estimated that 80% of the 300,000 ha cultivated are plowed by oxen.

Surveys showed that of the total of about 130,000 households in the region 58% own oxen and 6% own donkeys (Kilemwa, 1993). Households that possess animals have between one and 20 oxen and one and six donkeys. All ox-owners also own a mouldboard plow (known as the UFI plow) and a wooden sledge (see Table 1).

Mechanisation trends

With a good agricultural production environment, rising living standards and a high rate of popoulation increase, Rukwa farmers have been compelled for the past two decades to increase food crop production to feed the population and produce a surplus for sale to earn income. This has involved the use of animal power in agriculture. The number of oxen has risen by almost 400% over the last 20 years. For example, in 1975 there were an estimated 19,000 oxen in the region compared to 106,000 in 1995 (see Table 1).

Three factors are associated with the increased adoption rate of animal traction by farmers:

- in the 1970s, there were projects which advocated the use of oxen in farming activites, such as the National Maize Project, the Small Industries Development Organisation's common facility workshops

Table 1: Numbers of draft animals and animal-drawn implements in Rukwa Region

	1990/91	1991/92	1992/93	1993/94	1994/95
Oxen	27,750	36,000	58,870	76,300	106,000
Donkeys	3,800	4,200	4,800	5,000	5,400
Mouldboard plows	18,800	28,000	36,000	42,000	48,000
Harrows	220	340	720	560	520
Cultivators	102	213	250	230	220
Ridgers	100	450	460	390	350
Planters	6	10	12	8	8
Ox carts	120	250	230	250	260
Donkey carts	-	-	3	3	5

Source: Regional agricultural office

and Zana za Kilimo Mbeya, an agricultural implement factory in the Southern Highlands. These projects stimulated the use of animals as animal-drawn implements were readily available in the area

- the continuous influx of Wasukuma pastoralists from neighbouring Tabora and Shinyanga regions to Rukwa in search of pasture and water for their cattle. These people introduced and developed a substantial level of animal traction technology to the indigenous people, in particular in the Mpanda lowland and Rukwa valley

- the third factor was the villagisation programme which was carried out in 1974/75 by the Government with the intention of keeping scattered rural families in settlements to provide social services such as dispensaries, primary schools, roads and other collective infrastructures. Villagisation created a scarcity of suitable cultivable land near farmers' homesteads. Farming intensification then became a priority for most farmers which consequently called for the use of animal traction technology.

Appropriateness of animal traction

The potential contribution of draft animal power to agricultural development in Rukwa cannot be over-emphasised. Most farmers are now aware of how draft animals can reduce their workload and increase land productivity, whilst raising income, social satus and prestige among themselves.

Farming in Rukwa is done using three major system, the first system, the use of hand tools and manual labour, being the oldest from which the other two have evolved. It is estimated that currently 20% of the farming population in Rukwa, mostly resource-poor families, employ this system. Under this system, the area cultivated per family is usually low due to the difficulty and drudgery involved in performing farming operations.

The second system which is used widely is when farmers employ oxen for primary tillage only. Planting, weeding, harvesting and transportation are carried out by hand. In this system the area cultivated can be expanded and better land preparation is achieved. The bottleneck of this method is that weeding is done by the family, so it is usually done only once because of time limitations. This contributes to low yields.

Plowing, weeding and haulage of produce by oxen is the third system used in Rukwa. About 25% of the farming population use this system. In this system plowing, making furrows for sowing seeds, and weeding are done using ox-drawn mouldboard plows, while produce is transported from field to homestead using sledges or ox carts.

The advantages of this system are that the optimum crop density can be obtained as the seeds are planted in rows and that weeding can be carried out two or three times depending on weed intensity. The bottleneck of this system is during harvesting as it is usually done manually and the area planted is relatively large. However, the magnitude of this operation is reduced by using oxen to transport the crop home.

The use of tractors in Rukwa has not been encouraging. With 67 working tractors out of the 152 available, many farmers have not benefited from their use. Owners of tractors, usually a cooperative society, an institution, a settlement, or even individuals, hire the tractors out on a cash basis. The tractor is used for the first plowing, but planting is done without a second plowing (harrowing) because of inadequate financial resources at the farmer's disposal. Weeding and harvesting are done by hand. The area cultivated is in some cases expanded, though not necessarily so because the area correlates directly with the farmer's financial resources. Yield per unit area is mostly low as plowing is done unsatisfactorily and weed intensity and growth is accelerated on an unharrowed field. Delays in planting normally occur as farmers keep waiting to hire the tractor as the season progresses.

Environmental impact of animal traction

Since animal traction has been appreciated by many farmers to increase labour and land productivity, reduce drudgery and increase efficiency in farming practices, its effect on the environment has not be assessed and analysed critically. As in any situation the use of animals in agriculture also has negative effects. It has been associated with a number of environmental damages: soil degradation, vegetation depletion, water and soil moisture losses and other complementary effects brought by the technology. To most farmers environmental protection has not been their priority, rather their priority has been ways of increasing food production with minimum fatigue and low input use.

Ox plowing and soil structure

Plowing improves soil tilth, providing a good growing medium for crops, by increasing porosity and the water-holding capacity of the soil. However, plowing is damaging to soil as it involves cutting, loosening and inversion of a soil structure and therefore increased soil erodability. Research is needed to understand these conflicting effects of plowing. Researchers and extension agents have the role of guiding farmers to making decisions on what system of land preparation and cultivation does not cause excessive soil erosion and at the same time provides crops with a well-tilled seedbed.

Cultivation on hillsides

The most damaging impact of oxenisation on the environment is the widespread practice of plowing fields *along* the slope. Due to the undulating topography of Rukwa, farmers who use oxen for plowing demarcate their fields along the slopes in order to make longer runs for plowing. This practice opens up the soil to erosion by run-off. As a result many fields are now becoming difficult to plow as they are full of rills and gullies.

This practice not only subjects good soils to erosion but wastes farmers' resources as fertiliser is easily washed off the soil. This practice needs an urgent intervention by agriculturalists and leaders of Rukwa in educating farmers on either better ways of cultivating on slopes, or making topography a determining factor in the selection of farm equipment.

Use of sledges

A survey of animal traction in Rukwa Region revealed that most ox owners own sledges (Kilemwa; 1993). The sledges are simply V- or Y-shaped tree branches, and do not have a proper floor. They are used for transporting plows to and from fields, for hauling crops and, to a lesser extent, for domestic chores. The efficiency of sledges is hindered by their low carrying capacity and substantial grain losses due to the lack of a floor. They also cause environmental problems as they have a high frictional resistance which damages roads and footpaths and causes extensive soil erosion, especially on sandy and loose soils.

Cutting trees to make sledges is also a problem. Most farmers think cutting trees for sledges has a minimal effect on the environment, but experience has shown that farmers rarely plant new seedlings for the future.

Ox plowing and spread of weeds

In the villages there is a feeling that ox-plowed fields usually have many weeds and weeding is required regularly and earlier than on hand-tilled plots. This could be true, for two reasons. First, the finely tilled land that ox plowing produces encourages early and fast weed generation. Second, ox plows spread weed seeds during plowing, especially of vegetative weeds such as couch grass (*Digitaria* sp) and nut grass (*Cyperus* sp). It may be beneficial to advise farmers to clean their plows at every turn to reduce spread of weeds, or to perform two plowing operations before sowing, which has been found to check weed germination.

Sustainability of animal traction

One study in Rukwa showed that many farmers who do not own oxen would prefer to own oxen as they believe their neighbours' who own oxen and plow usually harvest more and have a higher standard of living (Kilemwa, 1993). From this observation it is safe to assume that the process of adoption of animal traction has been "farmers' personal involvement" after first being induced by extension agents and projects. Most farmers are aware that the overall increase in crop production depends greatly on bringing more land under cultivation, which is only possible through the improvement of labour efficiencies.

Farmers being aware of the benefits of animal traction as a source of farm power is a prerequisite for extension agents, researchers, manufacturers and distributors of agricultural equipment promoting sustainable use of animals in agricultural production. What is required is provision of all necessary technical know-how for ox and donkey training followed by ensuring the availability of all related implements, tools and other farm machinery for the smallholder farming sector. There is also a need to find ways to alleviate short-term constraints which slow the rate of adoption of animal traction in Rukwa.

Blacksmith training and development

This is an important component of animal traction since the use of animal-drawn equipment creates a demand for maintenance and spare parts. For sustainable use of animal traction there is a need to have locally-based production, distribution and sale of the parts needed. With training, this could be done by village blacksmiths. Centrally-located production centres, mostly in urban areas, have proved inadequate, as they force farmers to travel long distances in search of spares. This is a costly and inconvenient exercise, especially during the farming season.

Training programmes for farmers

On-site training programmes have proved to be better in terms of attendance and low cost. They also have the advantage of giving better understanding to farmers as they use similar conditions to the farmers' local environment.

Provision of credit

Draft animal power involves both large expenditures and high risk which could not be met by many subsistence farmers who want to introduce animal traction in their farming systems. No clear credit policy has been formulated by the few exisitng credit institutions and not enough credit channels have been established by the government to enable subsistence farmers to obtain credit for animal traction. The problem of securing credit is repsonsible for the slow adoption of animal draft power for farming.

Conclusion

Labour saving and area expansion are the main motivating forces in the adoption of animal traction in Rukwa. Indeed, an increase in agricultural production is a direct result of expanding area cultivated and reducing the drudgery involved. Oxenisation has been a key factor influencing crop production in the region despite some negative environmental impacts it created. These negative effects can be minimised by choosing appropriate equipment and by application of better farming techniques. However, it is important to keep environmental concerns in perspective by relating environmental damage costs to crop production benefits.

References

BRALUP, 1977. *Rural integrated development plan for Rukwa Region.* Bureau of Resource Assessment and Land Use Planning, University of Dar-es-Salaam, Dar-es-Salaam, Tanzania.

Kilemwa A M, 1993. *Animal traction and rural technology - Rukwa Region.* Integrated Food Security Programme - Rukwa Region, Sumbawanga, Tanzania.

Animal traction and sustainable soil productivity in Kenya

by

Isaiah I C Wakindiki

Appropriate Technology Centre, Kenyatta University, PO Box 43844, Nairobi, Kenya

Abstract

In the process of developing previously uncultivated land into permanent cultivation systems it is necessary to consider soil and water conservation measures. This paper describes the role of vegetative strips, trash lines, stone bunds and terraces as soil and water conservation techniques, and concludes that there is vast potential in Kenya to use animal traction in such indigenous soil and water conservation measures, especially on smallholdings in moisture-deficient areas.

Background

Land degradation is a widespread problem in Kenya, and a problem which is becoming worse as marginal lands continue to absorb the growing population (Stahl, 1994). Land that previously was able to sustain its inhabitants, and even produce a surplus, is now derelict.

Environmental assessment by the World Bank (1992), the United Nations Environment Programme (UNEP, 1992), the World Watch Institute (World Resources Institute, 1992) and the World Conservation Union (IUCN, 1992) questions Africa's ability to halt its environmental degradation and sustain livelihoods by the end of this century. This paper discusses land degradation in the context of indigenous soil and water conservation for sustainable soil productivity, and explores the possibility of employing the animal traction in this respect.

Traditionally, soil conservation has been perceived as a physical problem caused by inappropriate farming practices. Conservation projects did not analyse the problem from the farmers' perspective, so proposed solutions were often socially unacceptable, economically not viable or ecologically unsound (Fones-Sundell, 1989). Indigenous conservation is now gaining popularity among researchers and policy makers alike, who concede that the farmers themselves have a better understanding of the processes of ecological change, slope dynamics and biological regeneration (Zurick, 1990).

Unfortunately, the change in tillage practices from the traditional hoe to mechanised (tractor) plowing has meant that most indigenous soil conservation techniques are no longer used. However, agriculture in Kenya is currently characterised by an increase in the number of impoverished smallholder farms which are experiencing increased soil erosion and decreasing soil productivity (Kiome and Stocking, 1993). Tractor use is no longer an option for small-scale resource-poor farmers. Animal traction therefore remains a suitable technology if it can be developed alongside the indigenous soil conservation techniques to alleviate labour drudgery.

Indigenous soil and water conservation techniques

Vegetative strips

Vegetative strips are usually narrow grass strips grown across the slope. The grass acts as a barrier to run-off, and encourages the deposition of sediment, eventually leading to terrace development. *Imperata cylindrica* is commonly planted: other popular grasses include *Vetiveria zizanoides* and *Pennisetum purpureum*. Species such as *P. purpureum* may also be fed as fodder to draft animals and other livestock.

A modification of vegetative strips is live fencing surrounding cultivated fields (Thomas, 1988).

Trash lines

Trash lines range from simple bunds of cereal and legume stover (as applied in Embu, Kenya) to more sophisticated pegged brush lines (Gichuki, 1992). Apart from impeding run-off and enhancing infiltration, trash lines may also improve soil fertility and increase soil organic matter if they are incorporated into soil during plowing. Trash lines

---

are extensively used in the Tharaka area of Kenya (DAREP, 1994).

Stone bunds

Stone bunding is the most common indigenous soil and water conservation technique in areas of Kenya where stones are abundant, especially in the eastern drylands. Lines of stone are laid out in parallel lines or a grind pattern on compacted denuded land to increase infiltration and capture soil blown by the wind. Stone bunds present a semi-permeable barrier which allows the passage of excess run-off while trapping sediment. In Embu Region, Altshul (1995) reported quite sophisticated stone bund systems where cultivation on the inter-bund areas led, over time, to the formation of natural ridges.

Terraces

Levelled bench terraces and earth bunding on existing slopes are the two main types of earth terrace structure common in Kenya. Sometimes, especially in the highlands, steps are constructed across the hillside when strips of crop residue are covered with soil dug from above. The resulting incorporation of organic matter increases soil fertility and enhances infiltration (Thomas, 1988).

The *fanya juu* earth bunding system in Kenya has become a "modern tradition" (Critchley, Reij and Willcocks, 1994). The *fanya juu* terrace is a back-slope bench terrace designed to trap run-off and suspend sediment. It is formed by digging the trench and throwing the soil up-slope to form an embankment. These terraces have a reputation for being very effective, and there is evidence that crop production is increased. However, their construction and maintenance demand considerable labour input (Kiome and Stocking, 1993).

The role of animal traction in indigenous soil and water conservation

Earth-moving animal-drawn equipment can be used to construct terraces and earth bunds, thus saving on human labour. The mouldboard plow is common in Kenya, but it is rarely, if ever, used for this purpose.

Gathering stones to construct stone bunds is equally time-consuming and labour-intensive. Animal-drawn raking equipment that can gather and move surface stones can greatly promote this approach to soil and water conservation, but nowhere in Kenya is this opportunity being taken.

Vegetative strips (and modifications of this technique) are compatible with livestock systems because of their dual role: as a conservation measure and as a source of fodder. However, animal-operated forage harvesters and choppers are not well developed. Some grasses used in the strips may also pose a weed problem. Therefore, it is urgently necessary to develop specific animal-drawn weeding equipment. Planters that can be used to plant both the grasses and the crops also need to be developed.

References

Altshul H, 1995. Literature review on indigenous soil and water conservation prepared for the Dryland Applied Research Project planning workshop, 21–24 February 1995. DAREP (Dryland Applied Research Project), Embu, Kenya. 6p.

Critchley W R S, Reij C and Willcocks T J, 1994. Indigenous soil and water conservation. A review of the state of knowledge and prospects for building on traditional. *Land Degradation and Rehabilitation* 5:293–314.

DAREP, 1994. *Tharaka diagnostic survey.* DAREP (Dryland Applied Research Project), Embu, Kenya.

Fones-Sundell M, 1989. *Perspective on soil erosion in Africa: whose problem?* IIED Gatekeeper Series SA 14. IIED (International Institute for Environment and Development), London, UK.

Gichuki F N, 1992. *Indigenous land husbandry practices.* Overseas Division Report OD/92/2. Silsoe Research Institute, Silsoe, Bedford, UK.

IUCN, 1992. *Caring for the Earth. A strategy for sustainable living.* IUCN, Gland, Switzerland.

Kiome R M and Stocking M A, 1993. *Soil and water conservation in semi-arid Kenya.* NRI Bulletin 61. Natural Resource Institute (NRI), Chatham, Kent, UK.

Stahl M, 1994. Land degradation in East Africa. *Desertification Control Bulletin* 25:48–53.

Thomas D B, 1988. Conservation of cropland on steep slopes in Eastern Africa. pp. 140–149 in: Molderhauere W C and Hudson N W (eds), *Working with farmers for better land husbandry.* IT Publications, London, UK.

UNEP, 1992. *Saving our planet: challenges and hopes.* UNEP (United Nations Environment Programme). Chapman and Hall, London, UK.

World Bank, 1992. *World development report 1992: Development and environment.* Oxford University Press, Oxford, UK.

World Resources Institute, 1992. *World Resources 1990–91.* Oxford University Press, Oxford, UK.

Zurick D N, 1990. Traditional knowledge and conservation as a basis for development in a West Nepal village. *Mountain Research and Development* 10(1):23–33.

Meeting the challenges of animal traction

Gender challenges

Gender and animal traction: a challenging perspective

by

Juliana Rwelamira[1] and Lotta Sylwander[2]

1) University of Stellenbosch, PO Box 3060, 7602 Coetzenburg, South Africa
2) Sida, Birger Jarlsgatan 61, S-105 25 Stockholm, Sweden

Abstract

In sub-Saharan Africa migration of males from rural to urban areas and increased schooling of children has placed an increasing burden of agricultural labour on women's shoulders. Women in eastern and southern Africa constitute 60–70% of farmers and produce about 80% of the food. However, the contribution of women to the economic welfare of the peasant family remains a neglected topic in the analysis of agricultural production. Animal traction technology introduced to women is often inappropriate because women have not been consulted during design and planning.

In general, women's daily lives are dominated by the need to acquire basic necessities first. For women to benefit and participate fully in an animal traction project, these constraining factors have to be recognised explicitly: a gender-sensitive planning approach has to be adopted. Project planning should incorporate a detailed gender analysis and gender awareness needs to be emphasised in all levels and activities of a project or programme. Involvement of women as a group independent from men can be an effective method of ensuring women's full participation. However, care should be taken not to define women's advancement only as a concern of women, but as one which particularly requires cooperation and a change of attitudes by both men and women. This paper suggests practical steps which can be taken to achieve this and discusses the related policy issues.

Historical perspective

Sub-Saharan Africa has undergone tremendous changes during the last century: colonialism is gone, market economies progress, subsistence agriculture has made room for cash-crop production and advanced technology is being introduced into all sectors of society. Despite this, some things seem to remain the same. The division of labour in rural areas has not changed much since the beginning of the century. In about 1900, W S Bazeley, Native Comissioner to Umtali in Southern Rhodesia said "Women are required under the present system to grow food for the greater part of the native population" (Schmidt, 1992). This shows that as soon as men left the rural areas to enter wage employment and a growing number of children spent the better part of the day in school, an increasing burden of agricultural labour fell on women's shoulders. Tasks that had previously been accomplished with the help of other household members, men or children, were done by women alone. Even in 1944 it was estimated that women did as much as 80% of agricultural work (Schmidt, 1992). This pattern has not altered despite all the other changes in the region.

Animal traction is not an end in itself, but a means for intensified agriculture and increased production. With a rapid rate of population increase and high pressure on arable land, increased food production is one of the main concerns and priorities of the governments and peoples of sub-Saharan Africa. Improved technology and better management have been suggested as mechanisms to reach increased production. Improved timeliness when cultivating with animal power and the larger amount of land that can be cultivated greatly increases yields. The use of draft power also alleviates farmers from some of the hard manual labour tasks. These are the two main reasons for using animal traction in agricultural production.

Then who are the farmers and the producers of food and cash crops in eastern and southern Africa? Women in eastern and southern Africa produce about 80% of the food. However, men produce a larger proportion of the so-called cash crops and produce more, relative to the area, than women. Women constitute 60–70% of farmers, but cultivate a smaller area than men. The technology that women use is often simple, such as hand hoes; men may own or have access to draft animals.

Invisible female farmers

The role and the contribution of women to the economic welfare of the peasant family remains a neglected topic in the analysis of agricultural production. Women are in many ways the invisible agricultural producers. Invisible to 'developers', invisible to extensionists, invisible to those who disseminate agricultural technologies like animal draft power. Development programmes and the promotion of animal draft technology have been directed mostly at male farmers, forgetting that the majority of farmers are women.

Women, along with men, are subjects of development, but development so far has had a different impact on women relative to men. Although women are not a homogenous group, their weak economic positions result from an overwhelming concentration of their economic activity in the unrecorded domestic sector, the unpaid labour on the farm or other family enterprises (of their husbands') and labour done with patron-client relationship. Thus, the traditional division of labour between sexes has been one of the basic factors causing the unequal share of women in the development process, because it restricts women to the domestic sphere and to tasks that are traditionally 'female'.

Many of the problems with development planning and programming arise from traditional theories of development, which overemphasise economic concerns as opposed to the human resources and well-being. National development planning, within which animal traction programmes are implemented, has also often been ineffective due to lack of consciousness of the interdependence of the social and economic aspects of development. The situation has been compounded by gender inequalities that are manifested in all aspects of life.

This is especially important because women are not an isolated group in society, and the well-being of men and children is closely linked and dependent upon womens' well-being. Therefore, it is imperative that women contribute effectively to and benefit fully from development for both their own well-being and that of men and children. A number of researchers in the area of gender issues stress the fact that unless, and until, women are given wider access to economic and political power, food insecurity and malnutrition in sub-Saharan Africa will continue to increase. This is partly attributed to lack of access to and control over factors of production, including technologies such as animal traction (Jazairy, Alamgir and Panuccio, 1992).

The point at issue is the need to be aware that women operate within wider social and economic processes. In practice, these processes have the potential to constrain women from adequate performance even in their readily visible role as mothers. For example, poverty, the state of want or of lack of access to the basic necessities of life is, first and foremost, an economic and political problem. Thus, at the domestic level, a poor woman who is faced with the hungry eyes of her children cannot be expected to participate effectively in a project that does not address the poverty of her economic situation and priorities even if she recognises the intrinsic benefits of that particular project. In reality, her daily life is dominated by the need to acquire basic necessities first. This means that for women to benefit and participate fully in an animal traction project, such constraining factors have to be taken into consideration.

Gender planning

Thus, for animal traction projects with specific gender components to have long-term impact, women have to be perceived both as active agents of change as well as beneficiaries in the development process. Project impact should be measured both in terms of the division, by gender, of labour and resources within households as well as by division of returns to labour.

Methodologies for surveying and data analysis for purposes of planning and project design tend to be gender biased. For example, Jazairy, Alamgir and Panuccio (1992), point out the general assumption made by development planners that household heads are male, regardless of who is supporting the family. As a result, government and development agencies, which administer primarily to men have failed to make substantial investments that would increase female productivity. For example, concentration of animal traction use on cash and export crops, which are dominated by men, as opposed to traditional food crops grown by women, mainly for home consumption.

One strategy for reaching out to women to ensure full participation in animal traction projects specifically and in development generally as proposed by Makwanda (1994), is to involve women as a group independent from men. The same strategy is advocated by Marshall and Sizya (1994). Through organising women into groups they gained access to and were able to control draft animal technology.

On one hand, we have to recognise that 'women only' projects in one sense demonstrate the underprivileged status of women in society. It is because women have been neglected in development that they have to be given special attention in order to address the imbalance. On the other hand, care should be taken not to define women's advancement only as a concern of women, but as one which particularly requires cooperation and a change of attitudes by both men and women. In some circumstances, for example in sex-segregated environments, based on either tradition or religion, women-only projects are the only option.

Typical animal traction projects are introduced as a means of improving smallholder farmers' productivity and improvement on agricultural output and income. It has, however, become increasingly clear that targeting project benefits to the rural population generally and hoping that women within the communities in question will get their share, simply does not work. Moreover, the design of such projects does not take into consideration the nature of tasks to be performed by each gender within households. The time and labour requirements of such projects do not take into account the already overburdened schedules for women.

Women's participation in animal traction projects

In a number of cases, animal traction projects have tended to allow men to expand the amount of land under cultivation for cash crops and reduce their workload in land preparation, while increasing women's workloads in transplanting, weeding, harvesting and transporting produce from the field. This point cannot be overemphasised as it has been elaborated upon by a number of researchers (Rwelamira, 1993; Doran, 1994; Marshall and Sizya, 1994; Sylwander, 1994 and

Thrupp, Cliff and Estes, 1994). This, and similar negative effects of animal traction technology on women necessitate a reorientation of animal traction programmes to serve women better. Affirmative action type strategies aimed at simply providing oxen and animal traction implements will not suffice. An holistic approach which calls for the re-examination of the past and current socio-economic and political institutions within which women operate is essential.

A policy issue

Women are central to Africa's agricultural performance and food security. All the case studies that present data on the subject of women's roles in agriculture report that women are a key resource in food farming and provide a substantial part of agricultural labour for crops grown primarily for sale. Moreover, there are indications that with increasing male migration in search of waged work outside the food and agriculture sector, women's responsibilities in food and agriculture are expanding. However, despite the well-documented key role of women in food and agriculture, there is still a gap in the policy and technical support necessary for improving the value of women's labour in this sector. The evidence is that women still mostly carry out hoeing and other manual operations and even where agriculture is commercialised, the demand for labour is mostly for non-mechanised tasks (weeding, tea and coffee picking, etc).

The gender gap in agricultural policy and technical support derives in part from the historical policy bias in countries which have favoured the development of commercial agriculture aimed for the export sector (initially coffee, tea and later hybrid maize) and the neglect of the small farm sector which produces food for up to 80% of the population.

Within the above scenario, animal traction projects or project components addressing gender issues will make little headway in changing the status quo, if the national policies and institutional environment is not conducive. The International Fund for Agricultural Development's (IFAD) project experiences suggest, inter alia, a comprehensive approach for addressing issues related to gender and food production and security. Macro-economic and agricultural policies and programmes that will help rural women to make the best use of the resources available to them should be in-built into all rural development projects (Jazairy, Alamgir and Panuccio, 1992).

Agricultural extension and training

Traditionally, national extension and training institutions and curricula in developing countries are not particularly sensitive to gender issues. They are mainly based on western models and are staffed almost exclusively by men and offer services to men. Projects also suffer when information and training are given only to male heads of households. Advice on production, inputs and use of specific technologies like animal traction are often transmitted incorrectly from husbands to wives.

Sensitivity in animal traction training programmes is also essential for women with young families, who tend to be least flexible. For women between 18 and 45 years of age, child care and household demands limit the potential for learning new skills and new activities. To target this most productive age group of the female labour force, training programmes have to be based locally and for a short duration at a time. Hocking (1994) reported successes of mobile ox-plowing courses for women in the Western Province of Zambia. This strategy could be applied elsewhere to increase women's participation in animal traction courses.

Access to agricultural inputs and credit

Women need to save and borrow in the formal financial sector. However, formal credit systems are not geared to small farmers generally, and women who lack title deeds to their land or other assets to pledge as collateral are worse off. Several African countries have small-scale credit programmes under way to provide credit with less reliance on land as collateral and more on peer pressure and the ability to pay. For women to participate fully in animal traction projects they need credit to purchase equipment as well as hire or buy animals. Special loan schemes are necessary to circumvent the obstacles.

Female-headed households

The incidence of households headed by women and the growth in their number has emerged as an important indicator of poverty. It is estimated that 31% of rural households in Africa are female-headed. The capacity of such households to own and use animal traction and other resources effectively has significantly decreased over the years. Male migration has increased labour shortages, especially for land preparation, and has reduced productivity. The absence of male labour has increased women's reliance on child labour (boys) which has led to children being withdrawn from school in some instances.

In South Africa, an IFAD special programming mission reported that the use of oxen for plowing had been reduced because of male migration. Females who become de facto heads of households as a result of male migration are affected in several specific ways. First, they are often limited in their access to agricultural technical support and services because of the assumption in extension practice that men are the farmers. Second, men continue to hold overall authority. Third, since male migration is higher in the low or poor potential agricultural zones, female-headed households tend to be poor.

Female-headed households seem to require special programme emphasis because this phenomenon is an outcome of emerging economic circumstances which tend to work to the disadvantage of women.

Gender analysis and planning

To overcome the constraints and difficulties with involving both men and women in animal traction projects a gender-sensitive planning approach has to be adopted. The first step in sensitive project planning is a gender analysis. Various tools have been developed by ATNESA to assist planners to do this (see Sylwander and Mpande, 1995).

A gender analysis includes the following key issues:

- identification of the division of labour
- dentification of the resources available to men iand women and the benefits they derive from these activities
- analysis of the needs, the conditions and positions of women and men
- analysis of the relationship between the division of labour and the access to social, economic and environmental resources.

The first tool in gender analysis is to develop an activity profile for identification of gender roles. Some key questions for animal traction projects are:

- who is currently using animal draft power for which activities?
- which tasks could be done using animal power?
- which activities are most time-consuming and labour demanding? Who does this work?
- how can the project address the need for labour reduction in reproductive activities as well as productive activities?

The second tool to be used is an access and control profile. The analysis of the flow of the resources and benefits is essential in understanding how a project will affect women and men. The necessary differentiation between access and control of resources can be directly related to the control and access of benefits derived from project activities. Some key questions for animal traction projects are:

- do women and men have equal access to the animals and equipment used for traction?
- who controls draft animal power resources?
- can current patterns of access and control over draft animals be changed?
- who has access to credit for draft animal power?
- how has information and extension on animal draft power been disseminated in the community so far?

The third tool takes into consideration other factors that can influence the potential impact of a project, and presents opportunities and constraints to project goals and activities. The following factors have been suggested:

- *socio-cultural factors:* societal norms, societal organisation, traditions, religion etc.
- *economic factors:* poverty level, inflation rate, infrastructure, credit facilities etc
- *environmental factors:* quality and availability of land, climate, rainfall, availability of firewood etc
- *political factors:* power relations and control, government bureaucracy, legal system, land ownership etc

- *demographic factors:* migration, life expectancy, infant mortality, nutritional status, female-headed households etc
- *institutional factors:* health care, extension, education, veterinary services, hospitals etc
- *legal parameters:* right to ownership, right to vote, family rights, inheritance, the right to credit etc.

By using these tools, project planners may be able to develop a picture of gender roles and relations in a society.

The fourth and final relation in a society is to identify the specific gender needs that men and women have. *Practical gender needs* are the immediate and daily needs such as food, water, housing healthcare etc. Practical needs can often be met in a project context with specific inputs. *Strategic gender needs* refer to long term issues which are common to women and men. They relate to the disadvantaged position of women, lack of power, education, resources, decision making etc.

For animal traction projects it is crucial to identify both practical and strategic needs that can be met by the project and by the use of draft animal power. For further discussion refer to Sylwander and Mpande (1995).

Conclusion

Animal traction projects and programmes have to be reorientated to address issues of equity, human development, women's roles in society and sustainability of development. An holistic and integrated approach is needed to address gender issues in animal traction. This may not always be possible within a project context. Thus women need to be integrated thoroughly in all phases of development from the planning stage, as well as in their social and political surroundings. Animal traction technologies that can effectively reduce labour and time requirement for weeding, food processing, water and fuel fetching could alleviate women's drudgery substantially.

Gender awareness needs to be emphasised in all levels and activities of a project or programme. Women and men have to be part of all aspects of project design, implementation and evaluation. A strong recommendation is to ensure that the entire staff has been gender trained and gender sensitised.

To solve the inherent problems of the diverse and complex gender issues encountered in the introduction and promotion of animal traction programmes in eastern and southern Africa, animal traction research systems must be given a new direction. The emphasis should be on establishing a system which approaches research as a problem-solving process directly related to gender issues and evolving with changes in people's conditions of life, their resources, education, skills, family composition and the prevailing environment.

Most importantly, animal traction technology must be based on the active participation of the community for whom it is intended, from the definition of problems to the selection, application and evaluation of possible solutions. Too often, animal traction technology introduced to women is inappropriate because women have not been consulted during design and planning. Equally important is the fact that the development and diffusion of animal traction appropriate for women's major tasks require increased research funds to be allocated to food crops, food processing and transport projects.

Acknowledgement

The authors would like to thank Sida for its support.

References

Doran J, 1994. Transportation by women and their access to animal-drawn carts in Zimbabwe. pp 272–275 in Starkey P, Mwenya E and Stares J (eds), *Improving animal traction technology*. Proceedings of the first workshop of the Animal Traction Network for Eastern and Southern Africa (ATNESA) held 18–23 January 1992, Lusaka, Zambia. Technical Centre for Agriculture and Rural Co-operation, Wageningen, The Netherlands. 480 p. ISBN 92-9081-127-7

Hocking C, 1994. The impact of mobile ox plowing courses for women in the Western Province of Zambia. pp 288–291 in Starkey P, Mwenya E and Stares J (eds), *Improving animal traction technology*. Proceedings of the first workshop of the Animal Traction Network for Eastern and Southern Africa (ATNESA) held 18–23 January 1992, Lusaka, Zambia. Technical Centre for Agriculture and Rural Co-operation, Wageningen, The Netherlands. 480 p. ISBN 92-9081-127-7

Jazairy I, Alamgir M and Panuccio T, 1992. *The state of world poverty. An inquiry into its causes and consequences.* Publication for the International Fund for Agricultural Development by Intermediate Technology Publications, Southampton Row, London, UK. 481p.

Lombe M, Sikanyika C and Tembo A N, 1994. The importance of women's participation in animal traction in Zambia, pp 284–287 in Starkey P, Mwenya E and Stares J (eds), *Improving animal traction technology*. Proceedings of the first workshop of the Animal Traction Network for Eastern and Southern Africa (ATNESA) held 18–23 January 1992, Lusaka, Zambia. Technical Centre for Agriculture and Rural Co-operation, Wageningen, The Netherlands. 480 p. ISBN 92-9081-127-7

Makwanda AC, 1994. Women and animal traction technology: experiences of the Tanga draft animal project, Tanzania. pp 276–279 in Starkey P, Mwenya E and Stares J (eds), *Improving animal traction technology. Proceedings* of the first workshop of the Animal Traction Network for Eastern and Southern Africa (ATNESA) held 18–23 January 1992, Lusaka, Zambia. Technical Center for Agriculture and Rural Co-operation, Wageningen, The Netherlands. 480 p. ISBN 92-9081-127-7

Marshall K and Sizya M, 1994, Women and Animal traction in Mbeya Region of Tanzania: a gender and development approach, pp 266–271 in Starkey P, Mwenya E and Stares J (eds), *Improving animal traction technology*. Proceedings of the first workshop of the Animal Traction Network for Eastern and Southern Africa (ATNESA) held 18–23 January 1992, Lusaka, Zambia. Technical Centre for Agriculture and Rural Co-operation, Wageningen, The Netherlands. 480 p. ISBN 92-9081-127-7

Rwelamira J K, 1993. The social and economic aspects of animal traction in agricultural production among female-headed households of Lesotho and Swaziland, pp 227–229 in Lawrence P R, Lawrence K, Dijkman J T and Starkey P H (eds). *Research for development of animal traction in West Africa*. Proceedings of the fourth workshop of the West Africa Animal Traction Network held in Kano, Nigeria, 9–13 July 1990. Published on behalf of the West African Animal Traction Network by the International Livestock Centre for Africa (ILCA), Addis Ababa, Ethiopia, 306 p.

Schmidt E, 1992. *Peasants, traders and wives: Shona women in the history of Zimbabwe, 1870–1939.* Social History of Africa Series. Heinemann Educational Books, Portsmouth, UK.

Sylwander L, 1994. Women and animal traction technology, pp 260–265 in Starkey P, Mwenya E and Stares J (eds), *Improving animal traction technology*. Proceedings of the first workshop of the Animal Traction Network for Eastern and Southern Africa (ATNESA) held 18–23 January 1992, Lusaka, Zambia. Technical Centre for Agriculture and Rural Co-operation, Wageningen, The Netherlands, 480 p. ISBN 92-9081-127-7

Sylwander L and Mpande R (eds), 1995. Gender issues in animal traction: a handbook. Proceedings of a workshop of the Animal Traction Network for Eastern and Southern Africa (ATNESA), held Mbeya, Tanzania, 1-5 June 1992. ATNESA, Harare, Zimbabwe. 60p.

Thrupp A, Cliff E and Estes D, 1994. Women and sustainable development, pp 43–57 in *World Resources*. A Report of the World Resource Institute in collaboration with the United Nations Environment Programme (UNEP) and the United Nations Development Programme (UNDP). Oxford University Press, Oxford, UK.

Animal traction development and gender: experiences from Western Province, Zambia

by

G M Bwalya [1] and M Akombelwa [2]

1) Team Leader, 2) Animal Draught Power Promoter
Western Province Animal Draught Power Programme, PO Box 910067, Mongu, Zambia

Abstract

The use of oxen in Western Province has been dominated traditionally by men. Between 1989 and 1995 the Western Province Animal Draft Power Programme ran training courses targeting women farmers to increase their awareness, confidence and stimulate their interest in animal draft power technology. These were followed by credit schemes to increase access to oxen and implements. It was found that the initial inability of women to handle oxen and implements was due to traditions and cultural biases which had often excluded them from such tasks, believing that they were a 'man's job'. As a result of courses and demonstrations women farmers developed the confidence to approach oxen and use them for draft operations. After several courses and demonstrations, some women farmers are benefiting from the technology through reduced drudgery and increased independence from men. The paper describes the programme's methods and experiences and makes recommendations for future programmes, stressing the need to take into account the existing cultural aspects of animal use in the target area.

Introduction

The Western Province Animal Draught Power Programme was initiated in 1989 with the objective of contributing to the development of sustainable farming practices by increasing, intensifying and diversifying the use of animal draft power by the community in priority areas. The main objective was divided into seven immediate objectives, one of which was: "Introduction and implementation of proposals to ensure wider access to animal draft power considering the needs and abilities of various target groups and the existing inequality between male and female farmers" (Department of Agriculture, 1992).

The programme was implemented in two phases. The first phase began in 1989 and ended in 1992. During this period data collection and training of farmers and field workers were carried out throughout the province. Phase II began in January 1993 and ended in December 1995. During this phase the programme targeted areas with high potential for adoption of animal draft power technology.

Although Western Province has the second largest cattle population in Zambia, with about 541,000 head of cattle (Livestock census, 1994) the use of oxen and animal draft power technology has been limited. In the southern and central provinces of Zambia women use animal draft power technology, but in Western Province the technology has been confined mainly to men (personal observation). It has been recorded that women's fields are usually plowed late during the rainy season, mainly due to women's limited access to animal draft power (Vijfhuizen, 1992). Since women do most of the work even in male-owned fields the Western Province Animal Draught Power Programme deliberately targeted women with the intention of contributing to a reduction of farm drudgery, especially for weeding and harvesting.

To involve women, the programme first trained female farmers, to increase their awareness of the benefits of the technology, and second aimed to increase their access to oxen and implements. The extension and training activities during the first phase of the programme included both male and female farmers in the priority areas. In addition, mobile courses and demonstrations specifically for female farmers were carried out. These included courses on ox-plowing, weeding, ridging, groundnut lifting and animal management. The courses and demonstrations were popular; for example in Kaoma District 753 women attended. During Phase II of the programme extra attention was given to resource-poor female-headed households through oxenisation loans for female farmers.

Increasing access to oxen and implements

Early in the programme it was realised that lack of access to oxen and implements was a major constraint to adoption of draft animal power by women. In Western Province cattle have been kept traditionally by men, and women have no 'business' with cattle even if they own some (Beerling, 1986). It is impractical to maintain a kraal for only two oxen so even if a woman had a pair of oxen she would keep them with relatives, who are often men. In this case the men would often give her second priority on the use of the oxen and would ensure that they plowed their own fields first.

To make it possible for women farmers to own oxen and implements the programme introduced barter loans (Leeuwen & Siyambango, 1993). Farmers were given cash loans to buy oxen and implements, and repayments were made in kind, using maize. Nineteen loans were made, including ten to female farmers. Repayment in general was good and by women farmers was 100%. However, it is not known whether the good repayment was a result of benefits accruing through the use of oxen and implements or from other sources of income.

Since adoption of animal draft power technology was expensive for farmers the programme introduced an Agricultural Step-Up Programme as a way of increasing the adoption of draft animal technology among women (Department of Agriculture, 1994). In this programme female farmers were brought together, wealth ranked and assisted to come to a common decision of investment by a female 'Animal Draft Power Promoter'. From their savings, which were topped-up by the programme by 50%, the farmers were allowed to buy some fertiliser and seed for maize. It was hoped that timely maize planting followed by good rains would result in better yields, resulting in income that would enable farmers to invest in animal power technology if they wished. This programme was tried for the 1994/95 season and involved 250 women farmers, mainly from resource-poor female-headed

households. Unfortunately, it did not yield the results expected because of drought.

Gender experiences in animal traction development

The following are some of the experiences of the Western Province Animal Draught Power programme relating to the training of women.

It was found that the initial inability of women to handle oxen and implements was due only to traditions and cultural biases which had often excluded them from such tasks, believing that they were a 'man's job'. As a result of courses and demonstrations women farmers developed the confidence to approach oxen, span them, and guide them into the furrow to start plowing. However, after training they had little opportunity to practise on their own fields because of lack of access to oxen and implements.

Men did not want women to learn to plow because they thought that by doing so the status of men would be diminished. Traditionally, very few women are involved in ox-handling; they are therefore unfamiliar with oxen and tend to fear them. This makes it difficult for them to plow. Draft animals are regarded traditionally as wild animals which should be approached by men who seem to be stronger than women. However, it was observed that women who have plowed before are confident and more ready to use the skills. It is important to educate men about the benefits of women's involvement in animal draft power activities.

Few female farmers use their skills after training because of help from husbands or relatives. Women from female-headed households are more likely to utilise the skills acquired from the courses and demonstrations than those who are married or live with male relatives.

During plowing courses, women expressed their desire to learn how to yoke oxen, proper plow handling, guiding oxen into the furrow and at headlands, plow adjustments, function of plow parts and maintenance. Most women preferred these topics, as they contributed to making them more independent from their male counterparts (Hocking, 1991). Training in ox-management and health care was requested. Some women felt that since the opportunities for women to use oxen had been advanced by the courses and demonstrations, it was important that women should be aware of how they can best keep oxen. The farmers considered that diseases were contributing to the loss of productivity among oxen and as such their control or treatment would be of paramount significance to women farmers involved in animal traction development.

Most of the women who attended animal draft power courses were divorced, widowed or single.

Women tend to have greater freedom of expression and ability to respond positively when they are trained separately from male farmers. Mixed groups resulted in women letting men to do the heavier tasks.

Some women mentioned that they already had many activities to perform and did not have enough time to plow as this would increase their workload.

Besides being responsible for household tasks, women are involved in labour-intensive production tasks including field clearing, plowing, harrowing planting, weeding, ridging and harvesting. Weeding is believed traditionally to be a task for women and becomes a major labour burden to women farmers, especially in the absence of animal draft power.

Another reason is that women, especially, female-headed households have less access to oxenisation loan facilities due to lack of collateral. As a result this leads to limited animal draft power ownership and access. The barter loans to a large extent under-played the need for collateral.

In general, women complained that the courses were too short (eg one-week plowing course) for them to master the skills of plowing and ox handling (Simwinji, 1994). In addition lack of practice after training was perceived to be a problem (Simwinji, 1994). It was generally accepted that a refresher course just before the rainy season started would be helpful.

Benefits to women involved in animal traction technology

The benefits of animal power mentioned by women farmers included:

- women farmers acknowledged that yoking and plowing on their own helped them to become independent and enabled them to increase their crop production

- the demonstrations of reduced labour requirement through the use of animal draft power weeding and harvesting implements resulted in an increase in area cultivated and generally reduced farm labour demand for farmers using animal power

- women who mastered the skills did not need to hire someone to plow or handle oxen for them since they could do it themselves

- the skills taught could increase the income of women farmers through hiring by other farmers in the community. However, observations show that this is rare.

Conclusions and recommendations for future programmes

- Animal draft power technology programmes should target women to provide the opportunity for both women and men to benefit from the technology. This is important because whilst the majority of men already have cattle-related skills, women do not.

- Teaching women how to plow provides an opportunity for more effective use of animal draft power technology and increases the total input of plowing power into farming systems. With animal draft power, the potential for commercialisation can be better exploited. Increased production and the potential for cash crops could then lead to an increase in the standard of living of the households involved.

- Marital status influenced the impact of the courses and demonstrations. Single women made use of the opportunities to apply their skills more than married women, though married women seemed to have more access to oxen than single women.

- Women who have already used oxen are more likely to realise the advantages of improving their skills and thus are more likely to attend a course than women who have not used oxen before.

- Since labour-intensive operations such as weeding and groundnut growing are performed mainly by women, increased use of animal traction technology for these operations would reduce the labour production burden which they currently experience.

- Prior to training courses it should be made clear that participation is open so that farmers do not feel that they have to be selected before they are able to attend.

- Although specific courses should be targeted at women, men (particularly husbands) should also be involved during the sessions or meetings. By so doing more women are likely to attend, as the social stigma would have been removed.

- Course time and duration should be adjusted to suit the needs and the farming situation in anarea (eg plowing course should be set at the time of plowing operations and this could enable participants to correlate theory to practice of plowing).

- Female-headed households are in a worse situation than women from male-headed households in terms of need to plow on their own. Special attention could be given to them as a priority group requiring the knowledge. Access to credit and other agricultural inputs could improve their standards of living.

- Field officers associated with animal draft power programmes need to be receptive to female farmers. Cultural barriers can be overcome by adequately educating or informing men about the benefits of women's use of animal draft power.

- It is advisable to select a few women (head-of-households) from target groups to attend an intensive animal draft power course who could become trainers of other women in animal draft power gender programmes.

- Plans for introduction of use of draft animals by women should take into consideration the existing cultural aspects of animal use in the target area.

- Follow-up of training courses could be conducted shortly after the initial course programme to assess the results.

Acknowledgements

The Western Province - Animal Draught Power Programme (WP-ADPP) was sponsored jointly by the Government of The Netherlands and the Government of the Republic of Zambia. It was implemented in Zambia through the Agricultural Engineering Section of the Department of Agriculture. RDP Livestock Services BV of Zeist in the Netherlands implemented the programme on behalf of the Government of the Netherlands.

References

Beerling M L, 1986. *Acquisition and alienation of cattle in Western Province, Zambia.* Ministry of Agriculture and Water Dvelopment, Department of Veterinary and Tse-tse Control, Mongu, Zambia. 136p.

Department of Agriculture, 1992. *Western Province Animal Draught Power Programme, Second Phase Project Proposal.* Ministry of Agriculture, Food and Fisheries, Lusaka, Zambia. 33p.

Department of Agriculture, 1994. *Annual Report, Western Province Animal Draught Power Programme.* Ministry of Agriculture, Food and Fisheries, Lusaka, Zambia. 26p.

Hocking C, 1991. *The impact of mobile ox-plowing courses for women: A study in Kaoma East and in the areas surrounding the Lui river Valley, Western Province.* Msc Thesis. Department of Agriculture and Environmental Science, University on Newcastle-upon-Tyne, Newcastle-upon-Tyne, UK. 53p.

Leeuwen M and Siyambango N, 1993. *ADP loans in Kaoma District, a technical evaluation and assessment of credit for animal draught power in the district of Kaoma.* Department of Agriculture, Ministry of Agriculture, Food and Fisheries, Lusaka, Zambia. 15p.

Livestock Census, 1994. Working Paper 95/5, Livestock Development Project Phase II, Department of Veterinary and Tsetse Control Services, Ministry of Agriculture, Food and Fisheries, Lusaka, Zambia. 206p.

Simwinji N, 1994. *Report of the evaluation study of ox-plowing courses provided to women in rice groups by the Rice Promotion Programme.* Department of Agriculture, Ministry of Agriculture, Food and Fisheries, Lusaka, Zambia. 204p.

Vijfhuizen C, 1992. *Borrowers of oxen.* Livestock Development Project, Department of Veterinary and Tsetse Control Services, Ministry of Agriculture, Food and Fisheries, Lusaka, Zambia. 23p.

Gender and animal traction technology in eastern and southern Africa

by

Florence Lawan Tangka

Food and Resource Economics Department, Institute of Food and Agricultural Sciences
PO Box 110240, University of Florida, Gainesville FL 32611-0240, USA

Abstract

Animal traction technologies are not gender neutral. Men are generally involved in cash-earning activities and often have more access, control and ownership of improved technologies, such as animal traction, than women, though women do most of the domestic and agricultural activities.

Sustainable agricultural development is necessary to ensure the survival of the rapidly growing populations of eastern and southern African economies. This can only be achieved through the introduction of appropriate technology to boost the agricultural productivity of the majority of the farmers, most of whom are women. The time and effort put in by women to carry out their agricultural and domestic activities (such as weeding, harvesting, transportation of fuelwood, harvested crops and water) can be reduced greatly by the use of draft animals and appropriate animal traction technologies.

For animal traction technology to be developed and disseminated successfully, gender issues, such as differential access to land, credit, implements and extension services, need to be taken into consideration. A careful analysis of the potential constraints and benefits is important because the introduction of gender-blind animal traction technology can have different effects/impacts on male and female farmers, given the sexual division of labour and socio-cultural factors prevailing in the region. Women's lack of access to productive resources limits their ability to adopt animal traction technology and thus greatly reduces their production potential.

Introduction

Sustainable agricultural development in eastern and southern Africa can only be achieved through the introduction of appropriate technologies which will boost the agricultural productivity of the majority of the farmers, most of whom are women. Over the last twenty to twenty five years, there has been growing interest in the development and dissemination of appropriate technology for small-scale farmers in the region. Development of animal traction for smallholder farmers has long been identified as an appropriate technology. This is as a result of past failures of the more capital-intensive technologies, the costs of which are now very high. The spread of 'Green Revolution' technologies has been slow and uneven in the region due to

- lack of infrastructure for distribution of capital
- rising prices for chemical inputs
- lack of specific suitable packages for the highly variable soils, rainfall, and crops.

The need to increase the farm power available in the agricultural systems of eastern and southern Africa, where most of the farm activities are carried out manually with a hoe, is receiving particular attention.

Animal traction, used to supplement or replace human power, has proved to be of great benefit to farmers. It can allow farmers to increase labour productivity, overcome seasonal bottlenecks, increase yields because of improved timeliness of operations (particularly weeding), provide a source of manure, and, where appropriate, increase the area under cultivation. These benefits can only be realised fully if the introduction of animal traction technology is gender sensitive. Women play a dominant role in food production but face enormous social, cultural and economic constraints.

Intrahousehold activities

Farming households in eastern and southern Africa are complex institutions. The complexity arises from the numerous production systems and different accounts within the household, some of which are managed jointly by men and women, some by men only, and others by women only. In contrast with most parts of the world where households function as a single economic unit with common goals, resources and benefits, the practice in eastern and southern African households is different; family members have separate and sometimes competing own-account activities (Saito, Spurling and Mekonnen, 1994). Therefore the individual, rather than the household, constitutes the basic unit of production for this region.

Individuals are responsible for putting together factors of production for joint or own-account activities. Elaborate arrangements within the household determine what is to be produced, by whom, how it should be produced, and how the produce should be disposed – what quantity to consume, sell or store. A complex set of rights and obligations within the rural household, reflecting biological differences, social and religious norms, and customs dictates the gender division of labour, and land use. These generally place women at a disadvantage, resulting in differential access to factors of production, control of proceeds, and in more general terms, asymmetric economic and social relationships, thus distorting terms of exchange among household members.

The nature of the intrahousehold dynamics, the complexity and asymmetry of the rights and obligations between men and women in the farm household, and their evolution are important factors in animal traction adoption. The effective development and adoption of animal traction technology in eastern and southern Africa depends critically on the understanding of who does what, with what resources, and the process of decision-making within the household.

Women's role in agriculture

Several decades of studies have clearly demonstrated the pivotal roles women play in the agricultural sectors of developing countries. The changes in intrahousehold arrangements have particularly increased women's agricultural responsibilities. Data on the role of African women in agriculture are fragmented. The International Labour Organization (ILO) estimates that 78% of females in Africa are economically active in agriculture compared to only 64% of men (Buvinic and Lycette, 1988). Estimates from FAO reveal that women account for about three-quarters of the labour required to produce the food consumed in

eastern and southern Africa. Also, estimates of the time rural women spend on a variety of production activities in sub-Saharan Africa show that they contribute two-thirds of all hours spent in traditional African agriculture, three-fifths of all hours spent in marketing, and produce more than three-quarters of all eastern and southern Africa's basic food (Gittinger et al, 1990; Saito, Spurling and Mekonnen, 1994). Even though women contribute more labour to agriculture than men, men cultivate larger areas and produce more, due to their access to improved technology (Sylwander, 1992). Studies carried out by Tangka (1994); and Saito, Spurling and Mekonnen (1994) indicate that if male and female access to inputs were less unequal, substantial gains in agricultural output would occur, raising food production.

Aggregate data indicate that women account for more than 90% of the labour required for food processing, provision of household water and fuelwood, 80% of the labour for food storage and transportation, 90% of the labour for hoeing and weeding, and 60% of the work of harvesting and marketing (World Bank, 1989). These figures show the significant contribution of women to agricultural production despite unequal access to land, inputs, and information. They also highlight the need to incorporate gender issues in the development and implementation of animal traction technology in the region.

Official underestimation of women's labour contribution to agriculture is common and has been criticised for undervaluing female labour, especially efforts made towards unpaid work, as well as for undervaluing the worth of female-produced goods and services (Dixon-Mueller, 1985). The fact that women's involvement in agriculture is typically much greater than official figures suggest has enormous implications for the type and support of animal traction development and implementation as related to women farmers.

Potential benefits of animal traction use

The potential of animal traction is noted in a growing body of literature. Animal traction technology has been traditionally used in cash-crop production and tasks generally designated to men. Animal traction has been used mainly for land preparation. This has resulted in large areas being put under cultivation, with no corresponding use of animal traction for weeding. Animal traction use for land preparation alone can therefore increase the burden of work on women. Weeding is predominantly regarded as women's work and is the most constraining activity in eastern and southern Africa. Yield losses due to delayed weeding have been shown to average 8% per week for millet and sorghum in Northern Nigeria and 30% for a two week delay in cotton (Jaeger, 1986). The use of animal traction for weeding is yet to be fully realised in eastern and southern Africa. According to Jaeger (1986), weeding implements are used by less than 25% of animal traction farms in West Africa.

Women are responsible for weeding, food processing, harvesting, fetching and transportation of water and firewood – all of which take much time and effort. Use of animal traction in such activities can lead to delegation of women's duties to other members (especially young boys and men) within the households and hence reduction of the time and effort expended by women.

Determinants of women's adoption of animal traction technology

The number of female-headed households is growing with increasing migration of men from rural areas. Use of draft animals should help these women produce more and should reduce drudgery and labour bottlenecks. Animal use should also save women's time which will be of enormous benefit to communities in which they live, as more time will be spent on other communal activities.

However, use of animal traction requires time and cash investment, as well as access to land, credit, information, and implements. Though women play important roles in agricultural production, they usually have lower levels of physical and human capital than men (Tangka, 1994; Saito, Spurling and Mekonnen, 1994; Quisumbing et al, 1995). These differences arise because of legal, social, and institutional factors that create barriers for women to adopt new technology. Consequently, most women have less access to and higher effective costs for, land, information, technology, inputs and credit.

Access to land

Land, whether allocated, purchased, inherited, or seized, is the most basic resource in agricultural

production. In eastern and southern Africa, agricultural land reforms, the legal system and the patrilineal transmission of property have increasingly concentrated land rights and ownership in the hands of men, giving women little access to its ownership (Henn, 1983). Even in situations where civil law gives women the right to inherit land, local custom may rule otherwise (Quisumbing et al, 1995). In eastern and southern Africa where women do most of the agricultural activities, they are limited to use rights to land, and in most cases this is only with the consent of their male relatives or male community leaders. Women's social status impinges upon their access to land and consequently on their role in agricultural development. A positive relationship between individual land privileges and productivity has been established. Increased individualisation of land rights can improve a farmer's ability to gain returns from land investment, resulting in greater demand for land improvement and complementary inputs. Women's insecurity of land tenure reduces the likelihood of them investing much time and resources in usufruct land that they can be chased out of at any time. Clearly defined land rights improve the creditworthiness of farmers and augment the probability of obtaining formal credit for the purchase of animal traction implements. Women's inheritance of land was often discouraged traditionally by the widely-held assumption that a daughter marries and gains access to land in her husband's compound.

Access to credit

Cash flow, financial services or credit are important and necessary for the purchase of draft animals, implements and other farm inputs such as improved seeds. These are quite difficult for women to acquire, particularly those heading households with limited cash income. Property that is accepted as collateral for credit, (eg land) is normally held by men, and formal financial institutions often refuse the types of valuables (eg jewellery) held by women. The transaction cost involved in applying for credit is generally higher for women than for men owing to a higher opportunity cost for neglected activities. Lack of credit and limited income constrain women and female-headed households from investments in such productivity-increasing ventures. Women's low educational levels relative to men, their lack of familiarity with loan procedures, and social and cultural barriers limit their mobility and interaction with predominantly male moneylenders. Women who are not members of local groups such as farmers' groups, may be prevented from receiving not only extension advice, but also credit in the case where the extension worker plays an important role in obtaining it. Women are generally involved in production of relatively low-return crops that are not included in formal lending programs. Women with limited cash find it difficult to invest in animal traction because death, loss or injury of their animal can be very risky for them. It is therefore important to consider the needs and limitations of women in the development and dissemination of animal traction technology.

Access to agricultural extension

Extension has been shown to increase agricultural productivity and close the gap between technical knowledge and farmer's participation, by spreading information about new techniques. Literature shows that extension is cost-effective, and has a significant and positive impact on farmers' knowledge, adoption of new technology and hence farm productivity. Despite women's role in agriculture, they do not get a fair share of agricultural extension advice and other services (seeds and fertiliser provision). In eastern and southern Africa, where women and men may be responsible for different crops, it is very important that women get access to extension services. Unfortunately, men are still the main targets for modern extension services. Additionally, many societies place restrictions preventing male extension workers from meeting with female farmers and extension messages do not get passed on by men to women in the same household.

A relatively small number of extension agents are women, and it cannot be assumed that female extension agents are aware of the gender issues in agriculture and technology adoption. Few female extension agents have the relevant training in animal traction technology, and consequently have very little, if any knowledge to pass to women farmers (Sylwander, 1992). Domestic responsibilities sometimes render women immobile, making it impossible for them to attend meetings and courses away from home. Women

are less likely than men to speak the national language; this limits their participation in the extension meetings, as these often are not offered in the local languages. Thus unless timing, location, training and extension programs take into account women's multiple roles and responsibilities, particularly the severe constraints on their time and mobility, information and technology transfer may never reach them.

A note on methodology

The complexity of the gender issues in animal traction can be analysed using a linear programming model. Linear programming provides the framework within which issues such as the following can be examined:

- the appropriateness and comparative advantages of using specific combinations of draft animals and animal implements for male and female farmers – who have different access to capital, credit and grow crops for different purposes

- the evidence of complementarity that can be exploited from the practice of mixed farming (crop and animal production)

- the causes and implications of the under-utilisation of draft power

- the effects of farm size and economies of scale on the profitability of animal traction.

The linear programming model is a whole-farm approach which has the potential of providing a realistic assessment of the suitability and acceptability of animal traction technology. The complexity of the trade-offs of labour used for different tasks and different activities during a growing season can be difficult to sort out. Parametric variation of a linear programming model can be used to examine systematically the benefits from utilisation of animal traction, its effects on gender division of labour, potential scale economies, and the seasonality effects that are characteristic of the subsistence rainfed agriculture of eastern and southern Africa. Additionally, parametric variation of the model can provide insights about adjustments and responses to animal traction technology, that would otherwise require a much larger multi-year data collection effort to apply econometric techniques.

Linear programming is a computational technique for solving linear objective functions subject to a system of constraint equations. The most common farm models maximise profits (or cash income), after satisfying subsistence needs, for fixed input-output activities subject to constraints on resource availability by choosing the best cropping and/or animal production patterns.

Adoption of animals can be included as integer variables in the model. Each of these may be considered with the different sets of animal-drawn implements – the plow only, the weeder only, and plow and weeding implements together, as well as using animal traction for planting, harvesting and transportation. The model can be solved with each or a combination of these choices imposed.

The linear programming model is a whole-farm approach that can incorporate household objectives, risks, and constraints that other analysis techniques cannot handle. However, this requires detailed and reliable information about the objectives of the decision-maker, the farm-level constraints, and the input-output relationships.

Conclusion

Animal traction technology can provide the opportunity to increase agricultural and labour productivity for all farm activities (land preparation, planting, weeding, and harvesting), and to decrease drudgery and time spent in transportation of crops, water, fuelwood etc. Because women normally have a considerable workload, the direct and indirect benefits from the use of animal traction should increase agricultural and labour productivity and have a positive impact on the well-being of families and communities of which they are a part.

These benefits can only be realised if women, the majority of farmers in eastern and southern Africa, can make full use of animal traction technology. Full benefits can only be derived from the technology if development and implementation guidelines are made to ensure that women's needs,

opportunities or constraints be considered at all levels, and in all activities. Until training and extension programs include well-planned, active women's components, information and animal technology transfer is unlikely to reach female farmers. Thus, women's contributions and production potential are never fully realised.

References

Buvinic M and Lycette M, 1988.*Women, poverty, and development in the third world. Strengthening the poor: what have we learnt?* Third World Policy Perspective. No. 10. Overseas Development Council, Washington DC, USA.

Dixon-Mueller R, 1985. *Women's work in third world agriculture.* International Labour Organisation (ILO), Geneva, Switzerland.

Gittinger P J, Chernick S, Horenstein N R and Saito K, 1990. *Household food security and the role of women.* World Bank Discussion Paper Number 96. World Bank, Washington DC, USA

Henn J K, 1983. *Feeding the cities and feeding the peasants, what role for Africa's women farmers?* World Development Report 11(12). Pergamon Press

Jaeger W K, 1986. *Agricultural Mechanization. The Economics of Animal Draft Power in West Africa.* Westview Press / Boulder, Colorado, USA and London, UK.

Quisumbing A R, Brown L R, Feldstein H S, Haddad L, and Pena C, 1995. *Women: The Key to Food Security.* Food Policy Report. International Food Policy Research Institute, Washington DC, USA.

Saito K, Spurling D and Mekonnen H, 1994. *Raising the productivity of women farmers in sub-Saharan Africa,* World Bank Discussion Paper No. 230. World Bank, Washington DC, USA.

Sylwander L, 1992. Women and Animal Traction Technology. In: Starkey P, Mwenya E and Stares J (eds), 1992. *Improving animal traction technology.* Proceeding of the first workshop of the Animal Traction Network for Eastern and Southern Africa (ATNESA) held 18–23 January 1992, Lusaka, Zambia.Technical Centre for Agricultural and Rural Cooperation (CTA), Wageningen, The Netherlands. 480p.

Tangka F, 1994. *Analysis of male- and female-headed households' agricultural productivity: the case of Kenya.* Masters Thesis. Department of Agricultural Economics, Rutgers, New Jersey, USA

World Bank, 1989. *Women in development: issues for economic and sector analysis,* Policy, Planning, and Research Working Paper No. 269, World Bank, Washington DC, USA.

A note on gender issues in draft animal technology: experiences from Nyanza, Kenya

by

William Onyango Ochido

District Mechanisation Extension Officer, Kuria District, Kenya

Animal traction use diffused into Nyanza Province, Kenya, in the early 1930s. Women in Nyanza Province use animal power for plowing, planting, weeding and transport. Donkeys are used mainly as pack animals. The proportion of families using the technology varies from 30% to 80% within the province. In Kuria about 90% of women handle draft animals and they are involved in all aspects of animal power use. In Luo about 40% of women use animal power. However, although Luo women handle the animals they rarely operate the implements they pull. Kisii communities use little animal power as they have small farms that can easily be tilled by hand.

The main gender-constraints to the equal adoption of animal-drawn technology are:

- women do not have direct access to animal traction technology since they are in most cases reduced to assistants. Also, women's immediate problems such as transport of water, firewood and farm produce are given secondary attention by men

- women have many daily activites which prevent them from participating in animal traction programmes

- cultural constrints hinder the adoption of the technology by women as they are not official owners of the animals and hence all decision-making is done by men

- the animal-drawn equipment that is available is designed for use by men rather than women

- women lack access to credit facilities due to lack of collateral

- men are often cruel to their wives.

Many women are now becoming more involved in farming as a business, partly due to urban migration of (male) youths. In Nyanza Province many villages are now female-headed. Further studies and development of methodologies for participation of both sexes in animal traction adoption should focus on:

- understanding the position of women and the status and roles of women in the particular farming community

- analysis of the gender issues relating to the control of resources and access to resources

- involvement of the whole farming family in the activites of projects and programmes

- giving special attention to female-headed households in analysis, methodologies and implementation

- women should participate in research and development of appropriate animal traction technology

- training programmes should be gender-aware. A typical example is that in Kuria a course on animal traction was mostly attended by men even though it is not they who till the land.

Meeting the challenges of animal traction

Photo: Paul Starkey

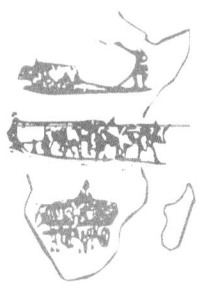

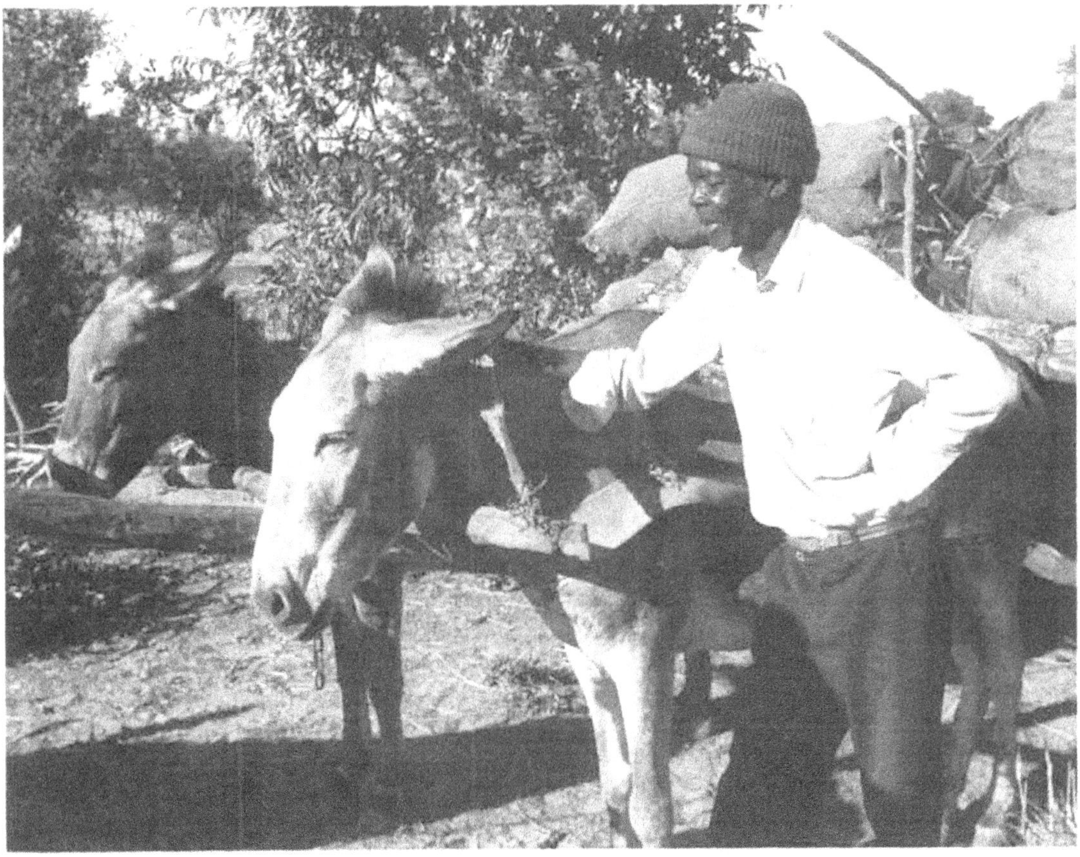

Farmer knowledge, extension and training

Farmers' informal knowledge in animal traction: case studies from the Southern Highlands of Tanzania

by

E Kwiligwa [1], J Rock [1], M Massunga [2] and M Sizya [2]

1) Ileje Food Crop Production Project (IFCPP), PO Box 160, Itumba, Ileje, Mbeya, Tanzania
2) Oxenisation Extension and Training Services (OXETS), PO Box 873, Mbeya, Tanzania

Abstract

The Ileje Food Crop Production Project (IFCPP) is a Coopibo (Belgian Organisation for International Cooperation) supported project operating in Ileje District of Tanzania. Oxenisation Extension and Training Services (OXETS) is a business organisation based in Mbeya but operating throughout Tanzania. The two organisations are working together to promote animal traction in Bulambya Division, Ileje District.

Both organisations believe in the participatory process as an important tool to improve farmers' self-confidence and strengthen their ability to take development into their own hands. The participatory rural appraisal (PRA) approach is used, whereby farmers are mobilised to analyse problems, seek possible solutions, experiment, and then evaluate the results according to their preferences. During the process valuable knowledge and experience of farmers are utilised as well as formal technical knowledge.

This paper describes two case studies that show how farmers are involved in developing solutions to their own problems. The problem in the first study is that ox carts are too expensive in the area; the solution has been to develop an improved four-wheeled sledge, with bush bearings made from used tins, a raised platform and wheels lined with rubber from used car tyres. In the second study farmers have identified some medicinal plants for treating draft animals. For example, utupa or fish poison (Tephrosia vogelii) is used to control ticks in livestock. The medicinal plants selected have been sent to the Tropical Pest Research Institute (TPRI) for identification and further work on classification, active ingredients, toxicity, dosage and alternative uses.

Characteristics of Ileje District

Ileje is one of eight districts in Mbeya Region in the Southern Highlands of Tanzania. It is located between latitudes 9° 15' and 9° 38' south and longitudes 32° 50' and 33° 45' east along the Tanzania–Malawi border. Ileje District has two divisions, with distinct agro-ecological zones. Bundali Division is in the highlands: it covers about 1000 km² in the south-west part of Tanzania, and much of the area is dissected escarpment with slopes generally exceeding 30% at altitudes between about 1200 and 2000 m above sea level (Kwiligwa, Ley and Mfanga, 1994). Bulambya Division is in the lowlands, at altitudes ranging from 1200 to 1700 m. The Songwe river passes through Bulambya Division, giving it a high potential for irrigation.

Bundali Division has a cool climate with annual rainfall of 1500–2000 mm: rain falls through most of the year (with only a two-month dry season). Bulambya Division is relatively dry with distinct dry and rainy seasons. Rainfall is erratic and rather unreliable, spread over six months (November to April). The low rainfall in the 1993/94 season led to drought and hence a serious food shortage

According to the 1988 census Ileje District had a population of 88,580 people (52% female) with a density of 40 persons/km² in Bulambya Division and 100 persons/km² in Bundali: the average population density over the whole country is only 26 persons/km². As a result land is scarce, especially in Bundali Division.

The farming systems

Agriculture (including livestock) is the chief source of income in the area: even the urban community is actively engaged in this sector. The major cash crops in Bundali Division are coffee, cardamom and bananas; the important food crops are maize, beans, sweet potatoes and bananas. Bulambya Division has no specific cash crop: the major food crops – maize, beans and sweet potatoes – are sold as cash crops when there is surplus production. Production of some traditional crops such as finger millet, groundnuts, sunflowers and cassava is increasing. Sunflowers and rice are potential cash crops in Bulambya Division and their production is increasing rapidly because there is a promising market. Ridge cultivation, mixed cropping and agropastoralism are important components of the farming system in Ileje district. In 1982 the Rural Integrated Development Project (RIDEP) in Mbeya showed that only about 13% of the total land area in Ileje was cultivated by oxen (RIDEP, 1982).

The district has the lowest animal population in Mbeya Region (about 3% of the regional population). The most common animals reared in the district are cattle (zebu), pigs, chicken, goats and sheep. Most of the animals, especially zebu cattle, can be found in Bulambya Division where animal traction is an important technology.

The objective of this paper is to highlight the animal traction activities carried out by the Ileje Food Crop Production Project in collaboration with OXETS (Oxenisation Extension and Training Services). Two case studies are described to show how the projects participate with farmers in technological development.

Ileje Food Crop Production Project

The Ileje Food Crop Production Project (IFCPP) is a tripartite organisation receiving personnel and financial support from Coopibo (the Belgian Organisation for International Cooperation), CDTF (Commnity Development Trust Fund) and Ileje District Council.

IFCPP was established in 1989 and has an overall objective of increasing the living standards of the smallholder farmers of Ileje through promoting food crop production. The intervention aims include: irrigation potential tapped, extension services improved, labour productivity increased through the application of animal traction and farmers organised so as to empower them. IFCPP uses Objective Oriented Planning in its planning processes. The project believes in the participatory process as an important tool to boost the self-confidence of farmers (both men and women), and strengthen their ability to take development into their own hands. The Participatory Rural Appraisal approach is used, whereby farmers are mobilised to analyse a problem, seek possible solutions, experiment and then evaluate the results according to their preferences.

During the process valuable knowledge of farmers is utilised either directly or with some modification. More often than not, animal traction activities at IFCPP are subcontracted to (ie done in collaboration with) OXETS, based in Mbeya. OXETS is a privately owned business organisation which originated in the Mbeya Oxenization Project with an idea of sustaining project services. The services provided or offered by OXETS include:

- three-week courses on draft animal technology suitable for rural development workers, coordinators, supervisors and farmer trainers
- one- to two-week problem-solving-oriented training courses/seminars for oxen users
- Participatory Rural Appraisal surveys related to animal traction
- animal traction related studies, for example, impact studies, feasibility studies, etc
- extension related animal traction activities, · such as giving demonstrations of different technologies (eg, ox weeding), organising competitions, and designing and producing leaflets, manuals and other promotional materials.

Animal traction programme

A monitoring exercise in 1995 estimated the number of oxen in Bulambya Division at 18,; most of them in Chitete Ward. There are nearly 250 oxen users, 65% of whom hire animals (mainly from Mbozi District). The cultivated area is about 2840 ha. Assuming that a pair of oxen can plow about 4 ha per season, only about 13% of the land is worked by oxen.

The original strategy of the project was to establish a farm service centre where ox mechanisation could be one of the services rendered by training farmers and their animals and distributing animal-drawn implements. However, very few farmers came to the centre for training. An impact study carried out in 1993 showed that even without major effort from the project (except for distribution of animal-drawn implements), the use of oxen for plowing had increased from 14% in 1989 to 20% in 1993 (IFCPP, 1993). It was then realised that there is substantial networking and knowledge exchange among farmers. The project decided to change its approach and started devoting more effort to catalysing and enhancing farmer-to-farmer exchange of knowledge.

Village ox mechanisation trainers

In each village ox users select two of their number to be trained as ox mechanisation trainers. These trainers are then supposed to train an agreed number of farmers and their animals each year. The terms and conditions of the training programme are agreed mutually by the trainer and trainees without any influence by the project. The training activities are monitored by IFCPP and OXETS to determine constraints and opportunities.

Demonstrations and competitions

Demonstrations and competitions are held annually in collaboration with OXETS in pilot villages. Winners in the competitions (gender specific) are rewarded with animal-drawn implements and spare parts. At these occasions farmers exchange knowledge and evaluate the performance of their trainers. They also identify and elect the traditional healers who receive training in modern veterinary techniques (Massunga and Sizya, 1995).

Case studies

A problem analysis exercise undertaken by a group of farmers in 1993 identified two serious constraints to ox mechanisation:

- they could not afford to buy ox carts, which posed a major transport problem
- they could not afford to buy modern veterinary medicines, which presented a major challenge to draft animal health.

Under the guidance of the project (through follow-up surveys and a workshop), the farmers themselves sought solutions to these problems.

Case study on traditional sledges

As a follow-up to the problem analysis exercise, IFCPP conducted a survey to investigate the extent to which the rural communities are tackling the problem of unaffordability of ox carts. The survey revealed that several designs of traditional sledges, with and without wheels, are being used (Mbepera and Rock, 1994).

In March 1995 the project organised a workshop for all farmers using sledges for farm transport. The aim was to cross-check the survey results, assess the limitations of sledge designs and propose improvements to make them a more efficient and reliable means of transport. The

farmers analysed the problem of sledges currently in use, and built a problem tree.

Solutions to the problems were discussed. It was agreed that the wheeled sledge is the best option, and that the following modifications would make sledges stronger and increase their reliability:

- provision of plain (bush) bearings (made of galvanised iron pipe or rolled sheet of tin)
- use of tapered scrap-iron bars to join the frame structure
- raising the platform to distribute the load evenly over the wheels
- lining the wheels with rubber from used car tyres to increase cushioning of the sledge
- using harder wood
- more frequent greasing
- avoiding overloading (maximum load 300 kg).

Two farmers were nominated to fabricate the new prototype, and IFCPP supplied the required tools and some of the materials. A farmer in Msia village has already made one of the improved sledges and is now assessing its performance.

Case study on ethnoveterinary medicine

The same 1993/94 problem analysis exercise, referred to above, identified unaffordability of modern veterinary medicines as a major challenge to draft animal health.

IFCPP's follow-up animal health survey investigated the traditional medicines used by farmers to control the various animal diseases and pests (ectoparasites and endoparasites) common in Bulambya Division. The objective of the March 1995 workshop (attended by users of traditional medicines) was to translate the outcome of the survey into a concrete plan of action (Kwiligwa, Mbepera and Rock, 1995). The basis for this plan was a problem tree developed from discussion of the problems faced by users of ethnoveterinary medicines in the area.

The traditional healers felt that most of the problems could be solved by enhancing knowledge among healers within and outside the project working area. Six of the most important and widely used medicinal plants (*Nanjoka, Nandete, Namatusi, Ikukulemba, Pembambuzi* and *Ntelembe*) were selected for further research. Samples of these have been sent to the Tropical Pests Research Institute (TPRI) in Arusha for classification, identification of active ingredients, and studies on toxicity, dosage and alternative uses.

In the meantime IFCPP is promoting the use of *utupa* or fish poison (*Tephrosia vogelli*) to control ticks on draft animals. To make a spraying solution, freshly picked green *utupa* leaves are dried in the sun for one day and then pounded and mixed with water (1 kg of dried leaves in 10 litres of water) and a small piece of neutral soap (to make the spray stick to the animal's skin). The mixture is stirred out overnight. Then it is filtered and ready for spraying. The oxen at the project have been sprayed once a week for more than one year, and no ticks can be found on them.

Conclusions

External knowledge and farmers' knowledge should complement each other in order to achieve a more sustainable agricultural development. The two case studies at IFCPP give a practical example of integrating formal and informal knowledge in a participatory development process. It is crucial that both the project workers and farmers involved should play their role effectively so as to avoid frustrating the other party.

References

Kwiligwa E, Ley G and Mfanga B, 1994. *A rapid rural appraisal of Bundali Division, Ileje District.* IFCPP (Ileje Food Crop Production Project), Itumba, Ileje, Mbeya, Tanzania.

Kwiligwa E, Mbepera A J and Rock J, 1995. *Ethnoveterinary medicines and improved traditional ox-carts in Ileje, Tanzania.* Proceedings of a workshop organized by Ileje FCPP 15–19 March 1994. IFCPP (Ileje Food Crop Production Project), Itumba, Ileje, Mbeya, Tanzania.

IFCPP, 1993. *Preliminary report of impact study: February to March 1993.* IFCPP (Ileje Food Crop Production Project), Itumba, Ileje, Mbeya, Tanzania.

Mbepera A J and Rock J, 1994. *Report on survey of traditional sledges and ethnoveterinary medicines in Chitete ward Ileje District.* IFCPP (Ileje Food Crop Production Project), Itumba, Ileje, Mbeya, Tanzania.

Massunga M and Sizya M A, 1995. *Consultancy report on ox weeding training for ox mechanisation trainers.* Report prepared for IFCPP, March 1995. OXTETS (Oxenisation Extension and Training Services), Mbeya, Tanzania.

RIDEP, 1982. *Proposal for an indicative development strategy for Mbeya Region, 1982 to 2000.* Volume 1. Report 49. RIDEP (Rural Integrated Development Project), Mbeya, Tanzania.

Towards privatised draft animal power extension in Zambia

by

Martin van Leeuwen

Advisor, Western Province Animal Draft Power Programme, PO Box 940007, Kaoma, Zambia

Abstract

The Western Province Animal Draught Programme (WP-ADPP) has been promoting draft animal power in Kaoma District in the Western Province of Zambia, through research, extension and provision of credit. The training-and-visit extension system assumes that improved agricultural technology offered to village extension groups, will trickle down to the rest of the farming households automatically, via a process of diffusion. However, critical research into diffusion has concluded that it only takes place within homogenous populations and areas; unlike Kaoma District. A second problem with the training-and-visit system is the quality of the information disseminated: there are few clear extension messages that could contribute substantially to improved production. WP-ADPP believes that high quality extension requires taking account of the relevant differences in client categories and farming systems in a manner similar to marketing and advertising strategies.

Following dramatic changes in the availability and costs of inputs and crop prices, WP-ADPP initiated an association of farmers and traders. The Kaoma Input & Marketing Investment Association (KIMIA) consists of 15 rural farmers/traders and offers farmers the opportunity to barter maize against fertiliser, seed, oxen, plows and spares and to subscribe early in the season for information and inputs for crops other than maize. Providing extension information is a major part of KIMIA's work. Within KIMIA, crop extension actors are paid according to results, which are established during monitoring of a sample of their clients. Farmer/traders are not paid for their extension work but work on the basis of expected commission on their own barter trade. KIMIA has provided a self-targeting exchange of information that is of most benefit to farmers who already own animald and were most affected by the decline of government-supported lending facilities.

Background and farming systems
Population

Kaoma District had a population of 113,000 people in 1990, with 88% living in rural areas. Roughly 60% of the population live up to 10–15 km from one of the four main roads, where the density varies between 10 and 1000 people/km². The remaining 40% live in more remote areas and consist of households which rely heavily on fishing, hunting, forestry, trade and agriculture for subsistence. The population grew by 5% during the last decade as the district attracts immigrant farmers (CSO,1990).

Cropping systems

Maize surplus production increased by a factor of 50 over the past 30 years, which is attributed to the intensive use of fertiliser due to loan provision for its purchase. Kaoma is 170% self-sufficient in staple food production and produces 90% of the marketable surplus of maize, groundnuts and soyabean in the Western Province (PPU,1991). Marketed surplus was produced by 25–50% of the rural households with average yields of 25 bags, about 2.2 tonnes/ha. During 1994 an enormous change took place in input supply and marketing when government support to lending institutions stopped. As a result utilisation of fertiliser and hybrid seed dropped from 25% to 10% of rural households (Kakwaba, 1995).

Livestock

The total livestock population grew faster than the human population over the past 30 years. The cattle population is three times larger than in 1963; this growth rate is the highest of all districts in Western Province (MacLean, 1965; DVTCS, 1992). The overall herd of 28,000 livestock in 1992 is 20% oxen, a much higher proportion than in the rest of the province. Livestock are owned by 22% of households, and 13% own oxen. Of the 22,000 rural households, 26% have access to animal draft power, through owning, lending and hiring (Kakwaba, 1995). Kaoma needs to import about 12% of its required quantity of oxen each year (van Leeuwen and Siyambango, 1995).

The Western Province Animal Draught Power Programme

The Western Province Animal Draught Power Programme (WP-ADPP) of the Department of Agriculture started in 1989 with an emphasis on the task of province-wide coordination. During the first phase (1990–92) more intensive activities were initiated in sub-areas of Kaoma District with a substantial surplus agricultural production. During the second phase (1993–95) these activities have been increased and targeted more precisely.

WP-ADPP Kaoma is part of the Agricultural Engineering Section of the Department of Agriculture. This section assists field staff in 36 agricultural camps in Kaoma District. Other sections are: Animal Husbandry, Home Economics, Land Management and Women and Youth. The main functions of the Department in the period 1993–95 have been:

- promotion of crop cultivation
- crop forecasting and organisation of agricultural shows
- land settlement
- staff training
- promotion of self-help projects
- promotion of animal draft power.

Research

Agricultural research by the Farming Systems Research Team follows farming systems research (FSR) methodology, with trial groups in one 'recommendation domain' in the district and is mainly formulated and evaluated at provincial level. Tests with technology aimed at alleviating farming systems constraints have led to a domain-wide formulation of extension messages regarding fertiliser use, and some useful specific crop memos for extension staff have been produced (Muwamba and Heemskerk, 1991). Research areas with high relevance to WP-ADPP iinclude:

- dry season feeding of animals
- groundnut cultivation and processing
- intercropping with legumes
- soil fertility management.

Useful information on animal draft power in Kaoma District has also been provided by external organisations. The Palabana Animal Draft Power Programme has tested animal-drawn equipment in Kaoma for the last five years, and has trained about 30 field staff. Tests with ridgers and groundnut lifters were successful and resulted in sales afterwards (Hoogmoed, 1992).

Some research into participation in, constraints to and impact of adoption of animal power has been undertaken in Kaoma District (van Agt, 1992; Hocking, 1994). The WP-ADPP has carried out research on credit for animal draft power (van Leeuwen and Siyambango, 1993), on ox hiring and borrowing (Kakwaba and van Leeuwen, pp301–305 in this volume), on oxcarts (van Leeuwen and Siyambango, 1995) and on the stability of use of draft animal power.

Group promotion

Since 1983, the Department of Agriculture has promoted formation of groups of resource-poor farmers for savings and credit schemes and small agricultural projects. The experiences of eight female group promoters within the 80 groups have been variable. Defunct groups contrast with a few groups which have expanded their activities and repaid their loans seriously. Although detailed monthly reporting is still carried out there has been no major analysis of the factors contributing to the successes and failures of the scheme at district level. Group promotion has been relevant to WP-ADPP, since many groups perceived lack of animal power as one of the most important constraints for further expansion of their agricultural production. Some groups acquired draft animals and implements through WP-ADPP.

Extension environment

Training-and-visit extension

The background information about Kaoma District given above shows clearly that that not all farmers in the District are the same. It would be wrong to consider the almost 400 medium-scale commercial farmers as progressive farmers who will soon be followed by all other farmers in Kaoma. This misconception is inherent in the so-called 'progressive farmer extension approach', which is at present implemented through the training-and-visit system. This approach assumes that improved agricultural technology offered to village extension groups, will trickle down to the rest of the farming households automatically, via a process of diffusion (Roling, Ascroft and Chege, 1976).

However, critical research into diffusion has concluded that it only takes place within homogenous populations and areas (Roling, 1988). Kaoma District does not host such an homogenous group of farmers. Apart from differences in ethnic and socio-economic background (ranging from retired officers to hunters), there is unequal access to fertile soils, water, labour and cattle for all rural households in Kaoma. A review of 33 training-and-visit extension projects concluded that their performance has generally been poor as inadequate understanding of the farming systems has often led to inappropriate recommendations (Farrington, 1994). It is the opinion of WP-ADPP that high quality extension requires taking account of the relevant differences in client categories and farming systems in a manner similar to marketing and advertising strategies.

A second problem with the training-and-visit system is the quality of the information disseminated. Clear extension messages that could contribute substantially to improved production in the difficult agro-economic environment of Kaoma District are not plentiful (see also Farrington, 1994).

Most farmers in Kaoma do not perceive methods of crop cultivation to be the main constraints. In general, farmers in Kaoma perceive the lack of a reliable and timely supply of inputs and the lack of a reliable market and transport to be the major constraints (Kakwaba and van Leeuwen, pp301–305 in this volume; 18). In addition, it appears that farmers lack the organisation to overcome their own production constraints. This lack of organisation could, to a large extent, be a consequence of government interference in the recent past.

Extension could play a very useful part in mobilising farmers to reorganise themselves to gain greater benefits. This does not mean that extension should become a jack-of-all-trades again. The experience of training-and-visit extension has made that clear. Agricultural extension should not be involved in credit provision, nor in providing markets; it should definitely focus on exchange and organisation of information within different categories of farmers and on production-related factors. Training-and-visit extension does not incorporate techniques and strategies that can organise the different target categories of farmers to overcome their specific constraints.

Targeting animal draft power extension

The main objective of WP-ADPP is to contribute to increased access to draft animal power by different target groups in Kaoma. In 1993, WP-ADPP identified three general categories of farmers, each with its own constraints to improved access to draft animal power:

- **Category 1** (Owners): Owners of draft animal power, or farmers with enough surplus production to invest individually in draft animal power.Their main constraint is a lack of reliable dealers or insufficient self-organisation to supply oxen, implements and spares and to transport produce and seasonal inputs in time.

- **Category 2** (Potential owners): Farmers with some but not enough surplus production to invest in major components of draft animal power technology. Their major constaint is insufficient opportunities to save and/or borrow capital over 1–2 years for investment in the major components of animal power.

- **Category 3** (Resource-poor households): Farmers, mainly female-headed households, with insufficient resources (labour/capital and sometimes even land) to invest in animal draft power. Their main constraints are insufficient capital/resources to save and insufficient access to markets, inputs and ox-hiring.

Services provided to the target groups

As discussed above, extension services can only be effective if targeted properly, so WP-ADPP provides specific services to the target groups:

Category 1: Owners

The first category appeared to have responded to WP-ADPP's promotion of labour-saving animal powered technology for weeding, harvesting and post-harvesting operations (WD-ADPP, 1995a). The main services which WP-ADPP offers these farmers are detailed written information about local suppliers, prices and implements, and wholesale supply of implements, carts and essential spares to local cart manufacturers (van Leeuwen and Siyambango, pp 176–182 in this volume).

Category 2: Potential owners

The second target group appears to take a high interest in mobile courses on ox-training, plowing and adjustments of plows. Experiments with fifty draft animal power loans towards this target group showed very high repayment rates (over 95%; van Leeuwen and Siyambango, 1993). The Loan and Saving Groups, introduced by WP-ADPP required high down payments and repayment in maize rather than cash to counteract inflation.

Category 3: Resource-poor households

A tentative strategy, the 'Agricultural Step-Up Programme' (Bwalya and Akombelwa, pp85–88 this volume; WP-ADPP, 1994) towards a further integration of this category into animal power-based and surplus-producing agriculture was initiated in 1994. It has been critically analysed by Huisman (1995).

KIMIA: privatised extension

Liberalisation in 1994

During 1994 input and output relations of agricultural production in Kaoma changed dramatically. Firstly, seasonal input supply became erratic, when the Government of Zambia arrested its support to agricultural lending institutions. Secondly, the maize price was left to market forces and farmers had to organise and negotiate marketing of surplus production on their own.

In 1995, after a fruitless effort to support a town-based small-scale trader of agricultural inputs, WP-ADPP decided to initiate an association of farmers and farmer/traders in Kaoma, called KIMIA. (Kaoma Input & Marketing Investment Associations). KIMIA offered farmers the opportunity to barter maize against fertiliser, seed, oxen, plows and spares and to subscribe early in the season for information and inputs for crops other than maize.

Input and marketing associations

KIMIA consists of 15 rural trader/farmers who supply inputs to about 1,000 farmer members in their areas, upon consignment from WP-ADPP's ADP Promotion Fund. Through KIMIA, WP-ADPP hopes to accelerate the adaptation of farmers to the new market conditions. WP-ADPP's decision to initiate KIMIA and diversify its service supply was based on three assumptions (WP-ADPP, 1995b):

- that animal draft power in Kaoma can only increase and be sustained by sufficient marketable surplus production, which requires inputs and markets
- that farmers and rural traders must carry greater responsibility for the costs of lending, marketing and input acquisition
- that project activities should increase regional management capacity and accountability and pave the way for private sector involvement.

As of November 1995, KIMIA had supplied on a barter basis fertiliser and seed for 1000 hectares of maize, 40 oxen, 70 plows, 30 ox carts and spares to a total value of over US$2,000.

Self-targeting extension

Farmers within KIMIA paid a membership fee for extension services, especially on prices and marketing conditions. WP-ADPP uses this income to finance a bi-monthly newspaper called 'The Kaoma Farmer', and is able to facilitate transport and pay local experts for extension on cultivation techniques for new crops and processing techniques. Within KIMIA, crop extension actors are paid according to results, which are established during monitoring of a sample of their clients. As a part of three 'package programmes', WP-ADPP supplies sunflower, groundnut and castor seed as well as oil expellers and other processing and harvesting equipment to specific areas. The package programmes are an effort to integrate a number of criteria for more successful surplus production in distinct producers' areas. Among these criteria, such as financing, mechanisation, and marketing, agricultural extension still plays a major role. The information supply and farmer organisation needed to achieve 1000 clients investing into inputs this year was realised by KIMIA farmer/traders, who organised meetings in their areas, informing farmers about decisions and prices established during KIMIA traders' meetings with WP-ADPP. These farmer/traders are not paid for their extension work but work on the basis of expected commission on their own barter trade. This commission depends on the maize sale price at the end of the marketing season.

Agricultural extension within KIMIA differs from the more fashionable 'participatory extension approach'. It does not only listen to farmers, but expects farmers to pay for relevant information.

KIMIA extension does not enter an 'open-ended discussion' with farmers, but informs about costs and different options and expects decisions of the clients. While the concept of participatory extension allows the farmer to play a role, KIMIA extension negotiates with specific clients about their opportunities and eventually modifies its available services (Bijl, 1987).

KIMIA has provided a self-targeting exchange of information that is of most benefit to farmers who already own animals and were most affected by the decline of government-supported lending facilities. Substantial numbers of households in the other two categories benefit from KIMIA as well.

Conclusions

WP-ADPP focuses on identifying and experimenting with locally available (individual) management capacity to exchange consignments of inputs and to provide adequate information on prices, techniques and conditions on a privatised basis. Its longer term objective is not so much to build a new institution, but to make farmers less dependent on loans and more aware of changed price ratios. It is important to create the skills and power needed to counteract the exploitative actors in marketing and input supply who emerge in a liberalised agricultural economy.

References

van Agt, P A J , 1992.*Go ahead with oxen, priorities and possibilities of farm households in Kaoma District.* Western Province Animal Draft Power Programme, Zambezi Livestock and Lands, Mongu, Zambia and RDP-Livestock Services, Zeist, The Netherlands. 98p.

Bijl G, 1987. *Farming systems in a changing policy environment.* Project Planning Unit (PPU), Mongu, Zambia. 163p.

CSO, 1990. *Analysis of census supervisor areas of the 1990 population census.* Central Statistics Office, Lusaka, Zambia.

DVTCS, 1992. *Livestock census 1988-1992.* Department of Veterinary and Tse tse Control Services (DVTCS), Mongu, Zambia.

Farrington J, 1994. Public sector agricultural extension: is there life after structural adjustment?. *Natural Resource Perspectives* 2. Overseas Development Institute (ODI), London, UK.

Hocking C, 1994. The impact of mobile ox-plowing courses for women in the Western Province of Zambia. pp288-291 in: Starkey P, Mwenya E and Stares J (eds), 1994. *Improving animal traction technology.* Proceedings of the first workshop of the Animal Traction Network for Eastern and Southern Africa (ATNESA) held 18-23 January 1992, Lusaka, Zambia.

Technical Centre for Agricultural and Rural Cooperation (CTA), Wageningen, The Netherlands. 490p.

Hoogmoed W B, 1992. *Alternative ox-drawn tillage systems in Kaoma District.* Thesis research, Agricultural Univeristy of Wageningen, WAU-Wageningen, Wageningen, The Netherlands. 89p.

Huisman D, 1995. A socio-economic survey among female-headed households in Kaoma District. Western Province Animal Draft Power Programme (WP-ADPP), Department of Agriculture, Mongu, Zambia. 33p.

Kakwaba K, 1995. *Preliminary household enumeration report.* Department of Agriculture, Kaoma, Zambia. 9p.

Kakwaba K and van Leeuwen M, 1995. Ox hiring and borrowing in Kaoma District .Western Province Animal Draft Power Programme (WP-ADPP), Department of Agriculture, Mongu, Zambia and RDP Livestock Services, Zeist, The Netherlands.

van Leeuwen M and Siyambango N, 1993. *ADP loans in Kaoma District, A technical evaluation and assessment of credit for ADP.* Western Province Animal Draft Power Programme (WP-ADPP), Department of Agriculture, Mongu, Zambia and RDP Livestock Services, Zeist, The Netherlands. 15p.

van Leeuwen M and Siyambango N, 1995. *Ox carts in Kaoma District, an economic and technical assesment of ox carts in a rural district of Zambia.* Western Province Animal Draft Power Programme (WP-ADPP), Department of Agriculture, Mongu, Zambia and RDP Livestock Services, Zeist, The Netherlands.

MacLean H A M, 1965. *An agricultural stocktaking of Barotseland.* GRZ, Lusaka, 49p.

MAFF, 1989. *Draft recommendations for maize growers in Kaoma ARPT-EP.*Adaptive Research Planning Team (ARPT), and Ministry of Agriculture, Food and Fisheries, Mongu, Zambia. 38p.

Muwamba J and Heemskerk W, 1991. *Lima crop mimeo: Kaoma District.* ARPT/Department of Agriculture, Mongu, Zambia. 48p.

PPU, 1991. *Provincial medium term development plan 1991-1996* (Draft). Project Planning Unit (PPU), Mongu, Zambia.

Roling N, 1988. *Extension science, information systems in agricultural development.* Cambridge University Press, Cambridge, UK. 233p.

Roling N, Ascroft J and Chege F, 1976. The diffusion of innovations and the issue of equity in rural development. *Communication Research* 3(2) 155 170.

WP-ADPP, 1993. *Crop forecasts for Kaoma District 1983–1993.* Western Province Animal Draft Power Programme (WP-ADPP), Department of Agriculture, Mongu, Zambia.

WP-ADPP, 1994. *Agricultural Step-Up Programme Fund proposal to the WID/RD Sections of RNE Lusaka.* Western Province Animal Draft Power Programme (WP-ADPP), Department of Agriculture, Mongu, Zambia.

WP-ADPP, 1995a. *Annual Report 1994.* Western Province Animal Draft Power Programme (WP-ADPP), Department of Agriculture, Mongu, Zambia.

WP-ADPP, 1995b. *KIMIA's for food security through farmer organisation.* Western Province Animal Draft Power Programme (WP-ADPP), Department of Agriculture, Mongu, Zambia. 15p.

Animal draft power training in Zimbabwe: experiences and future challenges

by

B Mudamburi [1], B Chikwanda [1] and J Francis [2]

1) Institute of Agricultural Engineering, PO Box BW330, Borrowdale, Harare, Zimbabwe
2) Department of Animal Science, University of Zimbabwe, PO Box MP167
Mount Pleasant, Harare, Zimbabwe

Abstract

Training of draft animals is of paramount importance because it ensures efficient use of their potential draft capacity. This is only possible if the people training the animals are also properly trained.

Training programmes used in Zimbabwe should be reviewed, seeking opinions of all target groups and incorporating experiences from other countries. There is also a need to find ways of ensuring that more women receive this training.

Participants are assessed before they take a course to establish their knowledge base, and soon after training to find out how much they have gained. Continuous follow-up evaluation of the training programmes must be reinforced on-farm, even well after completion of courses, to establish adoption levels. However, this proposed programme is likely to be expensive, and probably unsustainable as well.

Introduction

More than 85% of the 1–1.2 million communal (smallholder) farming households in Zimbabwe use animal draft power in agricultural production (Francis and Mudamburi, in press). This power is mainly provided by oxen, although use of cows and donkeys is increasing.

Many of these animals are used for draft without having proper training. Some animals are put to work before they are mature enough to carry out such tasks (Francis, 1993). They are also poorly fed. Yet, temperament, physical development, training and feeding management are principal determinants of draft capacity.

This paper presents a summary of animal draft power training courses offered by the national Institute of Agricultural Engineering (IAE). A critical analysis of the current training programmes is also provided.

Animal draft power training in Zimbabwe

Historical perspective

Animal draft power training in Zimbabwe was launched as small-scale mechanisation at Domboshawa Farm Machinery Training Centre, near Salisbury (now Harare) in 1964. Training was done through on-station and on-farm courses and mobile training units. In 1983, with financial assistance of the German Agency for Technical Cooperation (GTZ), the training centre was moved to the IAE. The main reason for this change was to have agricultural training and research close to each other so that there could be better flow of research findings into training and extension.

Main elements of training courses

The overall objective of animal draft power training in Zimbabwe is to provide hands-on experience and instruction on efficient use of draft animals, animal-drawn and hand-operated machinery and equipment. Table 1 summarises the contents of the training courses offered at the IAE.

The main target group for these courses is extension staff. Training is also offered to:

- farmers, but only if they will train others later; knowledge on animal draft power training would then be transferred through horizontal farmer–farmer interaction
- agricultural teachers and instructors from agricultural institutes and colleges
- sales representatives from manufacturing companies
- agriculturalists from Eastern and Southern Africa
- any other needy animal draft power users.

Table 1: Main elements of animal draft power training courses offered at the Institute of Agricultural Engineering, Zimbabwe

Course code	Course content
AP 1	Selection, care and training of draft animals Selection criteria and procedure; feeding and health care; training procedures
AP 2	Plowing with draft animals Parts, operations and maintenance of the single furrow mouldboard plow; plowing systems
AP 3	Seedbed preparation and crop establishment Equipment used, including planters
AP 4	Crop maintenance Hand hoes, knapsack sprayers and ultra-low-volume sprayers, mist blowers
AP 5	Harvesting and processing equipment Hand-operated shellers and winnowers; animal-drawn groundnut lifters, cutter bar mowers and carts
AP 6	Donkeys in Zimbabwean agricultural production Selection, management and training
Regional animal draft power courses	All courses listed above, including yoke and donkey harness making

Future challenges for animal draft power training in Zimbabwe

Assessing training needs

Proper training of draft animals and animal draft power users is important, as evidenced by the current inefficient use of draft animals and related farm equipment by many farmers. To ensure that appropriate training is provided, appropriate solutions to several challenges should be found. The impact of current training programmes should be assessed. This implies that thorough follow-up studies should be conducted in areas where farmers and agricultural extensionists have been trained, to establish the levels of adoption of the IAE training programmes. The assessment should provide answers to such questions as:

• have farmers and other target groups modified the training programmes?

• if so, what has been altered and why?

Where there has been poor adoption, it is important to discover the reasons for this. After this assessment, the training programmes should then be modified such that they meet the needs of all the different target groups. Training manuals would then be developed based on these findings. Even though this approach holds a lot of promise in meeting the needs of all target animal draft power trainees, it appears to be very expensive.

Involvement of manufacturers and farmer organisations in training

Manufacturers of animal draft power equipment and machinery do not supply brochures on how to calibrate and use the numerous implements they produce. For this reason, they should be more actively involved in training participants during development and evaluation of animal draft power courses.

Farmers are recruited for animal draft power training courses by extension workers in the field. The extension workers also keep a record of farmers who receive this training. It is doubtful if this is an effective approach.

Farmers' organisations, such as the Zimbabwe Farmers Union, have not been used as effective communication channels between farmers, researchers and animal draft power trainers. There is great potential for strengthening the training programmes through close collaboration with such organisations. In addition, experiences of other countries in training should be sought and used to improve training programmes in the country. One forum where such ideas could be tapped would be meetings of ATNESA (the Animal Traction Network for Eastern and Southern Africa) or other regional networks.

Table 2: Division of labour between men and women in African agriculture

Task	Level of participation (%)	
	Men	*Women*
Land preparation	95	5
Crop planting	30	70
Weeding	20	80
Harvesting	30	70
Processing	10	90
Transporting harvested crops to homestead	20	80
Storing crops	5	95
Marketing surplus	40	60
Caring for domestic animals	40	60
Collecting water and fuel	10	90
Feeding and caring for the family	5	95

Source: Mwoyowehama (1995)

Gender considerations

Although many women are involved in agriculture (see Table 2), they rarely attend training courses. The reasons for this non-participation should be sought, so that ways can be found to ensure that they receive training. In Zimbabwe, for example, farmers attend animal draft power courses free of charge. It is possible that because women are always faced with heavy agricultural workloads and family responsibilities, they cannot be away from their homes for such long periods as one to six weeks (duration of IAE training courses). In fact, the training centre is situated inappropriately in the capital city. If it had been located in a smallholder farming area (close to farmers), more women farmers would probably be able to attend the courses offered. However, only one national training centre located in a rural area would not adequately cater for all women farmers. Establishment of sub-training centres would be a better alternative.

Agricultural education

The overemphasis of agricultural syllabuses used in educational institutions on tractor power (but not on animal draft power) tends to perpetuate the commonly held view in the country that animal draft power technology is backward. Almost 90% of people use this source of power, so there is need for a revolutionary approach towards developing more realistic curricula. This radical shift should start with making agriculture a compulsory subject from as early as primary school level. Syllabuses at colleges and universities should then be revamped to correct the imbalance between teaching tractor power and animal draft power, in favour of the latter.

References

Francis J, 1993. *Effects of strategic supplementation on work performance and physiological parameters of Mashona oxen.* MPhil Thesis. Department of Animal Science, University of Zimbabwe, Mount Pleasant, Harare, Zimbabwe. 169p.

Francis J and Mudamburi B, in press. Animal draft power in Zimbabwe: where are we? In: *Meeting the challenges of animal power in Zimbabwe.* Proceedings of a workshop held at Chibero Agricultural College, Norton, Zimbabwe, 30–31 August 1995. APNEZ (Animal Power Network for Zimbabwe), Harare, Zimbabwe.

Mwoyowehama I, 1995. Gender perspective. *Agritex News and Views* No 61, April–June 1995. Agritex, Harare, Zimbabwe.

Extension to improve the welfare of traction animals

by

Cheryl M E McCrindle and Limakatso E Moorosi

Department Production Animal Medicine, Faculty of Veterinary Science
Int Box 170, Medunsa 0204, South Africa

Abstract

The position of animals in resource-poor societies is of necessity utilitarian and the welfare issue is not one of animal rights, but rather prevention of avoidable suffering. This paper suggests a systems approach similar to farming systems research and extension methods, for use by animal welfare agencies working in resource-poor areas. A situational analysis using rapid appraisal, cost-benefit analysis and ranking of welfare objectives should be carried out prior to intervention. This intervention should be participatory and proactive rather than reactive and prescriptive in nature. Animal welfare agencies in South Africa have shifted from a legislation-based approach to a more participatory approach. The paper describes a case history of a participatory project in the Soweto area of South Africa. Extension messages should be designed specifically for their target group based on situational analysis and should be evaluated after the project to determine their success or failure. People at different socio-economic levels have different concerns; extension messages relating to the concerns of the target group are most likely to be successful.

Introduction

In South Africa, animal traction has been largely ignored by agricultural researchers. Yet a survey performed during 1994 showed that cattle, donkeys, horses and, to a lesser extent, mules have an important role as traction animals in both rural and urban areas (Starkey, 1995). Historically animal power was important to all sections of South African society. As a result there is knowledge and legislation is available to maintain the well being of traction animals. This knowledge is only being transmitted to low-income and resource-poor traction animal owners by animal welfare agencies.

In the past the intervention of animal welfare agencies has been reactive and prescriptive. Animal owners have been arrested and prosecuted under the Animal Protection Act (1962), section 2,

offences in respect of animals. The following contraventions, mentioned in the Act, may lead to a fine of R2000 or imprisonment of up to 12 months for any person who:

- "... cruelly overloads, overdrives, overrides, beats, kicks, goads, ill-treats, neglects [an animal].."

- "... being the owner of any animal, deliberately or negligently keeps an animal in a dirty or parasitic condition..."

- "... uses on or attaches to any animal any equipment, appliance or vehicle which causes or will cause injury to such an animal or which is loaded, used or attached in such a manner as will cause such an animal to be injured..."

- "...drives or uses any animal which is so diseased, or so injured or in such a physical condition that it is unfit to be driven or to do any work...".

The powers of inspectors from the Society for the Prevention of Cruelty to Animals (SPCA) include arrest without warrant of offenders and confiscation of animals if there are reasonable grounds that the ends of justice would be defeated by delays in obtaining a warrant of arrest. Despite the powers granted to them, the attitude of animal welfare agencies in resource-poor communities has undergone a transition and more participatory methods are being considered. The importance of education in preventing cruelty has been realised but targeting and evaluating extension messages has been difficult.

The aim of this paper is to propose a more participatory and proactive stance for animal welfare agencies interested in promoting the wellbeing of traction animals.

The design of a particpatory extension method

Any extension message must have a motivational aspect because the recipients of a message are more likely to listen to a message if it benefits them (McCrindle, 1995). Socio-economic factors also have a strong influence on the distribution, dynamics and significance of animal well-being in South Africa (Krecek, Cornelius and McCrindle, 1995). People at different socio-economic levels have different concerns and so will perceive the benefits of looking after their animals differently. Extension messages should be designed approriately for the concerns of the target group. For example very poor people are likely to be worried only about physiological benefits to them so a message such as "Improve the feeding of your donkey and it will be able to carry water for your family" is likely to be effective. At higher socio-economic levels other benefits such as increased security, increased social standing and increased esteem become important so more appropriate messages could include "Improve the health of your draft oxen and they will be strong if your family needs them for plowing", "Look after your cows and you will be able to give milk to your neighbours" or "Improve your cart and you can win a prize at the show".

The rapid appraisal method has been used successfully for situational analysis of the status of animal traction in both rural and urban communities in South Africa (Starkey, 1995). It can also be used to investigate the welfare and well-being of traction animals. The team should preferably include a veterinarian and a sociologist or socio-anthropologist as the socio-economic status and cultural characteristics of the target community are important. In practice, we have found liaising with the local social worker is simpler (McCrindle, 1995).

It is vital that the local community participates as partners in the process (Bembridge, 1991). To ensure this an animal welfare team visits a particular area and uses direct observation, key informants, semi-structured interviews, workshops and brainstorming, ranking, scoring and case-studies to gain knowledge rapidly and progressively with appropriate precision. Most important are listening skills - the team needs to listen to the needs and problems of the community as well as observing animal welfare problems in the traction animals. After the appraisal the team meets to discuss appropriate interventions. These are analysed, with help from an economist, on a cost-benefit basis (McCrindle, 1995) and an appropriate programme of intervention chosen.

Once a programme of intervention has been decided upon, an intervention team is mobilised and contacts are made with key persons in the community to finalise arrangements for the intervention. After the intervention the team meets again to discuss evaluation. This includes direct evaluation of the intervention as it takes place (making it more efficient) as well as long-term evaluation (rapid appraisal applied a few months or years afterwards to assess whether intervention had the desired results). The process is summarised in Figure 1, which is a systems approach adapted from farming systems research and evaluation methods.

Using this approach, animal welfare agencies may address the needs and perceptions of people in such a way that the wellbeing of animals is improved. This presupposes that the majority of animal abuse in low-income communities occurs through ignorance and neglect and there is no deliberate intent to inflict pain and suffering. In the case of animal abuse due to sadism, the approach would ideally involve punishment of the offender and prevention of further opportunities to abuse an animal.

Welfare objectives

An extension message has several components. The technical information must be up to date and accurate, it must meet the needs of the target community and it should be simple to understand (Bembridge 1991).

The following animal welfare objectives have been identified and extension messages formulated for promoting the wellbeing of traction animals in South Africa:

Improving the condition of animals

- correct feeding and watering
- teaching condition scoring
- bits which do not injure mouth
- shelter from elements

Improving the health of animals

- how to treat for external parasites

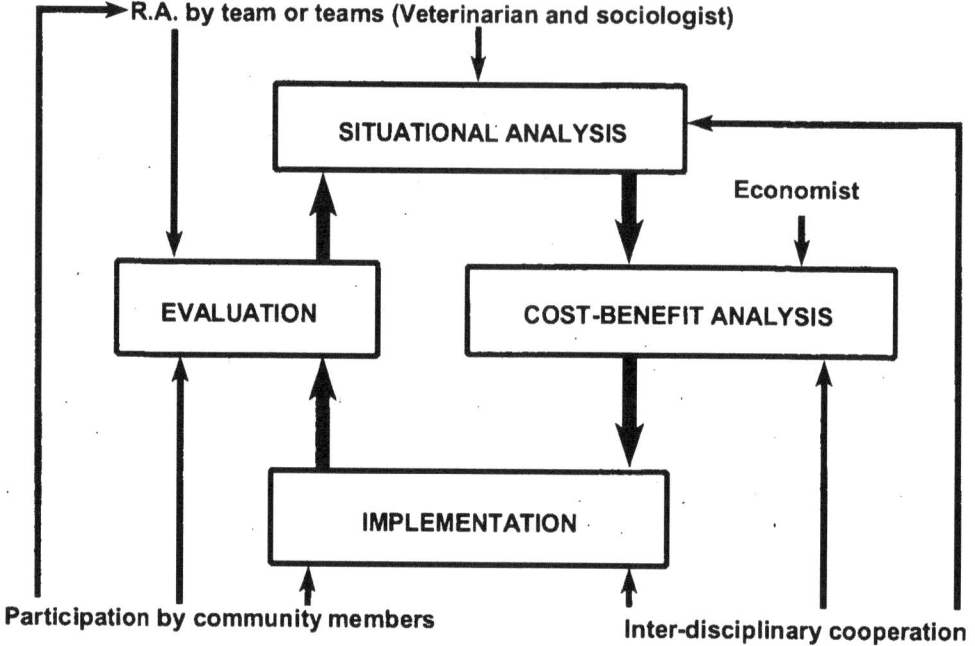

Figure 1: Diagram illustrating the systems approach used for selecting, ranking and evaluating animal welfare objectives (after McCrindle 1995)

- how to treat for internal parasites
- recognition of disease
- who to talk to if an animal is sick

Preventing injury to animals
- correct harnessing
- correct structure of carts
- correct loading of carts
- recognition of injuries
- treatment of minor injuries

Motivation to improve animal wellbeing
- 'Bring your animal to our show and win a prize if it is the best.'

Much of the technical information required for developing extension messages about correct harnessing, cart structure and cart loading is available in the publication by Starkey, Mwenya and Stares (1994), among others. Technical information for extension messages about nutrition, condition scoring, recognition of diseases and treatment of minor injuries should be formulated with the assistance of a veterinarian who has experience of the target species of traction animal (for example it is counterproductive to ask

a companion animal practitioner about condition scoring oxen).

The method in practice

This method was successfully used in 1994 to initiate a companion-animal welfare outreach clinic to low-income urban areas near Pretoria (McCrindle, 1994). A similar method was used successfully by the Soweto Society for Prevention of Cruelty to Animals (SPCA) in a settlement area known as Orange Farm, to improve the welfare of horses by exchanging new bits, donated by saddlery shops, for the rusty wire bits used previously.

A case study: Orange Farm, Soweto

The education department of Johannesburg SPCA launched an adult education programme for Soweto and Orange Farm in August 1994. This programme was extended in April 1995 to include the owners of working horses. Education officers meet these owners once a week at an established point where they bring the horses for shoeing, deworming, vaccination and treatment by the horse-care unit of the Soweto SPCA. Topics such as feeding, grooming, care of hooves, care of

Table 1: Achievements of the Soweto SPCA programme, March–October 1995

Operation	Number
Shoes replaced	2,825
Number of horses shod	749
Harnesses repaired	61
Harnesses replaced	8
Bits replaced	50

equipment (bits, harnesses), carts, paddocking and basic health care are discussed and reinforced with appropriate literature available in Zulu, Sotho and Shangaan. Suitability of the breed of horse and care in relation to its ability to work well, are emphasised. Education officers suggest means of caring for horse and equipment which are within the parameters of the owner's means and circumstances for example:

- wash and dry bit after use

- clean harness after use and lubricate with old cooking oil to keep it soft - this eliminates cracking and breaking, therefore fixing the harness with wire is prevented as are harness galls caused by stiff, hard leather

- keep horse in paddock free from broken glass, wire, tins etc - this prevents injuries which will keep the horse from working.

As a result of donations from two saddlery shops, old broken bits are exchanged for new ones. In most intances owners apply the knowledge they have gained form this programme and general improvement is seen. Statistics obtained from Soweto SPCA for the period March–October 1995 for Soweto, Orange Farm and the surrounding areas are shown in Table 1.

Conclusions

The position of animals in resource-poor societies is of necessity utilitarian and the welfare issue is not one of animal rights, but rather prevention of avoidable suffering. Proactive strategies with community participation are suggested. Extension messages selected should result in benefits which promote wellbeing for both animals and animal owners. By actively using, evaluating and modifying interventions using the suggested systems approach, it is felt that the welfare of traction animals in Africa could be considerably improved.

Acknowledgements

The assistance of Joanne Lombard of Soweto SPCA, who supplied the case history is gratefully acknowledged.

References

Bembridge T J, 1991. *Practical guidelines for extension workers - a field manual.* Development Bank of Southern Africa, Halfway House, Gauteng, South Africa.

Krecek T, Cornelius S, and McCrindle C M E, 1995. - Report on a workshop on the socio-economic aspects of animal disease in Southern Africa. *Journal of the South African Veterinary Association* **66** (3): 115–120.

McCrindle C M E, 1994. *Using community orientated veterinary extension techniques to facilitate companion-animal welfare.* (Abstract) Proceedings of the 19th World Congress of the World Small Animal Veterinary Association held 26–28 October 1994, Durban, South Africa.

McCrindle C M E, 1995. *The community development approach to animal welfare: an African perspective.* Paper presented at the World Veterinary Conference held 3–9 September, 1995, Yokohama, Japan.

Starkey P, Mwenya E and Stares J, (eds) 1994. *Improving animal traction technology.* Proceedings of the first workshop of the Animal Traction Network for Eastern and Southern Africa (ATNESA), held 18–23 January, Lusaka, Zambia. CTA, Postbus 380, 6700 AJ Wageningen, The Netherlands. 490p. ISBN92-9081-127-7

Starkey P, 1995. *Animal traction in South Africa: empowering rural communities.* Development Bank of Southern Africa, PO Box 1234, Halfway House, Gauteng, South Africa. 159p. ISBN 1-874878-67-6

Meeting the challenges of animal traction

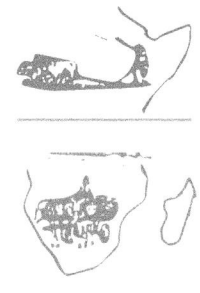

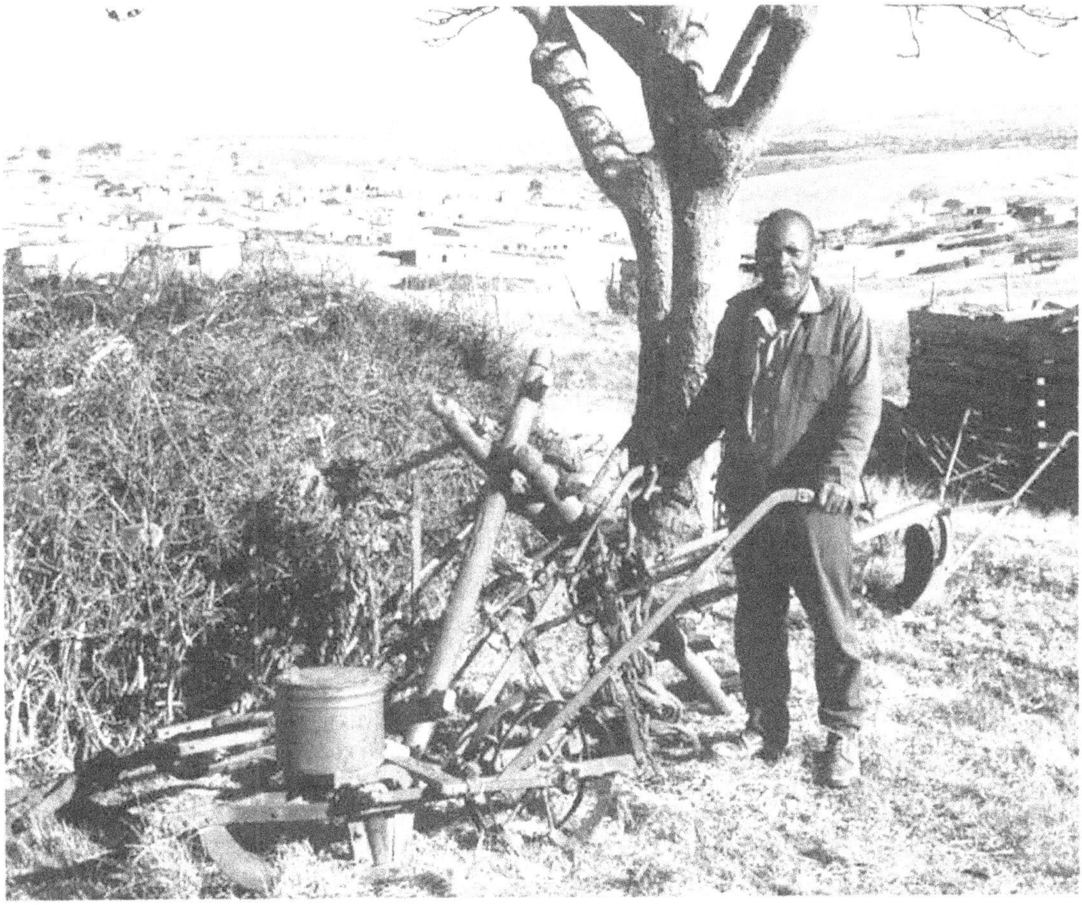

Animal-drawn equipment and harnessing

Conservation tillage for soil and water conservation using draft animal power in Zambia

Kenneth Chelemu and Peter Nindi

Farm Power and Mechanisation Group, Palabana Animal Draught Power Development Programme
Private Bag 173, Woodlands, Lusaka, Zambia

Abstract

Southern Africa is currently experiencing drought conditions characterised by late and scattered rains and insufficient total rainfall. Conventional plowing needs to be carried out after the onset of the rainy season. As a result, planting is often delayed. With the current short rainy season this often leads to partial crop failure. In these conditions, conventional plowing can also result in large losses of soil moisture due to inversion of the topsoil, in reduced infiltration in the furrows and in increased soil erosion. As a result farmers are ready to consider alternative 'conservation' tillage practices that maximise use of available water and reduce erosion. The Palabana Animal Draught Power Development Programme in Zambia has been promoting ripping and subsoiling as conservation tillage practices that are appropriate under certain conditions. Mechanical planting has also been promoted as a quicker method of seeding that can improve timeliness and make the best use of the short rains.

Ripping can be carried out before the beginning of the rains, facilitating dry planting and improved timeliness. It reduces soil disturbance and can channel soil moisture to the ripped areas. However, it can lead to serious weed problems if timely weeding is not carried out. Palabana has developed a ripper attachment for conventional I-beam plows that has been popular with farmers in demonstrations.

Whilst conventional animal-drawn mechanical planters are effective, they have not been widely accepted by farmers due to high initial costs, high maintenance costs and complicated operation. The Palabana programme has developed a new planter attachment for conventional plow frames that should be cheaper, easier to use and more reliable.

Subsoiling is only effective in a limited set of situations and it requires at least four strong oxen. Trials of a subsoiler developed at Palabana are being carried out and the response from farmers requiring this specialised tillage method have so far been positive.

Introduction

Depletion of soil moisture is influenced by many factors which can be categorised in two broad groups: climatic, and soil physical conditions. Most factors are natural and can only be controlled indirectly. The most critical factor is the amount of moisture received to be conserved. During the recent past most countries in the southern part of Africa and particularly Zambia have experienced serious drought seasons for the past 5 years. As a result some agricultural research projects have focused on improving the catchment, conservation and efficient use of available water.

One human factor influencing the rate of water loss is the method and extent of soil manipulation. The most common tillage method is conventional plowing using mouldboard plows. This tillage technique completely disturbs and inverts the soil. Since this buries weeds and leaves the loose surface bare, sheet erosion becomes more probable. The compaction and smearing effect at the base of furrows creates a compact pan which retards water infiltration. This reduces depressional storage and encourages runoff and consequently increases soil degradation due to erosion. Moreover, conventional plowing can only be performed when soils are substantially wet. As a result plowing is often delayed due to the late and scattered rainfall currently being experienced. Consequently subsequent activities, especially planting, are delayed; this frequently leads to partial crop failure.

Many farmers, especially in the drier most southern part of Zambia, have realised these problems and are ready to try promising new techniques aimed at maximising use of the available precipitation. A number of tillage systems aimed at increasing depressional storage and reducing soil loss, collectively referred to as 'conservation tillage', are being promoted but not all are feasible in all situations. The concepts behind these techniques are well understood but most of them have a problem with weed infestation since weeds are not buried during primary tillage.

In some instances the conservation tillage techniques do not fit easily into the existing farming system. In the recent past the Palabana Animal Draught Power Development Programme has introduced ripping and subsoiling. These are conservation tillage operations that farmers, particularly those using animal draft power, have perceived as promising to address the problem of timeliness of field operations and efficient moisture utilisation during prevailing drought periods. The limitations of these methods are generally accepted, for example farmers in the sandy Western province of the country are unlikely to appreciate operations such as subsoiling. Similarly ripping would give farmers of the high-rainfall northern part of the country a massive weed problem if care is not taken to include a good preventative weed control measure.

This paper describes the conservation tillage methods of ripping, planting and subsoiling being promoted by the Palabana programme and its experiences with them.

Ripping

Ripping is the breaking-up or loosening/disturbing of soil in the arable layer at regular intervals, in principle without inversion. Planting is done in the ripped strips. Because of the low infiltration in the undisturbed areas, these strips act as catchment areas for the ripped areas and increases the amount of water available within the ripped confines. Ripping is carried out with a simple narrow tined point to a desired depth that depends on the crop being planted. The quality of ripped lines is determined by the design and setting of the ripper. Ripping becomes beneficial if it can be carried out sufficiently before the rains to facilitate dry planting. The recommended time is before (dry ripping) or upon the onset of the rainy season. Experience indicates that serious weed problems result if ripping is carried out long after the onset of the rainy season.

Ripping as a primary tillage field operation has many advantages including:

o it is quite a light operation for oxen, ie requires a low draft force compared to plowing

o it results in minimal disturbance to the soils of the field worked

● since it enhances field preparation time, planting can be done early/timely

● the planting depth can be regulated, unlike planting behind the plow

o using well trained draft animals and the right harnessing (eg a long double-neck yoke), crop rows can be made very straight and well-spaced for easy weeding using animal draft power

● the sturdy chisel tine is reversible

o there is no need for a complete new implement since the attachment is cheap and suits all I-beam type plow frames.

The major disadvantage of ripping is the possibility of serious weed problems if it is carried out too late. Post-planting operations may therefore be more intensive. To minimise the problems timely weeding is essential.

The Magoye ripper

During the past few seasons the Palabana programme has developed a ripper, called the *magoye* ripper (Figure 1) in close liaison with farmers (through the local extension staff) on one hand and manufacturers on the other. It looks very much like the traditional ard plow. It was introduced in Niger with promising results (Kruit, 1994) and further developed into a simple and sturdy attachment in Zambia by the Palabana programme. In over 25 districts where it has been demonstrated it has been well appreciated by farmers. Major positive points for its acceptance are its ability to facilitate dry planting (prior to the rains) and the fast and light nature of the operation.

The special features of the *Magoye* ripper include a sturdy tine and two small wings to make a wide clear furrow most suitable for large seeds especially for a staple crop maize. This attachment fits on the I-shaped plow beams in common use in the region. Since many of these beams are left unused by farmers because of lack of spares for the original implement, there is a readily available

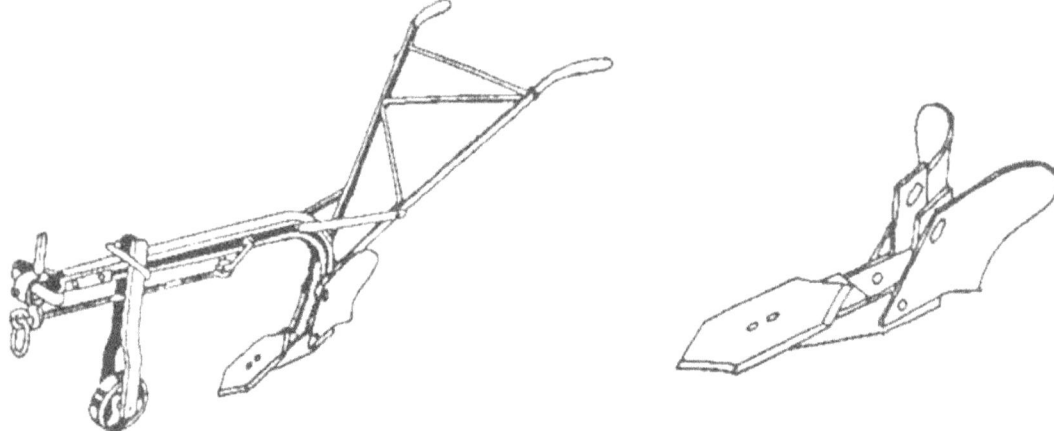

Figure 1: The Magoye ripper

supply. The working depth is easily influenced by soil conditions (whether wet/light or dry/heavy). This necessitates a good depth regulator. The distance between ripped lines is regulated by sequence of operation and the most practical way is to allow one draft animal to work in a previously ripped line. In this case the same yoke should be used for weeding. Figure 1 shows the major components of the ripper. The Palabana Animal Draught Power Development Programme has received encouraging reactions from soil and water conservation experts in Zambia, Zimbabwe and Tanzania who recognise the appropriateness of this tool to help farmers improve their tillage practices. For instance, a land management project in Tanzania recently ordered some *Magoye* ripper and subsoiler attachments to try under varying conditions.

The major components of the *Magoye* ripper attachment are:

- a reversible chisel tine for opening the furrows
- a subframe which attaches all components together and to the main beam
- left and right wings to keep the loosened soil from falling back into the furrow opened by the chisel tine

Harnessing for ripping

A harnessing system for ripping must space the draft animals at twice the desired inter-row spacing (Figure 2) unless a single animal is being used.

The same harnessing system should be used for weeding ripped fields using animal draft power.

Subsequent operations after ripping

Seed dropped into the furrow by hand may be covered by passing a cultivator between the planted furrows. This also serves as a preventative weed control measure (pre-emergence weeding). A harrow can be used instead. Conventional weeding can be carried out with either a cultivator or a ridger.

Status of Magoye ripper attachment

Prototype development of this implement started in 1992. The design is now acceptable to farmers and has already been taken up by some manufacturers. Jigs and templates to standardise its manufacture have been developed (and made on request or need) by the Palabana programme as a service to promote manufacture of *Magoye* rippers. These are intended for both centralised and small rural workshops. To mention but a few, MDM Engineering Company on the Copperbelt in the northern part of Zambia, SARO Agri-equipment Limited in Lusaka, and Kaleya Agricultural

Figure 2: A yoke for ripping

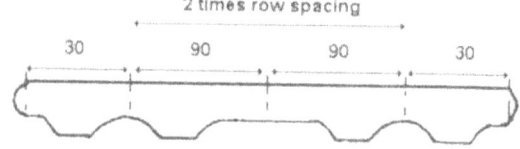

Engineering Company in the southern part of the country are some of the major companies that have taken up the challenge. Other smaller (rural) workshops have also been assisted and encouraged by Palabana to manufacture the attachment and supply their local farming clientele.

Palabana planter attachment

Mechanised planting using animal drawn planters has proved effective and efficient in some farming systems. However, this technique has often not been accepted by farmers. This has been because the available models of planter have been expensive, been complicated to calibrate and use properly and been vulnerable to breakdowns and so have a high maintenance cost. The common models are equipped with a blade-type coulter for furrow opening that requires moist and fine tilth. This delays planting because plowing must be carried out when the soil is moist.

With this background the Palabana Animal Draught Power Development Programme has developed a planter attachment (Figure 3) that makes use of the ripping concept to open up planting furrows. This speeds up planting and can even facilitate dry planting. The planter attachment consists of a seed hopper, a metering unit and two landwheels on either sides. The seed metering mechanism works using a cellwheel fixed on a horizontal shaft with a key and driven by the two wheels. The planter unit pivots in a 3-point linkage system behind the beam and thus follows the undulating surface neatly during planting. When used in combination with a ripper as a leading point the system has the same advantages as discussed for ripping alone.

The major components of the Palabana planter attachment are:

- a ripper subframe attaching the ripping chisel tine to the main beam
- the chisel tine itself acting, as usual, as a furrow opener
- planter side-plates connecting the planter unit to the main beam
- a seed pipe
- an aluminium seed wheel with 4 equidistant seed cells on its surface. This is fitted inside the seed pipe on an axle with drive wheels (100 cm circumference) on both ends

Figure 3: The Palabana planter/ripper attachment

- the axle itself – a 25 mm diameter shaft which drives the seed wheel by means of a key fitting in a key way in the bore of the seed wheel
- a rubber fitted by a bolt and wing nut to the front of the seed pipe to act as a brush preventing multi-drops of the seed
- an adjustable (by screw) metallic flap inside the back of the seed pipe to prevent seed from falling through the seed pipe backwards
- a hopper with a lid
- a stay that connects the seed pipe to the main beam
- a furrow covering assembly is optional.

Status of the Palabana planter attachment

The attachment has reached an advanced development stage. An extensive on-farm test is planned for this season, 1995/96. The design will be tested for technical suitability as well as acceptability by farmers. Most potential manufacturers in Zambia and the region have been informed about this project. It is hoped that the Palabana planter will be appreciated as an appropriate alternative to the regular type of planter. It will be less costly, less susceptible to technical failure, and will function in a wider variety of soil conditions. The possibility of planting without prior plowing and harrowing will be a most attractive advantage. However, weed control measures will need to be intensified. Several companies have already shown interest in its manufacture.

Harnessing for planting

Harnessing for planting depends on the desired inter-row spacing in the same way as harnessing for ripping.

Subsoiling

Subsoiling is the breaking up of the hard underlying pans beneath the plowing depth. The main purposes of subsoiling are to increase water infiltration and storage, and to allow easier penetration of the root system (especially tap roots) beneath the plowing depth (up to 25cm). However, this method is only useful in certain circumstances, and is hence expected to attract only a particular group of users, especially as the subsoiler requires at least two strong pairs of oxen. Subsoiling should be carried out in the dry season, well or just before the onset of the rainy season. The operation cannot be carried out in wet conditions.

Harnessing for subsoiling

Subsoiling is normally carried out on dry land to break hard pans so it requires high draft forces. Therefore a team of animals should be used, for example four draft oxen. The effective distance between two spanned animals moving abreast depends on how much subsoiling the operator wants in a particular field. If subsoiling is to be carried out on permanent crop rows similar harnessing to that described for ripping should be used.

Subsequent operations to subsoiling

Subsequent operations after subsoiling can be any primary tillage field operation. For conservation tillage, ripping or ripping and mechanical planting is recommended.

Like the ripper attachment, the subsoiler attachment fits onto any normal I-beam type plow frame. It consists of the following parts (see figure 4):

- a jumper bar which cuts deep into the subsoil layer

- a beam extension which holds the jumper bar and the main beam together and provides for adjustment of the jumper bar for deeper or shallower penetration

- a special hitch assembly goes with the attachment. It extends the main beam lengthwise to suit the reaction point geometry of the deep penetrating jumper bar.

Status of the attachment and its manufacture

The attachment is still in the development stage. In readiness for the forthcoming rainy season (1995/96), one engineering company was contracted by Palabana to produce a number of subsoiler attachments which a few selected farmers who requested and have a real need are currently using (November 1995). These farmers are mostly in the midlands and southern part of Zambia. An on-farm trial is being carried out as a follow-up of

Figure 4: Subsoiler attachment

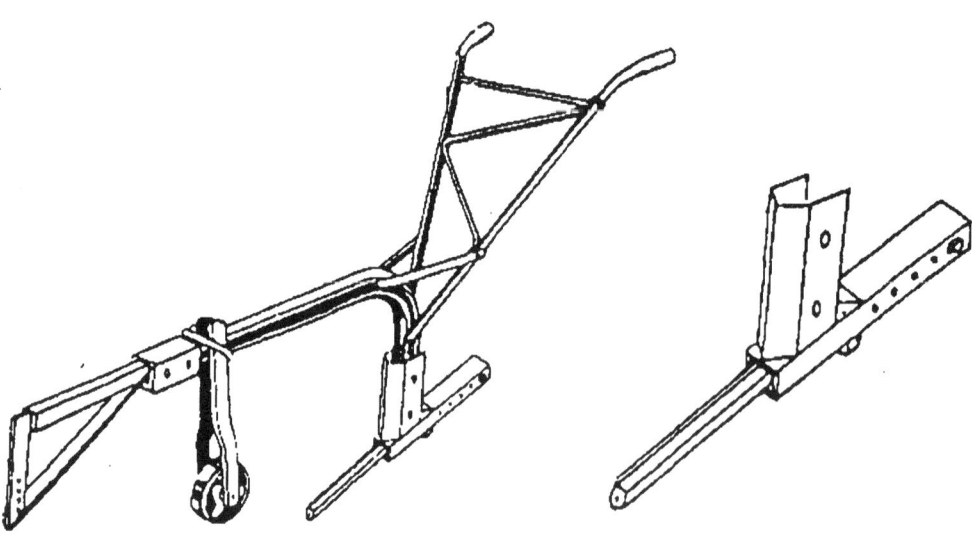

the subsoiler programme and so far farmers seem to appreciate the performance of the implement.

Discussion

The attachments decribed in this paper are not intended to turn the 'green' plow or ridger beam into a multipurpose toolbar. Multipurpose tool bars exist in French-speaking Africa and have their own shortcomings. However, the advantage seems obvious: only one frame is required to do many different jobs. It is only necessary to swap certain attachments instead of requiring a separate implement for each operation. This is helpful where finances are constraining. The main reason for Palabana to introduce the attachments is to combine the promotion of useful 'new' equipment on durable and well appreciated existing frames. There is simply no need for another frame. Moreover, many farmers using animal traction in Zambia have old 'green' beams. This is a result of the poor supply of spare parts, which has forced farmers to buy complete new plows or ridgers, whenever wear and tear of certain parts made the implement not functional. These unused old plows, currently scrap materia,l often have beams in good condition which could easily be re-used and turned into 'new' implements with these simple attachments.

The current short rainy seasons require timely planting. Conventional plowing delays planting so ripping and and direct planting seem to be realistic alternatives even though additional weeding is called for. An additional advantage of these methods is the efficient use of available soil moisture reducing erosion. As always, meeting the needs of farmers is the most important goal of agricultural projects. Assessing and working towards these needs without creating new problems is extremely difficult and calls for concerted efforts from all parties involved. To achieve this goal the Palabana Animal Draught Power Development Programme has joint the newly formed 'Minimum Tillage Lobby' in Zambia. The lobby, set in motion through the Zambia National Farmers Union and funded by the World Bank and European Union is expected to liaise monthly to monitor and analyse trends of the conservation tillage campaign in Zambia. The Palabana Animal Draught Power Development Programme welcomes comments and communications from people seeking more information, with suggestions and with questions about our activities.

Reference

Kruit F, 1994. Animal traction technology in Niger and some implications for Zambia., pp474–480 in Starkey P, Mwenya E and Stares J (eds), 1994. *Improving animal traction technology*. Proceedings of the first workshop of the Animal Traction Network for Eastern and Southern Africa (ATNESA) held 18–23 January 1992, Lusaka, Zambia. Technical Centre for Agricultural and Rural Cooperation (CTA), Wageningen, The Netherlands. 490p. ISBN 92-9081-127-7

The supply of animal-drawn implements in Tanzania

by

S Mkomwa and R M Shetto

SEAZ Agricultural Equipment Ltd, PO Box 2607, Mbeya, Tanzania

Abstract

Availability, affordability, and awareness are the major impediments to the widespread use of animal-drawn technology in Tanzania. While erratic availability and late deliveries cause frustration, availability of plows only instead of the whole animal-drawn technology package hinders adoption of animal power by precluding economic profitability.

Crop pricing that directly affects farmers' purchasing power is the most important factor for enhancing affordability of animal-drawn technology supplies by farmers. The current status of trade liberalisation and price de-confinement should be maintained. There should be no direct subsidies on animal-drawn equipment. However, indirect subsidies on particular equipment, for example weeders, for a specified amount of time, would be beneficial.

Efforts to inform and persuade farmers of the potential benefits of animal-drawn technology stand out as the most effective methods of increasing use of animal-drawn technology. Once a 'critical mass' of users is reached animal power becomes self-sustaining and spreads rapidly. To achieve this, donor support of private enterprises and provision of credit in key 'action zones' is necessary. The authors postulate that if effective promotion is adopted the number of animal power users in Tanzania could double within 5 years.

Introduction

Agriculture is the dominant sector in the economy of Tanzania, engaging 80–90% of the population. Smallholder cultivation accounts for 85% of the total area under cultivation. The hand hoe is the dominant tool and the estimated 3.6 million farm families cultivate only about 6 million hectares out of the 40 million hectares of potentially cultivatable land.

The objective of improving agricultural production in Tanzania by adopting animal draft power has not been smooth. To date only 20% of the land is cultivated by oxen despite a population of 12 million cattle. The 1 million draft oxen are mainly used for plowing, with a limited amount of transportation.

Efforts to promote the use of animal traction in Tanzania started about seventy years ago, with the missionaries and settlers playing a leading role during the colonial days. The initial spread of animal traction was more-or-less spontaneous, mainly through farmer-to-farmer contact and may have been associated with commercial production of cotton, rice, coffee, tea, maize and wheat.

Some of the major constraints which have been identified as limiting the widespread use of animal traction in the country are:

- low farm incomes which make the technology unaffordable to the majority
- unavailability of implements
- poor infrastructure and support services
- an ineffective extension system
- and social, cultural and gender issues.

Moreover, it should be noted that successful dissemination of any innovation requires that the price must be right, potential users must be knowledgeable in the use of the innovation and the necessary inputs must be available. Although farmers are a rational people, they are not aware of all the possibilities of animal power.

This paper reviews the supply of animal-drawn implements in Tanzania and its consequences for animal traction use. The supply of implements is observed as a function of availability, affordability, profitability and technological awareness.

Availability of animal-drawn implements in this context is defined as the timely physical presence at the farmer's place of residence of animal-drawn implements, harnesses, spares, repair services and technical know-how. Animal-drawn implements which are physically present in an area could still be perceived to be unavailable if farmers are not aware of their existence - a problem of education and promotion. Affordability can be referred to as the ability of farmers to procure animal-drawn implements, parts and services using income originating from farm produce where animal-drawn technology played a role.

Availability of animal-drawn implements

That availability of animal-drawn implements, spares and services has a fundamental influence on whether a farmer adopts animal power technology is obvious, in that it is impossible for farmers who cannot get implements to oxenise. In a survey carried out in Mbeya region, half of the farmers who responded indicated that they did not have the implements they needed because they were not available (Harder and Klassen-Harder 1988).

Parastatal monopoly of production

In Tanzania, unlike in many African countries, there has been a heavy involvement of the public sector in the supply of animal-drawn implements with little participation of the private sector. The main source of animal-drawn implements (up to 1994) has been Ubungo Farm Implements (UFI). UFI, a public parastatal factory, opened in 1970 with an annual installed capacity of 20,000 plows. It also produces hand tools, such as hoes, machetes, spades, forks, axes, etc and imports other animal-drawn implements such as weeders, harrows, ridgers and planters (Table 1).

Distribution points far from villages

The distribution of animal-drawn implements by UFI has been carried out through a network of three depots (Mwanza, Makambako and Kigoma), regional wholesalers like Regional Trading Companies (RTC), Cooperative Unions, the Agricultural and Industrial Supply Company (AISCO), the Tanganyika Farmers' Association (TFA) and a few private traders after the trade liberalisation in 1988.

The confinement policies in the 1980s severely restricted UFI's market, and the sale of farm implements to Regional Trading Companies on credit caused UFI severe financial problems. The wholesalers, being government controlled, were pushed through directives to sell animal-drawn implements in line with the government policy with disregard to profitability. The result of this antagonism was the accumulation of animal-drawn implements in town warehouses where nobody needed them while an acute shortage prevailed in the villages. The poor transport infrastructure further restricted accessibility of the implements to farmers. Over the years, animal-drawn implement needs (the market pull) were suffocated by bureaucratic central control over what was to be manufactured/imported, in what quantities and from where. Thus animal-drawn implement marketing became for the most part not consumer-driven, and the monopolistic tendency of UFI made it unresponsive to farmer requirements and wishes, such as an improved mouldboard curvature and angle of the frog piece so as to increase soil inversion. For example, farmers in Rukwa region find it necessary to send a newly-purchased UFI plow to a local blacksmith to work the mouldboard and frog piece to adjust the curvature and angle to increase soil inversion.

For many years UFI has concentrated only on ox plows. Other implements, such as weeders, have received little attention despite their importance in weed control which has been singled out as one of the major constraints to increased crop production (Croon, Deutsch and Temu, 1984; Rain, 1984). The use of the ox plow in tillage only shifts the labour bottleneck to weeding. This might be a disincentive to adoption where labour is scarce.

Table 1: Sales of farm implements by the parastatal Ubungo Farm Implements (UFI) in Tanzania 1978–1987

	1978	1979	1980	1981	1982	1983	1984	1985	1986	1987
Plows	7,464	7,254	15,098	5,676	16,000	30,829	80,435	2,027	24,846	22,023
Ridgers	131	111	20	0	289	1,226	1,168	195	42	81
Cultivators	38	172	60	15	40	1,236	116	1	9	60
Harrrows	48	50	0	0	240	152	82	7	0	41

Source: UFI annual reports 1978–1987

Similarly in areas where minimum/zero primary tillage is practised, for example in the semi-arid regions of Central Tanzania, labour is mainly critical in weeding, thus the introduction of the ox plow on its own would not address the farmers' needs and as such they might be reluctant to adopt the technology.

Affordability

Cash crops

Animal draft technology requires a relatively heavy investment which might not be possible for farmers with low farm incomes. To ensure sustainability, farms, like any other business venture, should be commercially viable and this can be achieved only when outputs have an excess (profit) over the inputs. Thus in most cases animal traction has been associated with cash income. It is therefore not surprising to note that the main area of animal traction use in the country is the cotton belt of Arusha, Mara, Mwanza, Shinyanga, Tabora and Singida which together account for 75% of Tanzania's work oxen (Starkey and Mutagubya, 1992).

Liberalisation of prices has brought in private companies like Cargil Tanzania Ltd, Chawe Transport and Co, Arusha Coffee, Douran Co ltd, Telae Winch Co.Ltd and many others, resulting to prompt cash payment and even better prices for farmers and thus enhancing animal traction use. For instance, in 1994 cotton prices nearly doubled from US$0.20 per kg to US$0.34. However, where there is no potential for cash crops farmers have benefited little from trade liberalisation due to the persistent government policy of cheap food. This has made it difficult for such farmers to afford the initial investment in animal traction.

Credit

The unaffordability of animal-drawn supplies by subsistence farmers requires 'unconventional' interventions to facilitate its introduction and/or adoption. While credit for farmers is classified as an 'unconventional' intervention in the Tanzanian context, most North American households and businesses cannot function without credit. Government-controlled banks, aid agencies and, more recently, private banks are all very wary of credit schemes because of the failure of earlier interventions. However, in most cases, loan default

has been due to organisational problems and not a 'boycott' by farmers.

Experiences from elsewhere in Africa including Mali, Togo, Benin, Cameroon and Côte d' Ivoire have shown that the provision of credit by crop marketing organisations and financial institutions has lead to widespread adoption of animal traction (Starkey and Faye, 1990). Unfortunately, there are few examples of this kind of intervention in Tanzania. A small number of donor-assisted projects and NGOs have extended credit to farmers successfully, but their overall impact has been low as they have worked in only a few areas.

In an effort to liquidate some slow-moving stock (*Cossul* cultivators and nylatron bushed axles), SEAZ, a small private company based in Mbeya, Tanzania managed to sell on credit in one season 77 *Cossul* cultivators and 20 complete carts at the full market price of Tsh 39,500 (US$ 79) and Tsh 150,000 (US$ 300), respectively, while their annual sales of the two implements on a cash basis averaged 11 and 3 units, respectively. The credit was repayable in one year at no interest. The loan repayment was 100%, presumably from the close contacts and follow-ups which were maintained. As a result of this credit arrangement, loan requests in 1995 stood at 160 cultivators, outstripping SEAZ's loan ability even with price increases of 20% for cultivators and 10% for carts to cover for devaluation. Only 40 farmers benefited from this arrangement before stocks of cultivators were depleted. This example suggests that the rate of adoption of animal traction technology can be increased significantly by the provision of credit and the loan repayment can be very high provided that the schemes are well organised with consistent and well-defined goals right from the beginning. Also, it shows that relatively inferior but workable technologies can still be accepted when credit is available as SEAZ's sales indicated farmer preference for other cultivator designs such as the *Agro Alfa* and *Mkombozi* over the *Cossul* and SEAZ *Nibebe* pneumatic roller bearings carts to the nylatron bushed axle carts. In the light of this, it is sad to note that US$ 295,000 set aside for credit by the International Fund for Agricultural Development's 'SHERFS' project in Mbeya Region has not been available to farmers for the last three years because of administrative wrangling.

Price controls and subsidies

It has been observed that the economic potential of any innovation, including animal-drawn technology could be raised through effective research, extension and marketing to attract more farmers to use it or through price controls and subsidies. The former is more difficult to implement and hence generally unpopular, so many governments opt for the latter approach. In Tanzania, in spite of the stringent price controls on farmers' produce, the government also subsidised animal-drawn implement prices until 1994. Mkomwa, Shetto and Mkoga (1994) report an indirect 53% price subsidy on the UFI plow compared to SADC region prices (Table 2). However, when the markets were liberalised the decrease in crop prices was much larger than the subsidies for implements.

The very large number of smallholder farmers who constitute 85% of the total employed population command little political power and are often seen as the group to be directly and indirectly taxed through price controls of their produce to subsidise food prices for the urban population and to support other, generally urban-based, state activities.

Massive state subsidy, much of it to support technological advancement, is in contrast to the approach used by developed countries for their minority of farmers. A comparison of the relationship between intervention through price controls/subsidies and technological advancement reveals exploited small-scale farmers producing well below potential, while rich large-scale farmers in developed countries produce an enormous food surplus which can only be sold at further subsidised prices.

Profitability of animal-drawn technology supply businesses

The authors believe that a businesslike approach to animal-drawn technology marketing is a prerequisite to serving the intended consumers and suppliers effectively. Animal-drawn technology supply must be profitable to justify involvement of suppliers, in this case manufacturers, importers, wholesalers and traders. Profit is also essential to ensure sustainable provision of quality implements and services.

It is sometimes argued that the price controls on implements were needed to enhance affordability by farmers whose crop prices were equally controlled. On the contrary, both farmers and the technology have suffered due to the following reasons:

- inadequate profit margins restricted the financial ability to effect sufficient availability of quality implements, spares and services. The inability by UFI to respond to feedback

Table 2: Prices of implements from UFI and other manufacturers in Africa in October 1992

Supplier	Country	*Implements and price in US$*				
		VS10 plow	S51 cultivator	Ridger	Zig-zag harrow	Planter
Zimplow	Zimbabwe	64	106	137	201	285
Bulawayo steel products	Zimbabwe	64	105	136	200	284
AgroAlfa	Mozambique	80	120	104	222	
Safim	South Africa	65	70			183
Cossul	India	86	55	114	75	
UFI	Tanzania	38	9 (from Cossul)	35 (from Cossul)	12	91 (from Zimplow)

Prices are ex-works in US$ in the host cities. UFI's prices include delivery.
Source: Mkomwa, Shetto and Mkoga, 1994

for improvement of their plow designs and erratic supplies are but a few examples. What is worse for UFI is that the government did not even reimburse the company for the subsidy between the market and controlled prices. Thus the recent liberalisation of markets has not fully benefited UFI because of their poor financial position and the inability to secure bank loans as a result of the poor financial performance of animal-drawn technology businesses over a number of years

• controlled pricing discourages individuals from investing in the sector. When competition is suffocated because of low profit margins, implement availability and repair services deteriorate as many intended suppliers are pushed out of business.

Economies of scale

Economies of scale demand a minimum number of units to be manufactured or traded to justify the capital invested. The implication is that manufacturing should be undertaken by large factories. But even these are vulnerable, as sales depend mainly on farmers' income and weather, and the underdeveloped infrastructure of the country limits the market to a few easily accessible areas. Ox carts are an exception, as their bulkiness and high capital tie-up render their production by small- to medium-sized workshops more economical.

Seasonality of animal-drawn technology business

Manufacture of and trading in animal-drawn implements and spares are not attractive as sales are highly seasonal, thus creating cash flow problems for most of the year, especially for smaller companies and traders. Mkomwa, Shetto and Mkoga (1994) reported that 62% of SEAZ's annual implement sales were realised in the four peak farming months of September to December in the Southern Highlands (Table 3). To ensure timely deliveries at the start of the farming season, suppliers have to build up their stocks off season when their sales are at the lowest level. This is difficult, especially where credit is not easily accessible. Seasonality of sales means that capital is tied up in stocks when it could give higher returns if invested elsewhere eg in consumer goods with more regular sales. In the 1994/95 season

SEAZ incurred a loss of US$ 7500 because late deliveries meant that they could not sell their implements in time and had to hold the stocks for 8 months. Due to the depreciation of the Tanzanian shilling during this period and the fact that it is unrealistic to pass on the cost of devaluation to resource-poor farmers who expect constant prices, this resulted in a loss to SEAZ.

Marketing and promotion

Marketing and promotion in this context is defined as communication between a manufacturer/trader and prospective farmers aimed to inform, persuade and influence the farmers to get them to buy animal-drawn implements, spares or repair services.

With the exception of donor-assisted projects like the Mbeya Oxenization Project, the Tanga Draft Animal Project, the Maswa Rural Development Programme, the Lake Zone Farming Systems Research Project and MIFIPRO, no significant animal power promotional activities have been made by either the parastatals or the few emerging private companies after the trade liberalisation.

The government-controlled extension systems have been ineffective and have little, if anything, to offer animal traction technology mainly due to the limited animal-power knowledge of the extensionists. Many farmers are not aware of the appropriate use and maintenance of other implements apart from the plow. For example, in one SEAZ marketing trip to Kahama District farmers were not aware of ox-powered weeding despite having used animals for plowing for over seven decades. Thus the advantages and profitability of animal traction have not yet been demonstrated fully under peasant farm conditions even in potential areas of widespread use and a conducive environment for adoption. A vigorous, well-coordinated extension programme has been credited for the accelerated rate of adoption in some parts of Gambia, Mali and Burkina Faso (Jaeger, 1984; Starkey and Faye 1990).

Strategies to alleviate animal-drawn technology supply constraints.

Marketing and promotion

Promotional efforts to inform, persuade and influence farmers about the availability,

Table 3: The seasonality of number of animal-drawn units sold illustrated by the sales of SEAZ Agricultural Equipment Ltd

Fiancial year	Implements	July–Sept	Oct–Dec	Jan–March	April–June	Total
1991–92	Plows	1	116	9	8	134
	Weeders	3	40	26	5	74
	Ox carts	59	43	11	16	129
	% of total	19	59	13	9	337
1992–93	Plows	68	222	59	40	389
	Weeders	11	103	52	42	208
	Ox carts	23	47	2	23	95
	% of total	15	54	16	15	692
1993–94	Plows	47	136	98	17	298
	Weeders	65	188	29	17	299
	Ox carts	35	12	10	32	89
	% of total	21	49	20	10	686

Source: SEAZ Quarterly Sales Reports

affordability and potential benefits stand out as the intervention with the best potential returns to the widespread use of animal draft power.

Conventional extension approaches as part of promotion are ineffective. Non-Governmental Organisations (NGOs), private companies and traders also need to be supported by donors to promote animal-drawn technology and stimulate demand.

Subsidies

No direct subsidies should be provided to parastatals or others to allow free market prices to determine the profitability and adoption of animal-drawn technology.

However, indirect subsidy on selected equipment such as weeders within a stipulated time-frame is justified and essential to promote sales and realisation of the potential of the technology. Rebate coupons for reimbursement to farmers by the Ministry of Finance are one option. In areas of animal-drawn technology introduction, or where farmers are poor with no access to cattle, provision of targeted subsidies is a prerequisite.

Credit

The provision of credit is desirable to enable the uptake of animal-drawn products with high investment costs such as weeders and ox carts. Increased production and trade and revenue from hire income would enable repayment of the loans. However, loan administration has to be controlled closely with special low lending rates, taking into consideration the low profitability of animal power and the long return period for the investment. The involvement of private companies and NGOs might be beneficial as they usually have closer contacts with the recipient farmers. The special lending rates could have a stipulated time-frame after which commercial or market interest rates could be imposed.

Action zones

It has been established that animal traction can spread rapidly and spontaneously once a 'critical mass' of people has adopted the technology (Starkey and Mutagubya, 1992). Hence particular geographical action zones should be established with the aim of meeting the critical mass.

The resulting economies of scale would also allow profitable provision of implements, spares and repair services. This is especially important because implements are heavy and bulky so transport is expensive even in areas with better-developed transport systems.

Conclusion

There is ample scope for expansion of animal power in Tanzania. With effective promotion, as outlined above, the authors postulate that the number of animal power users in Tanzania could double within 5 years.

References

Croon I, Deutsch J and Temu A E M, 1984. *Maize production in Tanzania's Southern Highlands: Current status and recommendations for the future.* Centro International de Mejoramiento de Maíz y Trigo (CIMMYT), Londres, Mexico.

Harder J M and Klassen Harder K, 1988. *Mbeya Oxenization Project inception report.* Mennonite Economic Development Associates, Winnipeg, Canada.

Jaeger, 1984. *Agricultural mechanization - the economics of animal traction in Burkina Faso.* PhD Thesis, Stanford University, Stanford, California, USA.

Mkomwa S, Shetto R M and Mkoga Z, 1994. *Experiences in animal draft technology implement business in Tanzania and opportunities for entrepreneurship.* Paper presented at the Tanzania Society of Agricultural Engineers (TSAE) 1994 Annual Scientific Conference held 3–6 October, 1994, Dar es Salaam, Tanzania.

Rain D K, 1984. *The constraints to smallholder peasant agricultural production in Mbeya and Mbozi Districts, Mbeya Region.* Report to Canadian International Development Agency (CIDA), Hull, Quebec, Canada.

Starkey P and Mutagubya W, 1992. *Animal traction in Tanzania: experience, trends and priorities.* Ministry of Agriculture, Dar es Salaam, Tanzania and Natural Resources Institute, Chatham, UK. 51p.

Starkey P and Faye A (eds), 1990. *Animal traction for agricultural development.* Proceedings of the Third Regional Workshop of the West Africa Animal Traction Network, held 7–12 July 1988, Saly, Senegal. Technical Centre for Agricultural and Rural Cooperation (CTA), Ede-Wageningen, The Netherlands. 475p.

Introducing animal-drawn cultivators in north Namibia: preliminary results and reasons for hope

by

Imalwa Veikko[1] and Carole Pitois[2]

1) Chief Agricultural Extension Officer, Department of Agriculture and Rural Development
P/Bag 5556, Oshakati, Namibia
2) Agronomist, North Namibia Rural Development Project (NNRDP), PO Box 498, Oshakati, Namibia

Abstract

The North Namibia Rural Development Project carried out on-farm tests of imported 'BS41' animal-drawn cultivators in northern Namibia with the aim of showing farmers that the cultivators could reduce the time required for soil preparation and weeding of plots of pearl millet. Four groups of farmers in each of two locations tested the cultivator for weeding and/or soil preparation compared to hand weeding and hand or no soil preparation. Using cultivators pulled by donkeys or oxen markedly reduced the time taken for weeding and soil preparation compared to performing these operations by hand or with a plow. On good soils yields were higher when a cultivator was used for weeding and/or land preparation. On poor soils the cultivator had no negative effect on yield. In general farmers were enthusiastic about the technology and when the cultivators were offered at a subsidised price all the farmers involved bought them. A few technical problems with the cultivators were noted. Other constraints included poorly trained animals and poor condition of the animals at the start of the rainy season. The paper gives a short description of proposed follow-up initiatives.

Introduction

This paper describes the animal traction extension work carried out by the North Namibia Rural Development Project (NNRDP) in collaboration with the extension services in two communities of North Namibia. The NNRDP is a "research-action" project funded by the French government whose role is to improve farming systems and to support extension services and local initiatives. It has worked in North Namibia for one and a half years. This paper is the result of pilot demonstrations run with Namibian extension officers and farmers in two communities (McKee and Pitois,1995).

Half the Namibian population, 800,000 people, live in scattered homesteads in the North-Central region of Namibia (Tötemeyer, Tonchi and du Pisani, 1993). Annual rainfall averages 200–400 mm but is highly variable. Northern Namibia is predominantly agricultural, but many farms depend on relatives working outside the villages (Durrand, 1994). The soils are mostly sandy and the main crop is pearl millet (known locally as *mahangu*). Livestock keeping is also important. There are no cash crops. Approximately 50% of the farmers own draft donkeys or oxen. These animals are used for plowing and in some cases for pulling carts. Currently, the only draft tillage implement used is the single-furrow mouldboard plow. A few farmers use either government or privately hired tractors for plowing. Durrand (1994), found three main classes of farmers:

- farmers who own no cattle, only a few goats (<20) and poultry. They usually have no off-farm income. As a consequence, they do not use animal traction and are not in a position to hire animals. They have little agricultural equipment and cannot afford a fence for their field

- farmers who own animals, with up to 20 cattle. They often have access to off-farm income which allows them to invest in fences for their fields. They use animal traction, but due to people working in towns, they have to hire workers, especially for weeding

- farmers with many animals, often over 40 cattle and 20 goats. They receive off-farm income, have fences around thier fields and have access to a cattle-post (veterinary services). They use animal traction and hire mechanised plowing services and workers for weeding.

Methodology

Diagnosis

An informal general survey of the area was carried out resulting in a better understanding of the agricultural systems and in the establishment of a typology of the farms. The typology (production system analysis) was based on how the farmers use their means (labour, land etc). Secondly, a series of informal interviews about the cropping systems was carried out with individuals from the chosen communities.

General meetings with the farmers gave extension officers and the project team the opportunity to carry the diagnosis a step further by discussing with them at length the external diagnosis and proposals aimed at alleviating the identified constraints. This stage was participatory: farmers were free to make proposals and to reject or to approve the findings presented by the extensionists.

The results of the surveys highlighted three main constraints:

- late plowing, due to a lack of draft power, leads to ripening problems and low yields of pearl millet
- only small areas of millet can be sown without prior soil preparation because of the time taken to weed the plots
- weeding is probably the most significant and arduous workload in the crop cycle.

The farmers confirmed this diagnosis.

Finding solutions

As a result of the diagnosis the project team formulated two proposals:

- reduce the time required for soil preparation so as to reduce the delays in sowing.
- reduce the time required for weeding and make it less exhausting.

The project proposed to achieve this by using an animal-drawn cultivator for soil preparation and weeding and a tracer for sowing in rows. The cultivator chosen was the "BS 41" from Zimbabwe (CTA, 1992). This implement was chosen mainly because it was easily available. The farmers involved agreed to test these implements. To maintain its principles of sustainability the project carried out the testing on the following basis:

- *self organisation:* people of the community interested in the trials were asked to organize themselves into four groups with a leading farmer for each group
- *grouping:* the project wanted to work with groups to minimize the workload and to facilitate the spreading of the information
- the project provided the material and demonstrated its use
- the lead farmer from each group provided a portion of their land and their animal power.

A member of the project team first demonstrated the use of the cultivator and tracer and then asked group members to try it themselves.

Implementation

Two trials comparing yield and time taken for weeding in hand-weeded and animal-power cultivated plots of millet were carried out by four farmers' groups at each of two localities (Eunda and Onamutanda). At the begining of the experiment there was an average of 15 farmers in each group. The average attendance for the different trials was 10 farmers for 6 groups, with a maximum of 20 in one group. The other two groups had poor attendance with only 5 farmers in each.

Trial 1 compared animal-powered cultivating and hand-weeding in plots with no prior preparation of the soil. The aims of this trial were:

- to show that the use of the cultivator for weeding alleviates the major constraint (the weeds) of the farmers on a soil which has not been prepared
- to show that weeding with the cultivator is less tiring and less time consuming
- to show that the cultivator has no negative impact on yields.

Each group prepared two adjacent 10x50m plots of millet according to the following scheme:

- Plot n°1 : no soil preparation, random sowing, hand weeding
- Plot n°2 : no soil preparation, sowing in lines with the tracer, weeding with the cultivator.

Trial 2 compared yields and weeding time in plots of millet with prior soil preparation using hand hoes or a plow and weeded by hand with plots prepared and weeded with the cultivator. The aims of this trial were:

- to show that soil preparation done with the cultivator is faster than with a hoe or a plow
- to show that weeding with the cultivator is less tiring and less time consuming than hand-hoe weeding
- to show that the cultivator has no negative impact on the yields.

Each group of farmers prepared two adjacent 10x50m plots of millet according to the following scheme:

- Piot n°1: soil preparation (either hoe or plow), random sowing and hand weeding
- Plot n°2: soil preparation with the cultivator, sowing in lines with the tracer and weeding with the cultivator.

For each plot weeding time was recorded by the team preparing it. The teams also harvested the plots and measured the yields. The main aim of the experiment was as a demonstration to test the acceptance of the cultivator, so the data were not analysed statistically.

Evaluation

The final evaluation meeting gave each farmer the opportunity to make his or her comments on the trials and the motivations. It was also a place for farmers to exchange their feelings about the cultivator with other farmers. The meeting was open to everybody: farmers who participated in the trials and farmers who had not participated.

Results

Time for soil preparation

Table 1 shows the time required for soil preparation by hand, using an animal-drawn plow or cultivator. On average, soil preparation using an animal-drawn cultivator was twice as fast as with a plow and 5 to 10 times faster than when using a hand hoe. These figures are averages of times from a number of locations, which were prepared by different farmers using different animals. In a particular situation the rate of land preparation will vary depending on a number of factors including the level of training of the animals and the animals' condition.

Time for weeding

On average, it took 60 hours for two people to weed a hectare by hand. Using a cultivator pulled by a pair of donkeys the average time was 6 hours and with a pair of oxen four hours. On average

Table 1: Comparison of time taken for soil preparation by hand, using a plow and using a cultivator

| Tool | *Time required (hours per hectare) with two people* | | |
	Hand	With donkeys	With oxen
Hoe	60		
Plow		20	7–10
Cultivator		13	4–5

weeding with a cultivator was 10 to 20 times faster than hand weeding.

Yields

Table 2 shows the yields from the experimental plots. In general, plots in which an animal-drawn cultivator had been used for preparation or weeding had higher yields. This was the case even when there was no prior soil preparation.

For poor soils (sandy soils, poor fertility, low retention capacity: group n°2 in Eunda and group n°2 in Onamutanda), the cultivator had no negative impact on the yields.

For richer soils (sandy-clayey soils, good fertility, high retention capacity: all the other groups in Eunda and Onamutenda), the cultivator had a positive impact on the yields. In the year of the study the rain pattern (concentration of the rains in February and March) favoured the soils with a high retention capacity.

Some technical problems and solutions

Tracer

The lines produced on the soil by the tracer are faint. A solution could be to adapt coca-cola cans on the teeth of the tracer to widen the traces. However, it is hoped that the use of the tracer will disappear in the medium run since after soil preparation, if the animals are well trained, the furrows should be parallel and a given number of furrows will correspond to a certain width.

Cultivator

The groups found a few technical problems with the cultivator:

- the mechanism to adjust the width gets stuck very easily when used in sandy soils

Table 2: Yields of millet from the experimental plots

	Yields (kg/ha)				
Soil preparation:	None	None	Plow	Hand	Cultivator
Weeding:	Hand	Cultivator	Hand	Hand	Cultivator
Eunda					
group n°1	240	440	480	x	480
group n°2	−[1]	−[1]	x	160	160
group n°3	560	920	x	600	1080
group n°4	−[2]	−[2]	−[2]	x	−[2]
Onamutanda					
group n°1	−[2]	360	x	360	560
group n°2	80	100	100	x	220
group n°3	360	440	460[4]	x	420[4]
group n°4	−[3]	−[3]	x	−[3]	−[3]

Notes:

X - no trial

− - trials damaged or destroyed: 1) chicken attack, 2) plots flooded, 3) plots flooded and/or attacked by army worms, 4) sticks delimitating the plots were removed so there is uncertainty about the yields.

- the bolt to adjust the depth (beside the wheel) does not fit in the last hole on some implements
- the BS41 is heavier than a plow and requires a higher draft power so it was difficult for donkeys to use, especially on larger plots
- when weeding, the arms supporting the duckfoot tine must be adjusted in such a way that the duckfoot tine move forward in the ground, almost parallel to the surface of the soil. This means the duckfoot tine cuts the roots of the weeds better and less draft power is required.
- the goosefeet must be sharpened before use.

Animals
- animals are weak at the beginning of the rainy season, the time when their strength is required for soil preparation
- in general, donkeys are not well trained (whatever the type of harness used) so several rows of millet are damaged in the process of weeding
- oxen are not used to the weeding yoke (not used to walk separated from each other in a straight line). They also require training.

Farmers' response

During the trials, six out of eight groups had a high level of attendance (8 to 10 people in each group). During the final meeting, the farmers expressed their views. In general, they were enthusiastic and recognised the advantages of the cultivators. At the end of the trials, the cultivators that had been used were offered to the farmers at a subsidised price (30% of the initial price). All the farmers who had hosted trials bought cultivators.

Plans for the future

To try to solve the technical problems with the cultivator the project plans to import different and lighter models. These will be tested using the methods described in this paper.

To try to solve the problem of animal nutrition the project will run trials of fodder improvement with urea. Intensive extension sessions on animal training will be organised with the aim of training farmers and extension officers in the use of animal-powered cultivators. After this they should be able to teach other farmers to train draft animals and use cultivators.

Trials will also be run with animals trained at this session to show farmers the difference between cultivating with trained and untrained animals. The aim will be to show that the tracer is not needed if the rows are straight.

The BS41 cultivator will be demonstrated to more farmers. BS41 cultivators will be on sale in the agricultural centres.

Pamphlets in local languages and English will be produced describing the cultivators, their use and associated benefits. Meetings with local businessmen and the farmers' cooperative are planned to raise their awareness of the purpose of the technology.

References

CTA, 1992. *Tools for agriculture, a guide to appropriate equipment for smallholder farmers.* Intermediate Technology Publications in association with CTA and GRET. London, UK. 237p.

Durrand O, 1994. *Typology of farms in the Omusati, Oshana, Ohangwena and Oshikoto regions.* North Namibia Rural Development Project (NNRDP), PO Box 498, Oshakati, Namibia. 10p.

McKee J and Pitois C, 1995. *The introduction of the cultivator in two communities in northern central Namibia: the reasons for hope.* North Namibia Rural Development Project (NNRDP), PO BOX 498, Oshakati, Namibia. 48p.

Tötemeyer G, Tonchi V and du Pisani A, 1993. *Namibian Resources Manual. Ministry of regional and local government and housing,* P Bag 13289, Windhoek, Namibia. 168p.

Dynamic steering response of animal-drawn plows

by

J M Mutua* and P A Cowell

Silsoe College (Cranfield University), Silsoe, Bedford MK45 4DT, UK

Abstract

Animal-drawn plows can be quite difficult to steer and place considerable physical demands on the operator. Swing plows and beam plows are steered by tilting them to one side and if the degree of tilt is excessive the quality of work may be reduced and extra physical effort may be required.

The paper addresses these problems by presenting the outline of a theory for predicting the dynamic steering behaviour of animal-drawn implements. It is shown that the implement steering behaviour is governed by a first order differential equation. The theory has been tested on an experimental plow whose design is based on the Victory plow, a type widely used in Kenya, and can be used to determine the effect of changes in design parameters and hitch attachment on the speed of response. For example the relationship between the length of the hitch, the position of attachment, the beam length and the angle of tilt and their effect on the speed of reaction to a steering input is clearly illustrated.

The understanding offered by the theory provides a rational basis for improving the performance of animal-drawn plows.

Introduction

The control of animal-drawn plows is a complex matter, involving control of depth, plow attitude and lateral position.

Operators of animal-drawn plows must walk behind the implement and control it by the handles. Steering control depends on the type of plow. Wheeled plows which operate with a pair of wheels running on the ground require careful setting after which they should be largely self steering. Swing plows and beam plows are mechanically simpler and normally operate without the aid of a wheel running on the ground surface. To steer, the operator must lean or tilt the plow to

---

* Subsequent address:
Dr Joseph Mutua
Kenya Agricultural Research Institute (KARI)
PO Box 30148, Nairobi, Kenya

one side. This action can affect the quality of work and may be awkward to manage.

Static force analysis of the equilibrium of both animal- and tractor-drawn implements is well established (Kepner, Bainer, Barger, 1972). It forms the basis on which implements and hitch systems are designed and provides the understanding necessary for setting up and operating plows in the field.

Steering a plow is a dynamic operation. When an operator attempts to steer a plow the speed with which the implement responds is important. An understanding of the factors determining the dynamic response of an implement to a given steering correction is required.

In this paper a theory for dynamic steering behaviour is presented which relates principally to swing plows but includes beam plows as a special case.

Previous work

Previous research on lateral dynamic behaviour of implements relates mainly to tractor-mounted plows. Reece, Gupta and Tayal (1966) showed that the lateral motion of a freely mounted plow is governed by a second order differential equation. Cowell and Makanjuola (1966) showed that for all practical purposes the motion can be described by a first order differential equation. When an implement is drawn from a real hitch point the spatial time constant (the forward distance travelled to achieve a 63% response to a disturbance) is the hitch length. This research was based on the hypothesis that, if an implement is constrained to move through the soil at a small angle to the direction in which it is pointing, then a side force reaction from the soil will be induced on it proportional to that angle.

Later Cowell and Sial (1976), working on the problem of implement penetration, used the simpler hypothesis that a directional implement such as a mouldboard plow always moves

---

List of symbols

x	lateral displacement of plow from central position, m
x_0	initial lateral displacement of plow from central position, m
x_{max}	maximum lateral displacement of plow, m
y	implement depth, m
y_0	initial implement depth, m
y_e	equilibrium depth of implement, m
y_h	vertical distance from centre of resistance to hake, m
s	forward distance travelled, m
v	forward velocity, m/s
l	horizontal distance from centre of resistance to attachment point on animal, m
l_p	horizontal distance from centre of resistance to hake, m
a	angle of lateral tilt of plow, rad
b	yaw angle induced by tilting the plow, rad

---

instantaneously in its preferred direction of travel, ie the direction in which it is pointing. This gave rise to a first order differential equation. Hence, starting from an initial depth y_0, the implement penetrates to its equilibrium depth y_e along an exponential path according to the equation:

$$y = y_e + (y_0 - y_e) e^{-s/l} \qquad (1)$$

Where l is the "spatial" time constant and governs the rate of penetration. This equation gave very good predictions for the penetration of mouldboard plows. *The significant point to note is that both theories indicate that the rate of response is governed only by the hitch length l.*

Steering theory for animal-drawn swing plows

Two theories have been developed: one based on the hypothesis used for lateral dynamic behaviour of tractor-mounted implements and the second based on that for plow penetration. In this paper only the latter theory will be described.

When a swing plow of the Victory type is in use, the support wheel is clear of the ground, so that the line of action of the traces passes through the centre of resistance of the plow. The plow is of the type which has a preferred direction of travel (known as the direction of pointing) and the side forces acting on it when in work are approximately balanced.

When an operator steers a swing plow he first tilts it to one side. This is illustrated in Figure 1 in

Figure 1: Isometric view of a Victory-type swing plow inclined at an angle a to the vertical. See box above for list of symbols

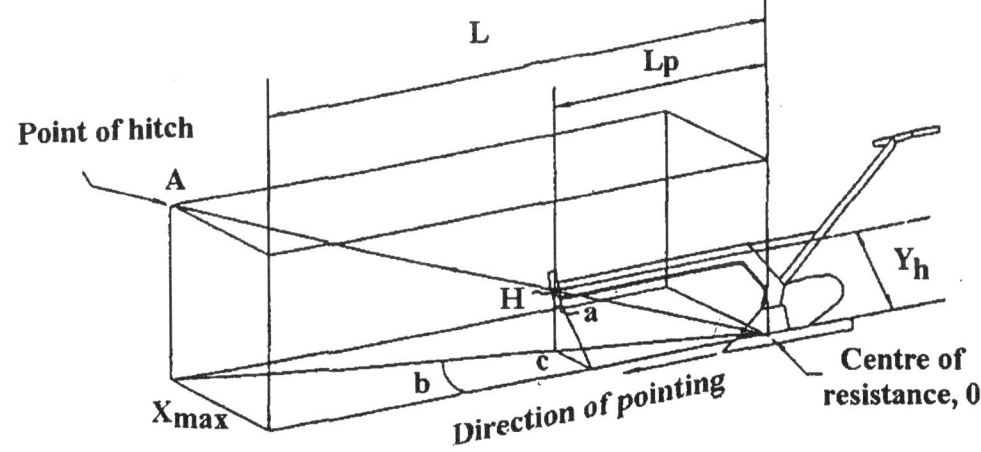

which the plow is leaned to the right. At this point the hitch point A, the hake attachment H and the centre of resistance of the plow O are not in line. The combined effect of the force in the traces AH and the soil force acting at the centre of resistance is to cause the hake point H to move into line, causing the plow to point to the left. As the plow moves forward it shifts to a new equilibrium position to the left where it is again pointing in the forward direction.

The hypothesis advanced is the same as that used by Cowell and Sial (1976) on plow penetration, namely that an implement with a preferred direction of travel (this time in the horizontal plane) will instantaneously move through the soil in the direction in which it is pointing. Consequently the implement will move (in this case) to the left until it reaches an equilibrium position where it is moving in the same direction as the animal is moving.

Based on the above hypothesis Mutua (1994) and Cowell have shown that the equation governing the lateral motion is:

$$\frac{dX}{dt} + \frac{V}{l} X = \frac{V}{l} X_{max} \qquad (2)$$

The response to leaning the plow sideways through angle a is

$$X = X_{max} (1 - e^{-vt/l}) \qquad (3)$$

If the forward velocity v is constant then the forward distance travelled $s = vt$

and $X = X_{max} (1 - e^{-s/l}) \qquad (4)$

Equation 4 shows that *the rate of response depends only on the hitch length l*. It can easily be shown that the implement achieves 63% of its total response after travelling forward a distance equal to the hitch length.

The maximum lateral displacement X_{max} obtained by tilting the plow through angle a can be understood from Figure 1 which shows the inclined plow at the equilibrium position.

Now $X_{max} = lb$

From the geometry it is readily shown that, for small angles

$$b = y_h \times a / l_p$$

and $X_{max} = a \times y_h \times l / l_p \qquad (5)$

Thus the maximum sideways displacement x increases in direct proportion to the lateral angle of tilt a , the height of the hake attachment y_h and the overall length l, but decreases in proportion to the frame length l_p.

Experimental work

Experiments were conducted using an experimental plow modelled on the Victory plow, a type widely used in Kenya.

Tests were conducted in the soil tank at Silsoe College. These consisted of a series of experiments in which the plow was held at a fixed angle to the vertical and the path taken along the ground was measured.

Good agreement was obtained between the theoretical and experimental paths.

Discussion.

Equation 4 indicates that the exponential rate of return to equilibrium is determined only by the hitch length l. Thus if the hitch length is doubled the implement must travel twice as far forward to achieve a given lateral movement.

Of particular interest is the sensitivity of the implement to a steering input. This is best described by the actual direction taken on the ground by the plow when a steering correction is initially made. From equation 4

$$\frac{dX}{ds} = -1 / l (X_0 - X_{max}) e^{-s/l}$$

The maximum rate of response is at the commencement of the correction, that is when $s = 0$.

If the implement is initially on the centre of the path, $X_0 = 0$, therefore

$$\frac{dX}{ds} max = X_{max} / l = a \times y_h / l_p \qquad (6)$$

The ratio y_h / l_p is also the slope or gradient of the traces of the plow to the horizontal.

This equation shows that the *initial rate of response (or sensitivity) increases in proportion to the lateral angle of tilt a and to the hitch point ratio y_h/l_p or the gradient of the traces of the hitch.*

These conclusions have been tested and confirmed.

Conclusion

It may be concluded that the parameters which determine the rate of response of an animal-drawn plow of the Victory type are the hitch length, the angle of the traces to the horizontal and the angle of lateral tilt.

In general, the plow makes 63% of its response after it has moved forward a distance equal to its hitch length - the greater the hitch length the further the plow travels to reach a new equilibrium position.

The sensitivity of the plow to a steering correction increases in direct proportion to the steepness of the traces and to the angle of lateral tilt.

It is significant to note that, provided the plow actually penetrates satisfactorily the theory indicates that the soil parameters have little, if any, effect on the steering performance.

References

Cowell P A and Makanjuola G A, 1966. The lateral dynamic behaviour of tractor mounted implements, with particularly reference to the three point linkage. *Journal of Agricultural Engineering Research* 11(2): 152–169

Cowell P A and Sial F S, 1976. A theory for the dynamic behaviour of mouldboard plows during penetration. *Journal of Agricultural Engineering Research* 21: 313–323

Kepner R A, Bainer R and Barger E L, 1972. *Principles of Farm Machinery* 2nd edition. Wiley (also AVI Publishing Company), London, UK.

Mutua J M, 1994. *Steering response analysis of an animal drawn mouldboard plow*. PhD thesis, Silsoe College, Cranfield University, UK (unpublished).

Reece A R, Gupta R and Tayal S S, 1966. The lateral stabilty of tractor implements. *Journal of Agricultural Engineering Research* 11(2): 80–88.

Developing suitable yokes for draft oxen in sub-Saharan Africa

by

Gérard Le Thiec and Michel Havard

Département des systèmes agro-alimentaires et ruraux (SAR)
Centre de coopération internationale en recherche agronomique pour le développement (CIRAD)
73 rue Jean-François Breton, BP 5035, 34032 Montpellier, France

Abstract

There are two main types of harnessing systems: yokes and collars. Collars are the most efficient for cattle, horses and donkeys, but are inappropriate for developing countries because of their complexity and cost. Yokes are therefore commonly used. However, in sub-Saharan Africa, where there has not been a long tradition of their use, lack of expertise and appropriate harnesses often leads to problems of injury to animals and lack of control over them. In sub-Saharan Africa, humped cattle, such as zebu, tend to be yoked with withers yokes, whilst for non-humped cattle, such as N'dama, neck yokes are used most often. However, the type of yoke used is less important than its quality, weight and proper use. In cooperation with the Cameroonian animal and veterinary research institute, CIRAD-SAR has developed an 'improved' double yoke and the 'ATECam' single-ox yoke. In trials of both of these yokes power output was improved. Their low cost makes them suitable for manufacture in developing countries.

Introduction

Although animal traction has been used for centuries in North Africa and Ethiopia, it has only recently been introduced to the rest of the continent. The Boers and French Huguenots brought it to South Africa in the 18th century. Around 1890, the French army, followed by the colonial agricultural services and trading companies, introduced animal traction to Madagascar and subsequently to West Africa in around 1920 (Bigot, 1985).

But it was only in the 1960s that animal traction really developed in western Africa and Madagascar through the efforts of research organisations, agricultural development corporations, religious missions, and, more recently, non-governmental organisations. Their know-how was based on the practices followed in their region in Europe. As a result, a wide variety of harnessing systems and animal-drawn equipment were introduced to Africa.

Historical and geographical development of harnessing systems

In animal traction terminology, 'harness' denotes a set of items that serves to link the draft animals and implement, so that energy generated by the animals can be used to carry out a specific operation. There are two distinct methods for harnessing oxen: yokes and collars. For yokes the main traction component is placed either in front of the withers (withers yoke), on the neck, behind the horns (neck yoke), or in front of the breast (breast straps). With a collar harness, the main component allows the application of tractive forces at other points of contact. These harnesses are more suited to the structure of the animals and distribute effort evenly.

The collar is the most efficient harness for horses, oxen and donkeys. The large contact surface, comfortable padding and convenient positioning of the trace attachments in front of the shoulders but slightly above the shoulder points, all combine to allow the animal to apply maximum force. However, production of collars is a sophisticated and expensive process. For these reasons they cannot be used widely in developing countries. The three-padded collar, which was tested about 20 years ago, has seldom been adopted, despite promising experimental results.

Europe: coexistence of different types of harnessing systems

References to withers and neck yokes can be found even in ancient documents (Columelle, 1st century, quoted by Delamare, 1969). In Europe, ox-team drivers and farmers have always differed about the merits of the two systems. Critics of the head yoke believe that it is too tight and can injure the animal. Nevertheless, this yoke was widely used and even replaced the withers yoke in some regions (particularly in Spain) from the 16th century onwards. The withers yoke is still used, however, in certain parts of Spain (Galicia), Portugal, France, and Italy (Le Thiec, 1991). The two systems have, therefore, coexisted in Europe through the centuries.

The neck yoke was preferred for deep plowing and transport. It is comparatively better suited for braking carts while descending slopes and for reversing. Animal teams can also coordinate their efforts better as they are yoked closer together; they therefore do not need so much training. The production of head yokes required skills that only specialists had. The profession disappeared in the 1950s when ox-drawn implements were replaced by power-driven equipment.

Despite their apparent handicaps, withers yokes are used throughout the world because they are not costly and can be easily adapted to different animals (oxen, donkeys, mules).

Development of harnesses in Africa

Apart from the Mediterranean region and Ethiopia, there is no real tradition of animal traction in Africa. Proper use of harnesses has therefore not developed. The withers yoke is more widespread than the neck yoke. It is well suited to the *Bos indicus* or humped type of zebu, and is therefore common in the Sahelo-Sudanian zone, which has a high population of these animals. Draft cultivation is highly developed in this zone.

The neck yoke is recommended for taurine cattle. It is mainly seen in the Sudano-Guinean zone and in Guinea, Sierra Leone, and Côte d'Ivoire, where the taurine cattle predominate. However, an Italian cooperation project has successfully promoted a withers yoke of Piedmont origin in Lower Zaire, particularly in Luala, for N'dama oxen (Le Thiec, 1991).

In Guinea, farmers have preserved and transmitted the technique of manufacturing (correct shape of yoke bows) and utilisation (use of cushioned covering) of yokes (Le Thiec, 1985). This know-how has endured for more than 25 years, despite an unfavorable agricultural policy that neglected animal draft cultivation since Guinea attained independence; the farmers

therefore had no support from extension services and technicians.

In Senegal, the Bambey research center and development corporations supported large-scale draft cultivation between the 1960s and the 1980s. From the beginning, they opted for the neck yoke even though the animals were zebus and crossbreeds.

Apart from these instances, manufacture and utilisation of yokes have not developed adequately in Africa. Neck yokes are usually cut and shaped carelessly and not properly fitted on the neck of the animal. They are loosely fastened to the horns, often with thin ropes that abrade the horns when the animal tosses its head. Withers yokes are often simple wooden beams, whose sharp edges are not planed off. As the yoke bows are not properly curved, the angular edge of the beam presses on the withers when the animal is in motion and makes any effort painful. Not surprisingly, the animals are difficult to manage and train.

The hames of the withers yoke are made of metal (round bars) or wood. The thin ropes tied around the throat of the animal to join the hames at their lower ends compress the windpipe. The yoke rotates constantly as it aligns with the line of traction. It moves forwards and backwards between the throat and the dewlap and almost chokes the animal.

Both the withers and neck yokes tend to be extremely heavy, often weighing up to 15 kg. Normally, their weight should not exceed 10–12 kg. Use of some lightweight, sturdy woods could reduce the weight to only 5–6 kg. However, it is commonly believed that a more massive yoke is better for making holes to facilitate hitching. The design, quality, and utilisation of yokes are therefore more important than the type.

Guidelines for better results
Animal team

Certain guidelines should be followed to obtain maximum output from draft animals, to keep them physically fit, and to increase their longevity:

- average effort should be 10–12% of the harness mass
- paired animals should be similar in size and strength for proper coordination of their work

- the harness and all the elements that transfer energy should not hinder movement of the animals or injure them; the animal should be able to freely move its head and joints, especially shoulder joints. They should also not block respiration and blood circulation.

The withers yoke is recommended because it does not constrain the animal. Moreover, it is easier to produce and use than the neck yoke.

Yoking

The guidelines for yoking also apply to the different links (yoke-shaft, yoke-chain, yoke-implement) in addition to the yoke.

The plane of the contact surface (tangent of the yoke bow curve) should be perpendicular to the line of traction and the surface should be as large as possible. This prevents the yoke from rotating during traction and the animal from choking. The minimum width of the yoke bow should be 10–12 cm, approximately the diameter of the yoke.

The yoke should not have any sharp edges. The surface that rests on the animals should be suitably rounded. Careless shaping is more harmful that none. The wooden pole used for the main piece of the yoke is naturally rounded and should be left as it is, rather than be incorrectly shaped.

The attachment point on the yoke should be lowered to reduce the angle of the line of traction. The points of contact should thus be lowered toward the shoulder points. The ideal position for the hitching hook is at the point of equilibrium on the contact surface. Only yokes with padded openings in the form of collars and those covering the shoulders offer this possibility. The optimum position is two-thirds less than the distance between the withers and the shoulder point.

Many authors have explained the diagram of the different forces and discussed the influence of the angle of traction (Ringelmann, 1908). This point will not be discussed in this paper. In practice, it is more important to reduce wastage of energy. A steep angle of the line of traction increases the vertical component of the traction effort and the work loss translates as a waste of energy. The line of traction should be so inclined that part of the load is transferred to the vertical component, which prevents the yoke from slipping onto the back of the animal. This way, the animals have a better ground grip, especially when heavy effort is required. An angle of 15–16° offers the best compromise. Table 1 gives some typical values of the traction parameters of withers and neck yokes in Africa.

Recommendations for production of yokes

Certain rules must be observed when linking the yoke to the shaft or the chain:

- the yoke hook should be fixed so that it is within the extended line of traction and it does not turn when under pressure
- the axis of hames should be perpendicular to the line of traction to prevent choking of the animal
- the tangent to the yoke bow should be perpendicular to the line of traction.

The hitching hook is thus fixed on the yoke to a metal double ring or a double loop made of braided leather strips that are passed around the yoke and left to rotate freely (Figure 1). Lateral slipping of the double loop is prevented by cutting two grooves on the upper surface in which the ring can turn half a circle, or by using staple nails.

Table 1: Average values of traction parameters for withers and neck yokes in Africa

Parameter	Neck yoke (N'dama)	Withers yoke (Zebu)
Withers-ground height (m)	1.1	1.3
Implement hook-ground height (m)	0.3	0.3
Implement hool-withers height (m)	0.8	1.0
Angle for 2.5m chain	18	23
Angle for 3m chain	15	19

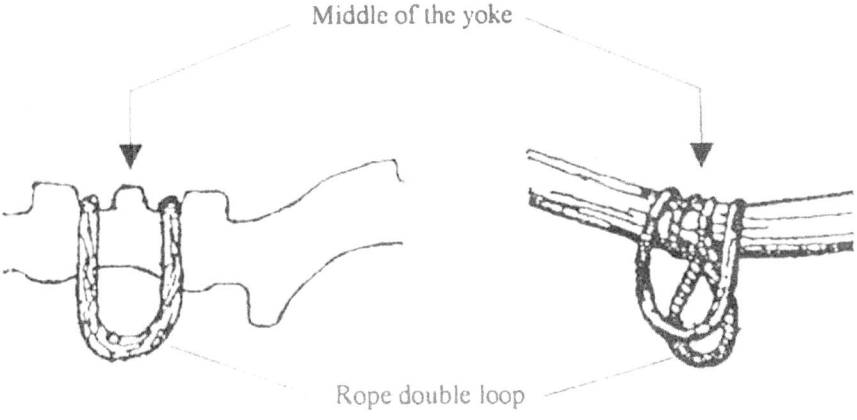

Middle of the yoke

Rope double loop

Figure 1: Centre attachment systems

Need for research on harnessing systems

Between the 1960s and 1980s, much research focused on the equipment, animals, and farming systems for draft cultivation. However, little research was carried out on harnessing systems.

The force generated by the animal depends on the type of harness, its condition, and the links that connect it to the implement. The output of the animal team depends on that of all the elements linked to it. We therefore attempted to identify harnessing systems and related elements that ensured optimum output. We compared harnesses adopted elsewhere – particularly in Europe – over the years. The original processes and know-how were analysed for a possible transfer of the technology to developing countries. We were aware that we were dealing with only one of the constraints to optimum output. Nutrition, care, and training of the animals and the farmer's ability to handle them also contribute significantly to better output.

CIRAD-SAR sought to increase the capacity of the traditional yokes through certain necessary improvements at a minimum cost. The result was the 'improved' double yoke and the single-head ATECam yoke.

The 'improved' double yoke

The design of the 'improved' double yoke (Figure 2) aims at a maximum lowering of the point where the chain is attached, the angle of the line of traction is reduced and rotation of the yoke

is stopped. The contact surface is increased for a better distribution of effort by adding prefabricated padding. In addition, the design easily allows local production of the yoke.

Trials were conducted in Burkina Faso from 1989–1991 to compare the performance of the improved double yoke and a traditional yoke. They were conducted jointly by the Burkinabe Agricultural Research Institute INERA; the Silsoe Research Institute, UK; and CIRAD-SAR. The trial protocol aimed to collect data on the total effort produced during two hours of uninterrupted work (Le Thiec, O'Neill and Bano, 1992).

The animals were made to pull a sledge to ensure maximum uniformity and control of experimental conditions. The load on the sledge was varied to obtain a range of data on efforts obtained on track and in closed circuit. The results showed that, compared with the traditional yoke, the double yoke improved energy output, which is particularly advantageous as the load is increased (Table 2).

ATECam single yoke

In 1993–1994 CIRAD-SAR developed a single yoke (Figure 3) for a research-for-development project in northern Cameroon in collaboration with CIRAD's livestock production department CIRAD-EMVT, and the Cameroonian animal and veterinary research institute IRZV.

The design specifications were: adequate contact area on the withers, no slipping of yoke when the

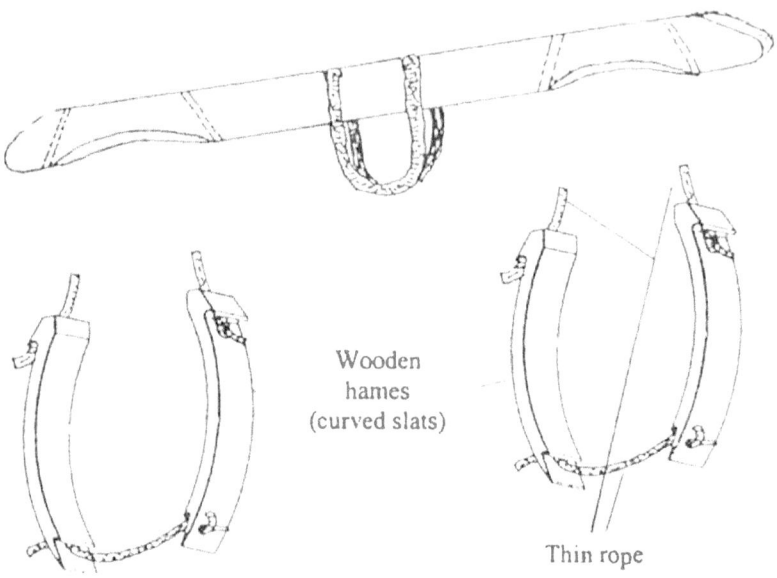

Wooden
hames
(curved slats)

Thin rope

Figure 2: The 'improved' double yoke

animal is stationary, automatic alignment with the line of traction during work, weight close to 5 kg, and low-cost production by artisans.

The prototype with the best results was called ATECam:

- A for its general shape
- TE (in French) for equilateral triangle; the two hames measure 60 cm each and as does the distance between the two points where the traces are fixed. The measurements can be reduced to 50 cm for smaller animals.

- Cam for Cameroon, where the preliminary trials were conducted (Garoua in northern Cameroon).

The two beveled hames are joined face together with a wooden bolt (TRcc 10 x 120 mm) at the top of the triangle. The wooden contact area on the withers remains in place because of it is designed to fit closely. The yoke opening is 120 mm wide near the withers, but it can range between 100 mm and 150 mm depending on the size of the animal. During work, the hames are automatically aligned with the line of traction. The strap around the throat and under the dewlap should be sufficiently

Table 2: Comparison of the increase in energy output of the 'improved' double yoke and the traditional yoke

Yoke	Average effort (% of the yoke weight)	% increase in energy
Traditional yoke	10	+4.8
	12	+8.4
'Improved' double yoke	9	+10.6
	15	+12.4

Source: Le Thiec, O'Neill and Bano, 1992

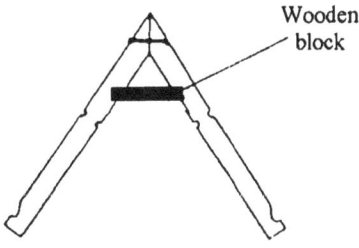

Figure 3: Cross section of the 'ATECam' yoke

long to allow this alignment and prevent the yoke from slipping on the hump when zebus are used.

The yoke weighs 6 kg. The traces are made from light material, either synthetic straps or a nylon rope. Metal chains are not recommended. There are no traction hooks. Semicircular grooves are cut at both ends of the swingle tree to fix the traces. The hitching hook ends in a ring which is gripped around the middle of the swingle tree and avoids its piercing it.

In trials with an animal in excellent condition the traction effort was 20% of the animal's weight for more than one hour (Le Thiec, 1994).

Conclusion

The CIRAD-SAR study highlights the importance of quality, proper design and production, and correct use of yokes.

The use of the withers yoke in Europe for many hundreds of years has shaken the traditional rigid dichotomy of the humped ox–withers yoke and ox without hump–neck yoke associations.

In Africa, harnesses are often considered of secondary importance in draft cultivation. The know-how acquired over the centuries in northern countries obviously cannot be transferred directly to the south. The socio-cultural context has a profound influence on the attitude, and sometimes affection, of the farmer towards the animals.

These aspects should not, however, hinder research on the design and development of harnesses. Although quality is perceived differently in Africa and may not be up to European standards, the techniques used in Europe or tested elsewhere need to be proposed to the African farmers so that they assimilate the basic principles of this know-how. These principles ensure

optimum utilisation of the potential draft power. The application of these principles enables us to develop improved yokes for these regions.

The experiments conducted by INERA, SRI, and CIRAD-SAR demonstrate that better finish and greater comfort can translate into a substantial increase in output. Qualitative improvement of harnesses becomes even more significant during intensive work periods, such as at the start of the cropping season in the Sahelo-Sudanian zone.

Single ox traction is not common in Africa, probably because of the absence of an appropriate single yoke. The development of the ATECam single yoke could fill the gap.

The low technology level of the local craftsmen and limited financial capacity of the farmers are important considerations for the application of the findings. The proposed improvements therefore focus on the fundamental principles; they are simple, low-cost solutions that ensure maximum efficiency. More sophisticated improvements requiring precise production methods can be introduced once this generation of farmers has acquired solid experience in animal traction.

References

Bigot Y, 1985. Quelques aspects historiques des échecs et des succès de l'introduction et du développement de la traction animale en Afrique Sub Saharienne. *Machinisme Agricole Tropical* 91: 4–10.

Delamare J., 1969. *Géographie et ethnologie de l'attelage au joug en France du XVIIème siècle à nos jours*. Musée des Arts et Traditions Populaires, Paris, France. 92p.

Le Thiec G, 1985. *Rapport de mission en Guinée*. Food and Agriculture Organisation (FAO), Rome, Italy.

Le Thiec G, 1991. *Les harnachements pour bovins, de "l'Europe à l'Afrique"*. Centre d'Etudes et d'Expérimentation du Machinisme Agricole Tropical (CEEMAT), Montpellier, France. 41p.

Le Thiec G, 1994. *Etude sur le travail animal et l'optimisation de l'efficacité du couple animal-outil. Amélioration des harnachements*. Rapport de mission au Cameroun. Centre de coopération internationale en recherche agronomique pour le développement, Département des systèmes agro-alimentaires et ruraux (CIRAD-SAR), Montpellier, France. 11p.

Le Thiec G, O'Neill, Barro A, 1992. *Expérimentation et mise au point d'outils à traction animale en zones semi-arides et amélioration des dispositifs d'attelage*. Centre de coopération internationale en recherche agronomique pour le développement, Département des systèmes agro-alimentaires et ruraux (CIRAD-SAR), Montpellier, France. 35p.

Ringelmann M, 1908. *Génie rural appliqué aux colonies*. Institut National Agronomique, Paris, France. 698p.

Design and manufacture of a withers yoke for zebu oxen and cows

by

Wells F Kumwenda

Ministry of Agriculture and Livestock Development, Farm Machinery Unit
Chitedze Research Station, PO Box 158, Lilongwe, Malawi

Abstract

Poorly designed yokes are commonly used in Malawi, often resulting in injuries to, and poor performance of, draft animals. This paper discusses the design criteria for effective yokes and concludes that the most important criteria are that: the design and materials should be simple, permitting artisanal manufacture; the cost be within the financial means of the intended users and the yoke should be comfortable for the animal. The different lengths of yoke required for plowing, ridging and transport are discussed. The paper details the design of a wooden yoke for zebu cattle that fulfills these design criteria.

Intoduction

The word yoke means a wooden cross-piece fastened over necks of two oxen or other draft animals and attached to the implement or cart that they are to draw. A yoke links two draft animals to each other more rigidly than single (collar) harnesses. There are many types of yoke depending on the animals for which they are designed and the materials available for making them. The wide range of yoke types can be divided into two main categories, those tied to the horns of the animal (head yokes) and those taking power mainly from the withers (withers yokes). The withers of an animal refers to the part of the back that is over the shoulders directly above the first thoracic vetebra. The zebu cattle (*Bos indicus*) used in Malawi have their withers immediately in front of the hump (Starkey, 1989). Both types of yokes are used in Malawi. The most common yoke is the withers yoke but in forestry operations the head yoke is used extensively. Head yokes that are securely fastened can facilitate braking when going down a slope and when reversing. However, head yokes restrict the freedom of movement of the neck and head so that the animal has difficulty in warding off flies (Vieberg, 1982) and the string

used for tying the yokes may create sores around the horns. It has been argued that yokes which definitely restrict side-to-side movement of the animals are undesirable.

In Malawi there are many examples of poorly designed yokes. Common problems include:

- yokes that are too short, making the animals bellies touch when hitched
- too narrow a space between the skeis so the animals' necks are squeezed
- skeis too short so the animals are almost strangled
- clamp holes drilled wrongly so the yoke rolls backward making the skeis point forward or the yoke rolls forward making the skeis point backward.

This paper describes the manufacture of a good withers yoke for two animals.

Design criteria for yokes

The common yoke used at present in Malawi consists of a round pole, four skeis and strings (also known as strops). According to Vieberg (1982), a good harnessing system should:

- not hinder the animals' natural movements, breathing and blood circulation, this being of particular importance for efficient power transmission
- permit reliable control and directing of the animals
- not injure animals or cause pain
- be as simple as possible to fit and remove
- be easily adjustable to fit the animals
- permit braking
- be of simple design and permit local manufacture.

- be inexpensive to enable farmers to obtain several yokes suitable for various types of work.

The most critical parameters for a good yoke include:

- the design of the yoke should be simple enough to enable farmers and rural artisans to make them in rural areas from materials available locally
- the cost of the yoke should be within the financial means of the users
- the yoke should be comfortable to the animals but at the same time durable to withstand the forces and the weather
- the yoke should have a large and smooth contact area with the animals because a small or uneven area concentrates the force, often resulting in sores at the contact point.

The withers yoke is better for zebu cattle because they are not sufficiently strong necked to use head yokes. Vieberg (1982) recommended that if a withers yoke is used each animal should always be harnessed on the same side so they become adjusted to work on their side through training and corresponding muscular development. When plowing, the right hand animal will become used to moving in the furrow and the left hand animal will become used to walking on the hard ground. The stronger animal should be harnessed on the right for plowing because it has to walk on freshly plowed ground with soft soil and thus has to work harder to provide the same amount of traction as the animal moving on the hard ground.

Neilson (1970) tried to develop several improved yoke types for oxen at Bunda College, University of Malawi. Unfortunately he used metal pipes and chains which are expensive and not readily available in rural areas. The same author later concluded that steel materials should be replaced by wood to reduce the cost and enable manufacture by rural artisans.

Methods of yoking

There are various types of yokes and individual harnesses and their regional distribution in Africa is probably based on several factors including:

- influence of colonial settlers

- availability of animals for example where there are many animals more tend to be yoked at once (eg Botswana)
- soil type
- type of equipment used
- the traditions that have been passed from generation to generation.

Yokes have major control on the output of draft animals. In Malawi three types of yoke are common: the plowing yoke, the ridging yoke and the transport yoke.

Plowing yoke

To reach the optimum depth and width of cut when plowing high draft power is required so the animals should be yoked close together. On a plowing yoke the animals should be placed 90 cm apart (centre–centre) on the yoke shaft. The traction chain should be at least 2.9 m long. If the yoke is too long (animals >90cm apart) a strip of unplowed land will be left at the centre during plowing, and if the yoke is too short (animals <90cm apart) the share cuts a narrow slice of unplowed land. This means that it will take more time to plow a given area of land.

Ridging yoke

To produce evenly spaced ridges the distance between the centres of the animals on the yoke shaft should be twice the desired ridge spacing. The recommended ridge spacing for most crops in Malawi is 90 cm so draft animals should be spaced 180 cm apart (centre–centre) on the yoke shaft. If other ridge spacings are required the yoking position should be adjusted appropriately. For example, if one requires a ridge spacing of 120 cm for tobacco then the draft animals should be 240 cm apart on the yoke shaft.

Transport yoke

The design of a transport yoke is based on the type of carrier to be pulled. The distance between the animals depends on the wheel track of the cart to be used. The wheel track is the distance between the centre of one wheel to the centre of the opposite wheel. This means that animals pulling a cart using a correctly designed yoke will walk in line with the wheels. A person coming from behind should not be able to see the hoof marks of the animals pulling the cart. As an example, a yoke for a standard one tonne oxcart

(1.8 m long and 1.2 m wide) should be 1.7 m long. The draft animals should be placed 110 cm apart (centre–centre) on the yoke shaft.

Manufacture of the yoke

The yoke shaft

The yoke shaft should be strong but not too heavy or too rough. Many farmers in Malawi use bluegum (*Eucalyptus*) poles. Freshly cut trees should be left to dry in the shade before constructing the yoke, otherwise the pole will crack and bend. A smooth finish is desirable to protect the skin of the animal. The diameter of the yoke shaft should be 8–10 cm.

The skeis

Each animal requires a pair of skeis, one on each side of its neck. The skeis in each pair should be 20 cm apart to accommodate the neck of the animal. If the animal is large the space should be increased to 25 cm. Too narrow a space results in the skeis damaging the neck of the animal. However, if the space between the skeis is too big the animal will be able to remove the yoke by pointing its head downwards, especially if it has small or no horns. Skeis with a shallow curve inward are best.

Each skei should be 45 cm long, 7 cm wide and 4 cm thick. The longest side has three notches 1.5 cm deep. The notches are made on the skei to fit three different neck sizes of draft animals. The notches assist in holding the string (also called strops) that goes round the bottom of the animal's neck after adjustment.

To make the notches on the skei, measure from the bottom of one side of the skei upwards and mark at 7 cm, 3 cm, 4 cm, 3 cm, 4 cm and 3 cm. Cut 1.5 cm deep into the skei at the bottom of all the 3 cm gaps then cut again at an angle from the top of the 3cm to the inside end of the 1.5 cm at an angle. The skei head should be 7 cm x 4 cm x 4 cm, the skei body should be 7 cm x 2 cm and the skei hole should be 7.5 cm x 2.5 cm. The skey

hole will allow the skei body to go through but not the head.

To make the skei holes, first mark the required distance from the centre of the yoke to the centre of the withers of one animal. From this position measure 10 cm towards the middle of the yoke and measure 10 cm towards the outside of the yoke. These measurements will enable you to mark 20 cm on each end of the yoke that will accommodate the withers of the animals. If the animals are large increase these measurements from 10 cm to 12.25 cm on each side so that the distance between skeis for each animal is 25 cm. On the top of the yoke measure a rectangle 7.5 cm long (away from the centre of the animal) and 2.5 cm wide. This should be made into a hole by chiselling.

Clamps

U-shaped clamps at the centre of the yoke are used for hitching to the implement through a chain or boom. Metal clamps are not good because over time they enlarge the size of the holes in the wooden yoke and they are rigid. The best clamps are made of plastic ropes; these are strong and durable but also flexible. It is possible to compensate for animals of different strengths when fixing clamps on the yoke shaft: the stronger animal should be closer to the clamps than the weaker one.

Strings

Strings (strops) go under the neck of animal from one skei to the other. These should be of adequate length, smooth and soft to avoid hurting the animal.

References

Neilson H, 1970. *Improvement of ox yokes.* FAO Technical Assistance Report No. AGS/TA/MLW/68/3. University of Malawi, Bunda College, Lilongwe, Malawi.

Starkey PH, 1989. *Harnessing and implements for animal traction: an animal traction resource book for Africa.* GTZ, Eschborn, Germany. 243p. ISBN 3-528-2053-9

Vieberg U, 1982. Basic aspects of harnessing and the use of implements. pp 135–221 in Munzinger P (editor). *Animal traction in Africa.* GTZ, Eschborn, Germany. 490p.

Harnessing techniques and work performance of draft horses in Ethiopia

by

Mengistu Geza

Assistant Research Officer, Institute of Agricultural Research, Nazareth Research Center
PO Box 436, Nazareth, Ethiopia

Abstract

A study was carried out to select suitable harnesses for horses and to generate indicative data on draft performance of horses. Three types of harness were used: collar harness, breastband harness and local neck yoke modified for a single horse. The weight of the horse used was 275 kg. A sledge-type loading device was used to apply three levels of pull: 25%, 30% and 35% of the body weight of the horse. The study was carried out on an oval test track. Data were collected on pull exerted, speed of work and on changes in body temperature, pulse rate and respiration rate of the horse.

The collar harness was found to be the most suitable. It offered more contact area for efficient utilisation of the strength of the horse and performed better in terms of speed, energy and power output compared to breastband and yoke harnesses.

The average working speed of the horse was 0.75 to 1.07 m/s for the collar harness,which is higher than that of Ethiopian oxen (0.4 to 0.5 m/s). With this range of speed the horse was able to generate up to 25 to 35% of its body weight (275 kg),which is 68 to 96 kgf. The working speed of the horse while pulling a maresha *ard plow was 0.86 m/s and the actual field capacity was 0.1 ha/h.*

Introduction

Despite the great progress of motorised agriculture, manual workers and draft animals will still continue to provide the main source of power for the farmers of many regions where the use of tractors and tractor equipment does not yet pay for itself. Different draft animals such as oxen, male buffaloes, camels, donkeys, mules and horses are used for draft purposes (Hopfen, 1969)

Oxen are the main source of draft animal power in Ethiopia. Of the 60 million hectares that are cultivable, only 6 million hectares are cultivated in any one year. Ninety-five percent of this is cultivated by small-scale farmers with five million oxen. The distribution of the oxen is not uniform: less than one third of the farmers own two or more oxen, about one third of them own only one ox and the rest do not own oxen. Thus farmers face acute lack of draft power for tillage which has limited the area of land cultivated and crop management practices.

There are 1.6 million horses, 1.5 million mules and 3.9 million donkeys in the country (Fielding and Pearson, 1991). These mules and donkeys are used mostly for pack work. Only a small percentage of horses are utilised for draft, mainly in the Northern parts of the country: Gojjam, Wello, Gondar and Northern Shoa (Pathak, 1988). In Inewari area (Northern Shoa), the use of horses is becoming very important. This shift from oxen to horses is attributed to the fact that horses can be grazed on herbage types which are not grazable by cattle, horses are multipurpose animals, on flat plots power output of horses is considerably higher than that of oxen and the cost of acquiring or replacing a horse is less than that required for an ox.

Oxen are harnessed in pairs using a yoke that is known locally as *kenber*. The yokes are made of light wood and are 140 cm long with a diameter of about 7.5–8 cm. Horses are commonly hitched with the same yoke using an artificial hump. Yokes are used on horses for convenience and simplicity as equine harnesses are not easily obtainable and yokes for oxen are already available (Starkey, 1989). However, yokes are unsuitable for horses, mules and donkeys since these animals are built differently from cattle. Their main strength is in their shoulders and chests. To improve their work output it is necessary to improve or select the harness and thereby improve the work efficiency of the animals. The design of the harness can help

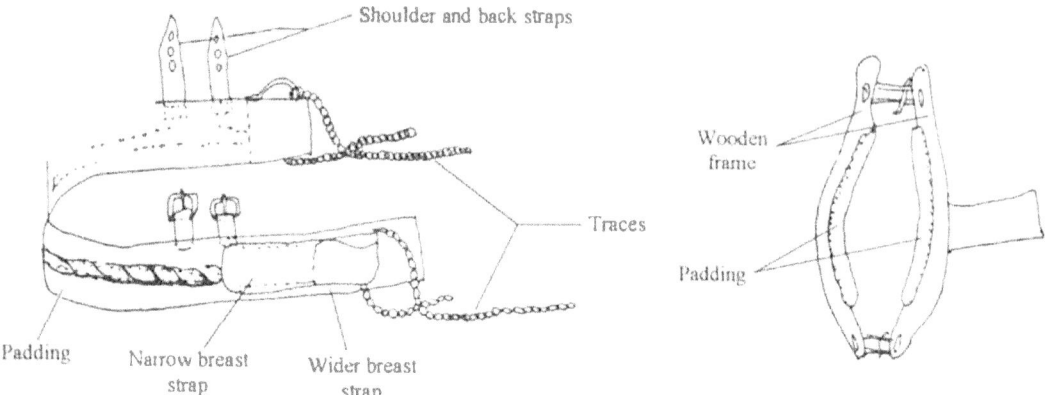

Figure 1: Breast band harness

Figure 2: Collar harness

in tapping the full power of the animals, making them more economical and useful to the farmers.

On this basis a study was carried out at Agricultural Implements Research and Improvement Center to select suitable horse harnesses and to generate indicative data on draft performance of horses and the permissible draft requirement of horse-drawn implements.

Methodology

Three types of harnesses: collar harness, breast band harness and local neck yoke modified for single horse with an artificial hump 'Embineger' (Figures 1, 2 and 3) were used. A horse weighing 275 kg was used to test the three harnesses. The study was carried out on an oval test track. Prior to testing, the horse was trained with each harness on the test track. Observations made prior to testing indicated that pull levels below 25% of body weight allowed the horse to move faster than needed for work animals, whilst at pull levels above 35% of body weight the speed of the horse was too low and the horse showed signs of fatigue in a short time. Therefore three levels of pull: 25%, 30% and 35% of the horse's bodyweight (68, 82 and 96 kgf) were applied using a sledge-type loading device. For each of the three pull levels each harness was used for three consecutive days in a week. The horse was allowed to rest for the remaining four days. During the non-work days the horse was allowed free grazing. On the working days supplementary feed (grain) was given to the horse.

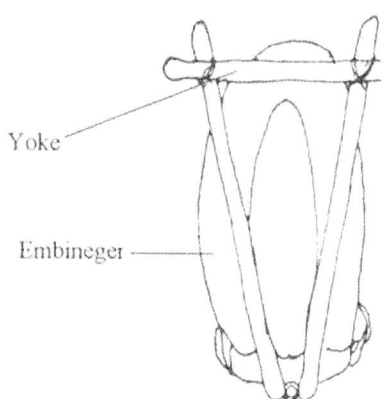

Figure 3: Yoke with embineger artificial hump

The horse was made to work continuously for three hours. An electrical load cell of 0–5 KN capacity was used to measure the pull. The variation of friction between the sledge and the track from one point to another was negligible. The angles of pull for all the harnesses were measured and were kept constant for the three levels of pull.

Speed of the horse was measured by recording the time taken to complete five rounds of the track throughout the working period. Body temperature was measured using a rectal thermometer before starting work, every hour during work and at the end of work. Pulse rate was measured at the throat latch of the animal at the same time as temperature was measured. Respiration rate was also measured at the same time by keeping the flat of the palm on the animal's flank. Data on ambient atmospheric

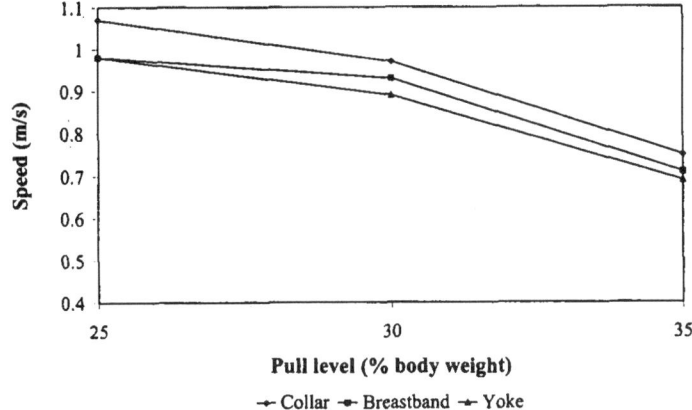

Figure 4: Speed at different pull levels for the three harnesses

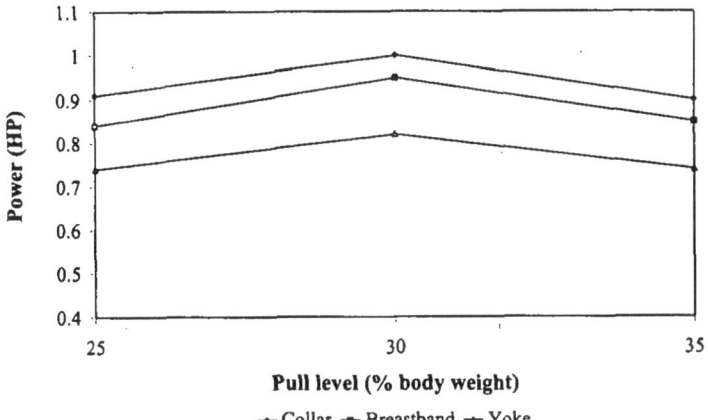

Figure 5: Power output at different pull levels for the three harnesses

beating was required to move the horse while working with the yoke harness. The yoke harness had poor stability against the animal's neck and was found to cause sores at higher pull levels, despite the padding. The breast band harness needed accurate positioning, otherwise it interfered with the movement of the animal resulting in sores around its chest.

The power output increased up to a pull level of 82 kgf, 30% of body weight of the horse, after which it started to decline as shown in Figure 5. The average speed of the horse at the lighter load, 64 kgf, was 1.1 m/s using the collar harness. With an increase in draft the average speed decreased to 0.75 m/s at a load of 91 kgf. Because of the decrease in speed, the power output started to decline as the draft was increased.

The change in speed during the work is shown in Figure 6. The horse's body temperature, respiration rate and pulse rate increased considerably during the first hour of work and then tended to stabilise. After the 1st and 2nd hour of work the respiration rate and pulse rate in most cases started to decrease slightly for all the harnesses. This can be attributed to the decrease in the speed of the horse, especially after the 1st and 2nd hour, indicating that the physiological responses are more dependent on the speed of the horse than the pull exerted.

Implement modification and testing

The results of the study have shown that the draft power output of a horse is sufficient to perform tillage operations, provided appropriate implements are available. In addition, the collar harness was found to be the most suitable harness for the horse. Therefore the local plow was modified for a single horse as shown in Figure 7

conditions during the study period were obtained from the Nazareth Research Center weather station. The mean temperature and relative humidity for the test duration were 25°C and 49.5% respectively.

Results and discussion

The results are shown in Table 1 and Figures 4 to 6. Figures 4 and 5 show the relationship between draft, speed and power for each of the harnesses. As the draft was increased, the speed of the horse decreased for all types of harness. Although the differences in the speed and power output of the horse at different levels of pull were not significant among the three harness, the collar harness performed relatively better followed by the breast band harness. As the draft was increased there was frequent stoppage and so continuous

Table 1: Work performance of a horse with three different harnesses

Pull level (% body weight)	Collar harness			Breastband harness			Neck yoke		
	Draft (kgf)	Speed (m/s)	Power (HP)	Draft (kgf)	Speed (m/s)	Power (HP)	Draft (kgf)	Speed (m/s)	Power (HP)
25	64	1.1	0.9	65	1.0	0.8	57	1.0	0.7
30	78	1.0	1.0	78	0.9	1.0	70	0.9	0.8
35	91	0.8	0.9	91	0.7	0.9	82	0.7	0.7

and it was tested at Inewari (central Ethiopia) on heavy soil with a moisture content of 13% and bulk density 1.1 g/cm^3. The test results are shown in Table 2. The working speed of the horse while pulling the *maresha* ard plow was 0.86 m/s and the actual field capacity was 0.1 ha/h.

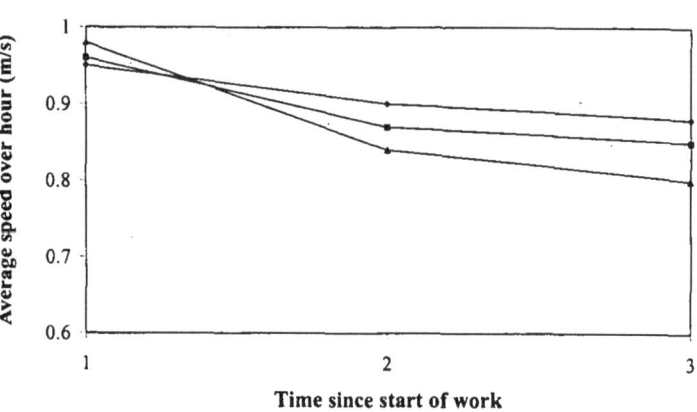

Figure 6: Change of speed with time since start of work. The figures for each harness are averaged over the three levels of pull

Conclusions

The results of the comparative test among the three harnesses indicate that the collar harness shown in Figure 2 is the most suitable. It provides more contact area for efficient utilisation of the strength of the horse compared to the breast band and yoke harness. It performed better in terms of speed, energy and power output compared to the breast band and yoke harnesses. Yoking is the most common practice in parts of Ethiopia where horse are used for draft purpose. However, yokes have a higher angle of pull, and are unsuitable and inefficient for horses, mules and donkeys. Therefore, for heavy draft it is advisable that the collar harness is used. For lighter work the breast band harness should be used since it is relatively cheap. In addition, no physical injury was observed using the collar harness, unlike the breast band and yoke.

Figure 7: The maresha *ard plow modified for use with a single horse*

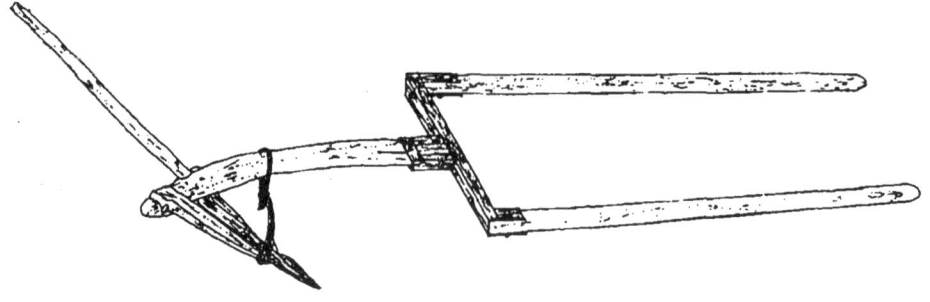

Table 2: Tests of a *maresha* ard plow modified for a single horse

Plot	Soil bulk density (g/cm³)	Soil moisture content (%)	Depth of work (cm)	Cross sectional area of work (cm²)	Capacity (ha/h)	Forward speed (m/s)	Draft (kgf)
1	1.1	14	11	175	0.1	0.8	51
2	1.1	14	13	179	0.1	0.9	49
3	1.1	12	13	183	0.1	0.9	58
Mean	**1.1**	**13**	**12**	**179**	**0.1**	**0.9**	**53**

The average working speeds of the horse were 0.75 to 1.07 m/s for the collar harness, which is significantly higher than that of Ethiopian oxen which range from 0.4 to 0.5 m/s. With this range of speed the horse was found to generate pull from 25 to 35% of its body weight (275 kg), which is 68 to 96 kgf. This is adequate to perform secondary tillage operations and to pull moderate loads under most soil and surface conditions provided appropriate implements are available. This range of draft output is also adequate to meet the draft requirement of primary tillage implements on light soils.

Performance tests of the *maresha* ard plow for primary tillage on light soil using a pair of oxen has shown the working speed of the oxen to be 0.66 m/s with an actual field capacity of 0.05 ha/h. Therefore, a single horse can work more area per unit time (0.1 ha/h) than a pair of oxen. However, a horse can work for a shorter duration per day compared to an ox. In areas where there is a shortage of draft animal power and where horses are available, it could be a good option to the farmer to use this implement modified for a single horse along with the collar harness.

References

Campbell J K, 1990. *Dibble sticks, donkeys, and diesels.* Philippines.

Fielding D and Pearson R A, 1991. *Donkeys, mules and horses in tropical agricultural development.* Proceedings of a colloquium held by the Edinburgh School of Agriculture and the Centre for Tropical Veterinary Medicine of the University of Edinburgh.

Hopfen H J, 1969. *Farm implements for arid and tropical regions.* Food and Agriculture Organisation of the United Nations (FAO), Rome, Italy.

Ministry of Agriculture, 1986. *Agricultural Statistics* (Amharic version). Ministry of Agriculture, Addis Ababa, Ethiopia.

Pathak B S, 1988. Survey of agricultural implements and crop production techniques.

Pathak B S, 1988. Draft performance of indigenous and cross bred oxen.

Starkey P, 1989. Harnessing and implements for animal traction. Vieweg for German Appropriate Technology Exchange, GTZ, Eschborn, Germany. 244p. ISBN 3-528-02034-2

Meeting the challenges of animal traction

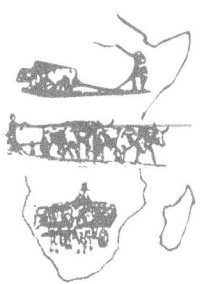

Animal-based transport

Meeting the challenge of animal-based transport

by

Ron Dennis

IT Transport Ltd, The Old Power Station, Ardington, nr Wantage, Oxon OX12 8QJ, UK

Abstract

Animal-based transport already plays a significant role in the economies of rural areas in eastern and southern Africa and in relieving the transport burden of rural households. However, it is considered that it is still greatly under-utilised and that there is considerable potential for it to make a major contribution to the economic and social development of rural communities. The aim of this paper is to identify the main constraints on wider use of animal-based transport in the region, to suggest actions which need to be taken to overcome these constraints and to propose an approach to implementing these actions. It is hoped that this will provide a theme for the working group on transport to develop and possibly to draw up a plan of action.

A questionnaire circulated to intending participants at the ATNESA conference in Kenya has been used to identify the constraints on animal-based transport. This has clearly identified the major constraint as unaffordability of carts due to a combination of high cost and lack of access to credit. Another strong constraint in some areas is a shortage of carts caused by a limited production capacity and capability. Constraints on animal ownership apply mainly to oxen and include disease, lack of grazing and high cost. There appears to be considerable potential for greater use of donkeys in transport work but this is restricted in several areas by lack of status. This is an area in which more promotion is needed and is justified.

Introduction

Access to an effective means of transport is an essential ingredient for promoting the economic and social development of rural people. This is particularly true for small-scale farmers. It enables them to provide greater inputs of fertiliser and manure to improve yields, allows them to move and market more produce, and reduces time and effort spent in household activities such as collecting firewood and water so releasing time for more productive activities.

One of the significant steps forward over the past few years has been the increasing recognition by major development agencies of the true nature of transport problems in rural areas and of the transport needs of rural people (for example, World Bank, 1995a). The need is not for primary roads for motorised vehicles, but for improvements to village level paths and tracks on which rural people can use appropriate and affordable means of transport to move relatively small loads over short distances.

Development agencies are also recognising the important role of intermediate means of transport (intermediate between head or back carrying and motor vehicles; Relf, 1995), such as wheelbarrows, handcarts, bicycles and animal-based transport, which are generally the only means of transport affordable to rural people. There are many rural people who cannot even afford to own this form of transport, although they may gain access through hiring or borrowing. Although simple and relatively low-cost, intermediate means of transport are a significant improvement for rural households and have a substantial potential to promote economic and social development.

Development agencies also consider intermediate means of transport appropriate because of their energy efficiency and low environmental impact. These features may not seem very relevant in terms of the present severe lack of transport in rural areas but they have important long-term implications. For example, there are estimated to be about 700,000 carts in Africa. If pack animals are added, assume the total (carts and pack) is equivalent to 1 million carts. Assuming these each carry an average of 500 kg for 1000 km per year, the saving in diesel fuel over trucks to do the same work is roughly 30 million litres.

Animal-based transport

Animal-based transport is a particularly appropriate mode of intemediate transport for rural areas since the animals are often already used for other farm activities. In these cases the costs of the animals for transport are marginal, as they cover their extra upkeep for the additional work. If animals are kept purely for transport work then obviously they will add more to the cost of transport. The modes of animal-based transport are:

- *pack animals,* where loads are carried in panniers or containers on the animal's back. The main pack animal used in eastern and southern Africa is the donkey. A few horses, mules and camels are also used. Pack animals are suited to operating on narrow tracks, particularly in hilly and rough terrain

- *animal-drawn sledges.* A sledge is a very low-cost form of transport, often carved directly from a tree. They are usually used where carts are not affordable but may also be used for particular activities such as carrying ploughs to the fields. They are usually pulled by oxen because of their large draft requirement

- *animal-drawn carts.* These are by far the most effective form of animal-based transport, carrying much higher loads than the other forms, but also the most costly. Most carts in the region are 2-wheel, although 4-wheel carts are also found, particularly in South Africa.

Carts are most commonly pulled by oxen or donkeys.

Animal-based transport is compared with other forms of intermediate transport in Table 1. The main advantage of animal-based transport over other forms of non-motorised intermediate transport is that the operating power is provided by the animal(s) rather than the user. In the case of carts, particularly ox carts, they can also carry significantly higher loads. Their overall cost depends on whether the animals are used only for transport work or also for other farm work. For example, if donkeys are kept only for transport, their cost should be added to that of the pannier or ccart. The cost of donkeys varies through the region – in South Africa it is quoted as about US$17 (Starkey, 1995). The cost of a single donkey and cart would in this case be roughly the same as a bicycle and trailer. However, it is advisable to keep a reserve donkey and often in West Africa the reserve travels with the cart so that the donkeys can be changed over when the working one becomes tired. The load of 300 kg carried in a cart could equally be carried by about 5 pack donkeys at roughly the same cost but this would necessitate the load being broken down into smaller portions. As shown in Table 1, the cost of animal-based transport is substantially lower than that of the cheapest means of motorised transport.

Table 1: Animal-based transport and other forms of intermediate transport

Mode of transport	Typical load (kg)	Average speed (km/h)	Daily range (km)	Transport capacity (tonne km/h)	Cost, compared to bicycle
Human carrying	30	4–5	15–20	0.12	–
Wheelbarrow	90	3–4	5–6	0.35	0.6
Handcart (1 person)	200	3–4	10–12	0.8	0.4–0.7
Cycle with carrier	40	12	40–50	0.48	1
Cycle trailer	125	10	30–40	1.25	0.6–0.8
Pack donkey	50–80	4–5	20	0.3	0.2
Ox-drawn sledge (2 oxen)	250	2–3	15	0.75	0.1
Donkey cart (1 donkey)	300	3–4	20	1.1	1.5–2
Ox cart (2 oxen)	900	3–4	20	3.5	2–3
Motorcycle	50	50	150	2.5	16
Motorcycle trailer	300	30	100	9	2

Note: Initial costs are given relative to the cost of a bicycle (roughly US$125). Costs of animal-based transport do not include the cost of the animal(s).

The role and issues of animal-based transport in the region have featured prominently in two previous ATNESA workshops. In Lusaka, January 1992, a paper by Anderson and Dennis (1994) reviewed both the technical and socio-economic issues. Since this was the inaugural meeting of ATNESA, the aim of the paper was to survey the important published literature on animal-based transport, covering experience from both Asia and Africa, to review the role and importance of animal-based transport in the region and to highlight the main issues and problems. It is believed that the paper provides a good background to the state of the art of animal-based transport at the time of the workshop.

One of the issues raised at the Lusaka meeting by the working group on transport was the need to collect and share information on all aspects of the design and production of animal-drawn carts. A follow-up workshop was therefore held in Harare in January 1993, 'The design, production and testing of animal-drawn carts'. The output of this workshop was a set of guidelines on the design, manufacture and marketing of animal-drawn carts (ATNESA, 1996).

Possibly the major issue arising since the previous workshops is that, in spite of the increasing availability of information on animal-based transport, it is still very much under-utilised in the region and there appear to be few signs of any significant increase. In fact, the last two to three years have been difficult for rural areas in the region because of poor harvests resulting from drought and the impact of structural adjustment policies. Rural economies have been depressed, resulting in low demand for transport devices.

Structural adjustment has influenced both the demand for and supply of carts. Evidence from Zimbabwe suggests that the real value of incomes has decreased, thus reducing the purchasing power of rural households and also the flow of returned money from migrant workers. On the supply side there has been a steady increase in the cost of materials, although the availability of materials has increased through relief of import restrictions. As a result the cost of carts in particular has risen whilst rural households have less money available to buy

them. It seems likely that this situation is fairly common in other countries in the region.

In the light of the above points it is felt that in terms of animal-based transport this workshop can best "meet the challenges of animal-traction" by attempting to establish and put into place a plan of action aimed at overcoming some of the barriers to wider use of animal-based transport in the region. The aim of this paper is to attempt to set the scene for the discussions of the working group by:

- identifying the main constraints on wider use of animal-based transport

- putting forward suggestions for ways in which these might be overcome and actions which could be taken.

The paper analyses information obtained from a questionnaire circulated to prospective participants at the workshop, and puts forward proposals for ways of tackling the constraints identified from the questionnaire.

Constraints on the wider use of animal-based transport

A questionnaire was sent out to members of ATNESA who had expressed a wish to attend the workshop. An extremely good response was achieved; 50 replies were received in time to prepare this paper. A copy of the questionnaire is contained in Annex 1 to this paper. It is in two parts – the first deals with constraints on the ownership of draft animals, and the second with constraints on the wider use of means of animal-based transport.

Constraints on ownership of draft animals

The results of the questionnaire are summarised in Figure 1. Brief profiles for each country from which more than one response was received are presented in Appendix 2. The first bar chart indicates the number of responses (out of 50) which identified the issue as a medium (M) or high (H) constraint. The second chart compares the strengths of the constraints, as identified by the responses, in terms of a weighted rating given by the number of high responses plus half the number of medium responses. This is a purely arbitrary weighting but gives a simple way of comparing the perceived importance of the constraints identified in the questionnaire.

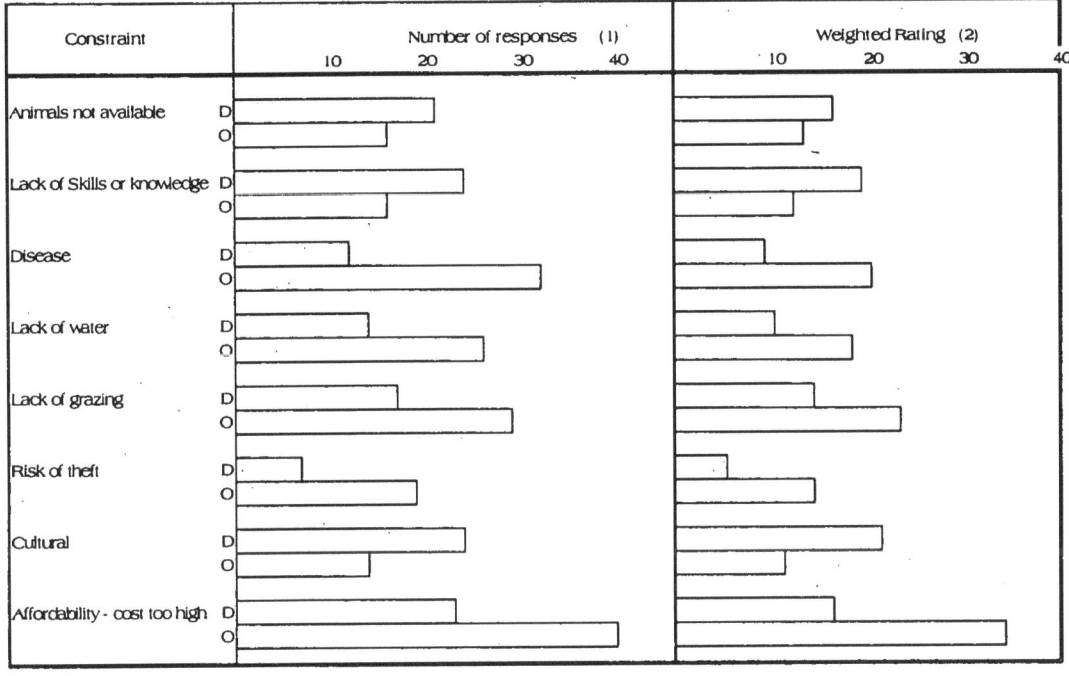

NOTES:

1. Number of responses of Medium, M, or High, H, constraint
2. Weight Rating: number of H responses + 0.5 x number of M responses
3. D - Donkey; O - Oxen

Figure 1: Constraints on wider ownership of animals

Although patterns of ownership vary across the region and some constraints are quite localised, the figure shows some trends which are common over much of eastern and southern Africa:

- the high cost of oxen is considered a major constraint across the region, even though their availability is less of a constraint than for donkeys. One of the reasons is the reduction in numbers in countries such as Zimbabwe due to drought

- lack of grazing is also identified as a significant constraint on ownership of oxen. This has also been identified as a constraint by a study of animal traction in South Africa (Starkey, 1995). This may be partly a result of the effects of drought over the past few years but it is also of serious concern for the future of animal traction in the region. As populations and population densities increase there will be increasing competition on land for food production and grazing. This is an

issue which needs careful study to see if it is a temporary constraint or more long term

- lack of water is also a serious issue but again this may be mainly a result of recent droughts. It needs further investigation

- donkeys are seen to have a number of important advantages over oxen–they are less prone to disease, lack of water and grazing are lesser constraints, they are less at risk of theft and they are cheaper. The main constraints on wider ownership are limited availability in some areas, lack of experience in their upkeep and handling, and cultural resistance. The latter varies considerably through the region but is a very difficult constraint to overcome as it involves changing people's attitudes. However, donkeys could make a major contribution to increasing animal-based transport in the region and I suggest that working to remove these constraints is an important area for action.

Technical constraints on wider use of animal-based transport

An overview of the technical constraints on wider use of animal-based transport is shown in Figure 2. The criteria used for comparison are similar to those in Figure 1. In the case of pack loading of donkeys the main constraint mentioned is the lack of knowledge of this form of transport, but this is a relatively low constraint. Lack of information on designs and methods of manufacture is not considered a serious constraint. However, it is apparent that some of the methods of loading used in the region are likely to cause discomfort to the animals and that there is a need to improve harnesses and distribution of loading. Users also need to be made more aware of the levels of loading that can be reasonably carried by donkeys without causing them distress.

Lack of awareness of the use of donkey carts is perceived as quite a strong constraint, more so than for ox carts. This probably relates largely to lack of knowledge on methods of hitching pairs of donkeys to a single drawpole by rural people who are more familiar with ox carts. This is clearly an area where development, testing and dissemination of suitable methods is needed. Also there is a need to increase awareness of carts for single donkeys.

Shortage of both donkey and ox carts is seen as a strong constraint. The main reasons are identified as lack of information on design and manufacture of carts, technical problems limiting production and shortage of critical components such as wheels, bearings and tyres. These problems can be partly overcome by improved dissemination of information and training of workshops, but the capacity and motivation of workshops may also be a serious constraint. This is discussed further below.

Figure 2: Technical constraints on the wider use of animal-based transport in the region

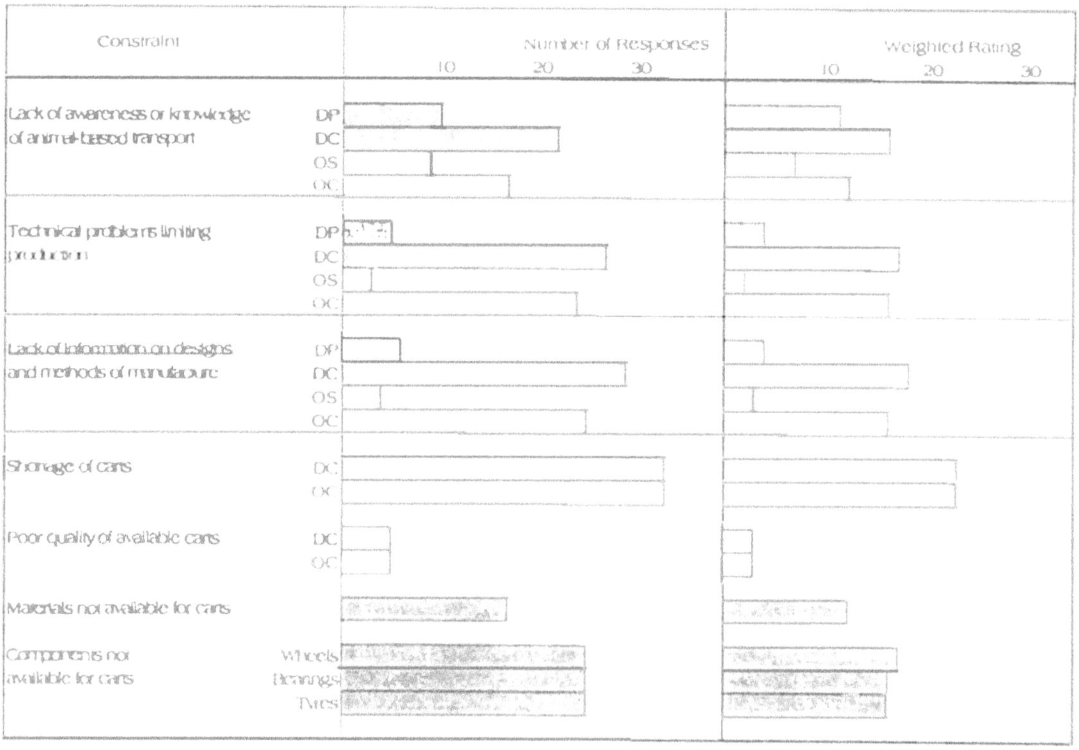

NOTES:

Pack donkey	DP	Ox Sledge	OS
Donkey Cart	DC	Ox-Cart	OC

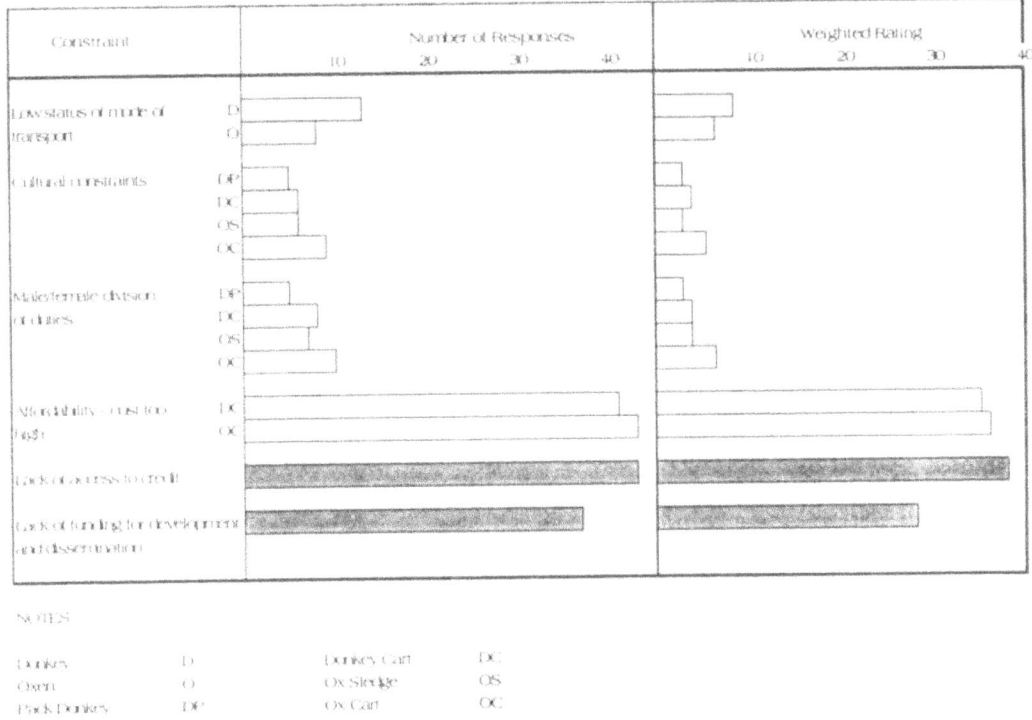

NOTES:

Donkey	D	Donkey Cart	DC
Oxen	O	Ox Sledge	OS
Pack Donkey	DP	Ox Cart	OC

Figure 3: Socio-economic constraints to wider adoption of animal-based transport in the region

Socio-economic constraints to wider use of animal-based transport

Socio-economic constraints are compared in Figure 3. The figure shows clearly that the major constraint is considered to be affordability due to the high cost of carts and the lack of access to credit to purchase them. Both donkey and ox carts are considered equally to cost too much. This is probably because many of the donkey carts used in the region are based on ox-cart designs with a single drawpole and need pairs or more of donkeys to pull them. The smaller, lighter and lower-cost carts for single donkeys which are widely used in West Africa are not commonly found in the region. Logically, it would be expected that they should have a major role to play in reducing cart costs in the region, but introducing an unfamiliar form of cart is likely to involve considerable time, effort and perseverance. This is reinforced by a significant response suggesting that donkey transport is seen as having a low status in several areas of the region.

Lack of access to credit is seen as a major constraint across the region apart from Zimbabwe, where the Agricultural Finance Corporation has played a significant role in promoting the relatively high number of ox carts found in rural areas, and Namibia where household incomes are possibly higher than in other countries. Other socio-economic issues are considered to have little impact on the level of use of animal-based transport in the region.

Unaffordability of animal-drawn carts

Unaffordability is identified clearly as the main constraint on increasing the use of animal-based transport in the region. It is a combination of the high cost of carts and the lack of access to credit. Affordability is also linked closely to household income. At low income levels there is little produce to move or excess crops to market and a means of transport has a relatively low priority for the household. Even though it might be very useful in reducing their transport burden, particularly collecting water and firewood, households would be unlikely to consider purchasing a means of

Table 2: Estimates of the cost of ox carts in various countries in the ATNESA region in 1993

Country	GNP/Capita (US$)	Estimated annual household income (US$)	Assumed cart cost (US$)	Number of months' income to buy cart
Ethiopia	100	100–200	350	21–42
Kenya	270	270–540	200	4.5–9
Malawi	200	200–400	550	16.5–33
Mozambique	90	90–180	350	23.5–47
Tanzania	90	90–180	200	13.5–27
Uganda	180	180–360	200	6.5–13
Zambia	380	380–760	350	5.5–11
Zimbabwe	520	520–1040	350	4–8

Notes:

1) Data for mid 1993 fromWorld Development Report, World Bank (1995b)

2) Estimated household income is based on data obtained from World Bank sub-Saharan Africa Transport Programme Surveys (Airey, Barwell and Strandberg, 1993; see also Table 3). These suggest for rural areas in the region typical household income is 1 to 2 x GNP/capita.

3) Assumed cart costs are based on following actual data from mid 1993.

Kenya: cost of ox cart made in small workshop US$180–220

Zimbabwe: ox cart produced by small/medium workshop US$300–400

*   ox cart produced by large manufacturer US$900*

Malawi: ox cart produced in informal sector US$550

Sudan: donkey cart produced by small/medium workshop US$340

Costs are assumed similar in neighbouring countries.

transport that would not pay for itself. As household income rises a means of transport may become essential to support the earning of income and may well generate enough extra income to pay off the loan needed to buy it.

It is not clear if there is a definable relationship between household income, affordability of a particular means of transport and hence ownership levels. However, this must clearly be an important criterion in assessing a household's ability to repay a loan for intermediate means of transport. Table 2 shows estimated data for household income and cart costs, and based on these, estimates of months of income needed to buy a cart. A number of points may be noted:

- actual cart costs are used for some countries, and it is assumed costs will be similar in neighbouring countries. There is quite a wide variation in costs across the region, being particularly low in Kenya and very high in Malawi. Different forms of construction and

costs of materials would be expected to give some variation, but the range seems excessive

- average household income is estimated from fairly limited data obtained from surveys carried out within the World Bank Sub-Saharan Africa Transport Programme (Airey, Barwell and Strandberg, 1993) and relating this to GNP per Capita. There could therefore be quite large variations from these estimates

- because ownership of carts varies considerably within a country it is not possible to draw any general trends regarding household incomes and level of ownership of carts. More specific data is needed on localised household incomes. It is possible that high levels of cart ownership in an area are associated with higher than average levels of income. Relationships can also be confused by income returned by migrant workers. However, the estimates for Zimbabwe suggest

Table 3: Bicycle ownership and rural household income

Country (District)	Months of household income needed to buy a bicycle	Bicycle ownership (% of households)
Zambia (Kasama)	4.7	28
Zambia (Lusaka Rural)	2.6	26
Uganda (Mbale)	8.6	15
Burkina Faso (Dedougou)	2.9	90

Source: Airey, Barwell and Strandberg, 1993)
Note: The high level of ownership is associated with widespread access to credit in Burkina Faso

that, in the presence of an effective credit scheme, quite high levels of ownership of carts become possible when their cost is equivalent to about 6 months' income. Information shown in Table 3 for bicycles suggests that even when only 3 to 4 months' income is needed, ownership is low unless credit is readily available

- This is clearly an area where more information would be useful to improve estimates of demand for carts and to differentiate between actual demand and apparent demand (when households say they would like a cart but they cannot afford it). This would lead to more effective strategies for dissemination of carts and other animal-based transport.

Actions to increase use of animal-based transport

Main constraints and needs for action

The main points from the responses to the questionnaire are summarised as follows:

- oxen are generally more popular for transport activities, but there are concerns regarding disease, availability of grazing, security and above all their high cost. Their main advantages are that their use in transport usually supplements their use in farming so that their transport cost is marginal, and they provide high draft

- donkeys have several advantages for transport use but they have a low status in a number of countries. It is not clear whether farmers can afford to use them purely for transport or whether their cost would have to be spread over other farming activities

- the main constraint on increased use of animal-drawn carts is unaffordability resulting from their high cost and the limited availability of credit

- there is considered to be a shortage of carts in many areas of the region. However, it is not clear whether in the light of lack of affordability there is an actual demand for more carts or if it is only apparent demand. The shortage is considered to be caused mainly by lack of information on design and manufacture, technical problems in production and limited supply of wheels, bearings and tyres

- lack of funds to promote the development and dissemination of animal-based transport is considered a major constraint. It is clear from the nature of the workshops producing carts that problems of manufacture, supply and marketing are unlikely to be resolved by purely commercial initiatives, and assistance will be needed in upgrading the capacities and capabilities of workshops.

There are clearly a number of issues relating to supply, upkeep and training of draft animals where action is needed but these are considered outside the scope of this paper.

The following needs for action on transport issues are suggested and discussed briefly:

- reducing the cost of carts
- improving the capacity and capability of workshops
- improving access to credit
- a coordinated approach to development and dissemination.

Although these actions are concerned mainly with supply, it is equally important to work with communities to assess more clearly their transport needs, estimate their priority and demand for means of transport and determine what means of transport are affordable and acceptable. Recent development experience suggests that new or improved technologies are more likely to be successfully accepted and assimilated through this kind of participative approach. Once initial surveys have indicated what forms of intermediate transport, if any, are likely to be affordable and appropriate then prototypes can be introduced to demonstrate their potential and to create an awareness in the area of the possible options. This needs to be done on a sufficiently large scale to stimulate interest and awareness.

Options for reducing the cost of animal-based transport

Carts are the most effective and acceptable form of animal-based transport. Sledges cost very little but they have limited applications and cannot be encouraged because of the damage and erosion they can cause. Improved sledges with wooden wheels and crude wooden carts ('flintstone') are used in a few areas and these forms of transport can play an important role in poorer communities. It may be worthwhile putting some resources into developing these but it is uncertain whether they are likely to achieve any wider acceptance. I suggest that further work is needed with poor communities to assess if transport is a priority and if there is real demand for very low-cost transport devices. Experience suggests that most rural households want carts with pneumatic tyres and rolling contact bearings and the main need identified by the survey is to make these more affordable.

A breakdown of the production costs of a range of cart designs is shown in Table 4. The table is in two parts:

- the costs of materials, components and consumables for a number of cart designs

Table 4: Breakdown of cart costs

i) Estimated costs of materials and components (based on costs in Zimbabwe in 1994)

Type of cart	Frame and body (US$)	Axle and tyres (US$)	Total cost of materials (US$)
Ox carts:			
Steel frame and body, pneumatic tyres, ball bearings	115	140	255
Steel frame, wood body, pneumatic tyres, ball bearings	85	140	225
Wooden cart, pneumatic tyres, ball bearings	57	140	197
Wooden cart, wood wheels, hardwood bearings	57	93	150
Single donkey cart:			
Steel frame, wood body, pneumatic tyres, ball bearings	35	105	150
Wooden cart, pneumatic tyres, ball bearings	25	105	130

ii) actual cost breakdown of ox carts for 3 countries in 1994

Item	Kenya	Malawi	Zimbabwe
Materials and consumables (US$)	160 (67%)	432 (72%)	280 (63%)
Labour (US$)	10 (4%)	8 (1.5%)	15 (3.4%)
Overhead and profit (US$)	70 (29%)	160 (26.5%)	150 (33.6%)
Selling price (US$)	240	600	445
Cost of axle assembly with tyres (US$)	70	280	155
Proportion of material cost	44%	65%	55%

presented in the ATNESA guidelines (ATNESA. 1996)

• a breakdown of the selling prices of carts produced by workshops in Kenya, Malawi and Zimbabwe.

The information in Table 4 shows that material costs make up a major part of the production cost and that the cost of the axle assembly, including tyres, usually makes up at least half of this and significantly more on the lower-cost carts. The large influence of the cost of the axle assembly on the selling price of the cart is seen by comparing the prices in Kenya and Malawi. Both prices are for carts with scrap axles and tyres but the costs of these components are much higher in Malawi, presumably reflecting their limited availability there. Where scrap axles are readily available, they provide a relatively simple and cheap base for making carts. However, the wheel bearings are often nearing the end of their life and are a common source of unreliability. If the cart breaks down it may be out of service for a considerable period because replacement bearings are not available or the owner cannot afford to repair it. The repair cost and loss of income while the cart is out of service can substantially increase the effective cost of the cart to the owner. A slightly more costly cart which is more reliable may often be better value for money.

Since material costs make up the major part of the cart cost, the approach to cutting cost must be either to reduce the amount of material used through smaller carts and/or better design, or to use cheaper materials. This suggests the following options for reducing cost:

• a cart with a wooden body on a steel frame is about 10% cheaper than a sheet steel body

• a wooden cart mounted on a steel axle assembly with pneumatic tyres and rolling element bearings reduces cost by a further 15%. This is simple to construct and will be adequately strong and durable provided that good quality wood is used and treated regularly with a preservative

• a smaller, lighter cart for a single donkey should cost little more than half the cost of a standard ox cart. It would have a much lower load-carrying capacity but it is felt that ox carts are seldom used to their full capacity and a smaller cart may suffice for many of their needs. For example, in recent trials of a small handcart/donkey cart in Zimbabwe, one of the participants claimed to use the small cart six

Figure 4: An intermediate design of 2-wheel cart - wooden body with steel axle frame. Carts like this are used widely in West Africa

times more frequently than his ox cart. Single donkey carts are used widely in West Africa (see Figure 4) and it is possible that a good quality, low-cost cart of this type has good potential in the region. It appears particularly appropriate for women.

- The other possibility for a smaller cart is a cart for a single ox. I do not know if any are used in eastern and southern Africa but they are quite widely used in Asia (see Figure 5). The cart has a pair of shafts and is hitched by a yoke/collar which is attached to the ends of the shafts and rests on the withers of the ox. The cart would have a capacity of about 500 kg and cost 60 to 70% of the price of a double ox cart.

I suggest that the best option for reducing cost is a cart for a single animal. The actual load-carrying requirements of rural households should be investigated to see if a smaller cart would meet most of their needs. If a large capacity is only needed once or twice a year, for example to carry grain to a depot, then this may not justify the cost of a larger cart. To help to overcome resistance to an unfamiliar type of cart it should be of good quality with a modern and attractive appearance. A particular selling feature, possibly a seat for the operator, may help to promote a market for this type of cart.

Figure 5: A single bullock cart in Sri Lanka Carts like this are common in Asia. A yoke is fixed between the ends of the shafts and rests on the withers of the animal.

Photo Paul Starkey

Upgrading workshops

Cart production can be classified generally into centralised production in factories and localised production in small- to medium-sized workshops. Centralised production has better access to materials and tools but incurs significant extra costs from higher overheads, distribution costs and often additional mark-ups for selling through agents in the rural areas. Large producers tend to produce one or two standard models which are at the top of the price range. Small- and medium-size enterprises tend to produce a range of models, varying from high quality carts which may be 20 to 25% cheaper than those from central producers, down to poor quality carts incorporating scrap materials which may be less than half the cost of the centrally-produced carts.

Localised production has a number of advantages: more direct contact for buyers, a better ability to respond to local needs and demand, and more effective back-up support for repair and maintenance. Disadvantages are lack of information on designs and manufacturing techniques, difficulties with access to materials and components, and limited finance so that cash flow often restricts manufacturing capacity. It is suggested that the advantages to rural communities of localised production outweigh the disadvantages and there is a strong case for putting resources into upgrading the capacity and capability of small- and medium-sized workshops to help overcome the disadvantages. This would not only help rural communities but also be of general benefit to rural economies. Three areas of action are suggested below.

Training of workshops

The following areas are important:

- training in manufacture of improved designs, particularly wheel/axle assemblies
- training in techniques to improve productivity and quality (see also Oram, 1995)
- the basics of business management, including keeping records, managing production, controlling cash flow and planning for expansion.

Promotion, marketing and customer relations

To maximise the use of resources some care is needed in the selection of workshops for support and training. Experience in Zimbabwe (see Box 1) suggests that artisans need to have a certain level of skill, business acumen and motivation to put training into productive use. It is likely that a successful small manufacturing sector will be built around workshop owners that are capable and have ambition to succeed and better themselves.

Improving access to tools and equipment

Many small- and medium-sized workshops only have access to basic tools, hand tools – a welding set and possibly a power drill and grinder. Manufacture of cart bodies is generally not a problem since, even if electricity and welding are not available, good quality bodies can be made from wood providing these are locally acceptable. The main problem for smaller workshops is access to good quality axles, particularly when, as is often the case, there is a shortage of scrap axles. It has been suggested (ATNESA Harare Workshop) that axles might be centrally produced and supplied to smaller enterprises. This appears to have a number of problems:

Box 1: A project at the Institute of Agricultural Engineering, Zimbabwe

This project arose out of a number of surveys in the rural areas of Zimbabwe which yielded two significant conclusions:

- ox carts are very important to the economy of small farms

- the supply of ox carts from rural workshops was severely restricted by a shortage of scrap axles which had traditionally been used by these workshops

A 3-year project was set up at the Institute of Agricultural Engineering (IAE) to overcome this bottleneck by training workshops to produce wheel/axle assemblies using a technology developed by Intermediate Technology. Support was provided by IT Zimbabwe and IT Transport. Some of the features of this project have been:

- Fifty workshops have been trained covering all areas of Zimbabwe. About half of these are now using the technology on a commercial basis. Over 400 carts and 150 axles have been produced by these workshops in spite of the depressed rural economy. A number of workshops are also using the technology to produce parts of axles, for instance rims for scrap hubs and hubs for scrap rims.

- As experience was gained in the selection of workshops for training it became apparent that to benefit from the training the workshops needed to have a certain level of business sense and ambition to develop. Some were content to carry on making a few carts on an irregular basis when they were able to find a scrap axle. Those that adopted the technology were able to produce carts on a regular and reliable basis without wasting time in searching for scrap axles.

- There was some initial resistance to carts incorporating a new technology. However, this was gradually broken down by the quality of carts produced by some workshops. The numbers sold have increased each year. Some workshops are offering 12-month guarantees on their carts and also delivering carts to buyers. These are strong selling points.

- Initially some workshops used scrap bearings but as supplies dried up costs increased and this became a bottleneck. A supply of low-cost bearings from China was therefore arranged through a local importer. These needed to be ordered in a batch of 2000 so that it was necessary to obtain estimates of needs from workshops and advise them of the source of supply. The arrangement has worked well to date. The bearings sell for about US$4 each.

IAE continues to monitor the progress of the workshops and provides back-up support as required. A number of regional training institutions are beginning to offer training in wheel making and providing support to local workshops. In a few places other workshops have seen the success of neighbouring workshops that are using the technology and have copied it themselves. The production tools needed are produced locally and sold commercially.

- axles are costly and many small workshops would have problems in buying them outright
- the income earned per cart by the smaller workshops would be substantially reduced and might not be compensated by increased sales
- the cost of the carts supplied by the smaller workshops would increase
- the scheme would require a good level of organisation.

It is possible that centrally producing axle components, rims, hubs and axles, may be more feasible, but even so there may be resistance from smaller enterprises. It is suggested that smaller-scale workshops may prefer to be independent of other suppliers and to make as much of the cart as possible themselves to maximise their income from the cart. A more appropriate and acceptable approach may be to upgrade the capabilities of small and medium-sized workshops by helping them gain access to a wider range of tools.

Workshops making carts would benefit from access to such tools as a power saw for cutting steel sections, a lathe for machining hubs and axles and wheel-making equipment. However, they may not be able to afford these or their degree of usage may not justify the investment. A possible solution may be to organise a better equipped workshop, possibly a mission workshop or technical training centre, to provide support to a group of smaller workshops. Support might include supervised use of machine tools, hiring out of portable tools and stocking of some standard materials and components. Training of the workshops might also be organised through the support centres.

Improved access to materials and components

Small workshops can only afford to buy in small quantities and so have low priority in access to materials and are unable to benefit from discount schemes. Bulk buying for groups of workshops could improve this situation but this requires outlays of money and may be difficult to organise. It may be possible through a local wholesaler. Relaxation of any government duties on materials used for cart manufacture would be a considerable boost for small workshops. Limited availability of wheels, bearings and tyres was identified as a strong constraint on cart production. Possibilities for overcoming this are:

- The manufacture of wheels is covered in detail in the ATNESA Guidelines on cart design (ATNESA, 1996). A number of practical methods for different levels of workshop are described.
- Bearings are a critical component for good performance of a cart. Scrap bearings are widely used but tend to be unreliable and are often in short supply. New bearings from major suppliers (SKF, NSK etc) are expensive but cheaper versions are available from some of the emerging industrial countries such as China, Korea and Eastern Europe. Although of lower quality, they are adequate for use on animal-drawn carts. It may be possible to obtain these at low cost by ordering in large batches of about 2000 (see Box 1). This involves a substantial investment and is probably best organised through an importer or wholesaler. However, the importer needs some guarantee of demand and considerable effort will be required to introduce a standard hub design and to obtain reliable pre-orders of bearings from workshops.
- New tyres are also expensive, particularly the larger sizes preferred for carts and scrap tyres are widely used. These are usually of poor quality, often with no tread, and are prone to punctures. It may be possible to import better quality scrap tyres from countries where regulations on the wear of vehicle tyres are stricter. For example, a preliminary study indicates that it would be possible to import 16" tyres with at least 1mm of remaining tread from the UK into Zimbabwe at a cost of about US$20 each.

Improving access to credit

Figure 6 illustrates the relationship between the important parameters of a credit scheme. Although simplified so that the interest rates shown are not exact, the graph does show accurate trends and illustrates the following important points:

- Operating costs of the scheme – vetting and setting up loans, administration and collecting payments – must be minimised. Typical rates seem to vary between about 5 and 30% of the total amount loaned out. It is best to work through existing schemes which are efficiently organised and have a reasonably large loan portfolio to offset their operating costs.

• Achieving high recovery rates is very important to keep interest rates down. For example, if the recovery rate drops from 95 to 90% the interest rate needed to keep the scheme sustainable must be increased from 14 to 20% (at C=5%). Strategies to encourage repayment are therefore very important. For instance loaning to groups to create peer pressure on members of the group that receive loans has proved quite successful (Devereux, Pares and Best, 1987). However, much of this success has been for relatively small loans over short periods of time and where members wish to keep taking out loans on a regular basis. There is less experience with larger one-off loans over longer periods of time where group pressure is more difficult to maintain, although the Agricultural Finance Corporation in Zimbabwe reports that about 40% of its loans for ox carts are now given to groups (IFRTD, 1994).Other strategies that

encourage repayment are to create a sense of involvement in the scheme – any sense of the loan being a grant must be dispelled – and maintaining firm and consistent action on defaults.

• Interest rates must be high enough to make the scheme sustainable taking into account the depreciation of capital caused by inflation. Since operating costs of rural credit schemes tend to be high – clientele are scattered over a wide area and recovery is affected by irregular and variable incomes – interest rates may need to be higher than commercial rates. For example, the AFC rate for cart loans in Zimbabwe is 21% and the general rate for rural credit in Malawi is 40%.

Box 2 gives a brief overview of the Agricultural Finance Corporation scheme in Zimbabwe. It would be very useful to obtain more detailed information on this and other schemes for relatively large loans in rural areas, including guidelines on household incomes needed to service the loans.

Figure 6: Illustration of the effects of recovery rate and operating costs on the interest rate needed to break-even on a credit fund. This graph represents a simplified approach and is used only to illustrate the relatively high interest rates needed to break even on a credit fund and the critical need to minimise operating costs and to maximise recovery of loans. Break even means that the fund sustains its value taking into account inflation

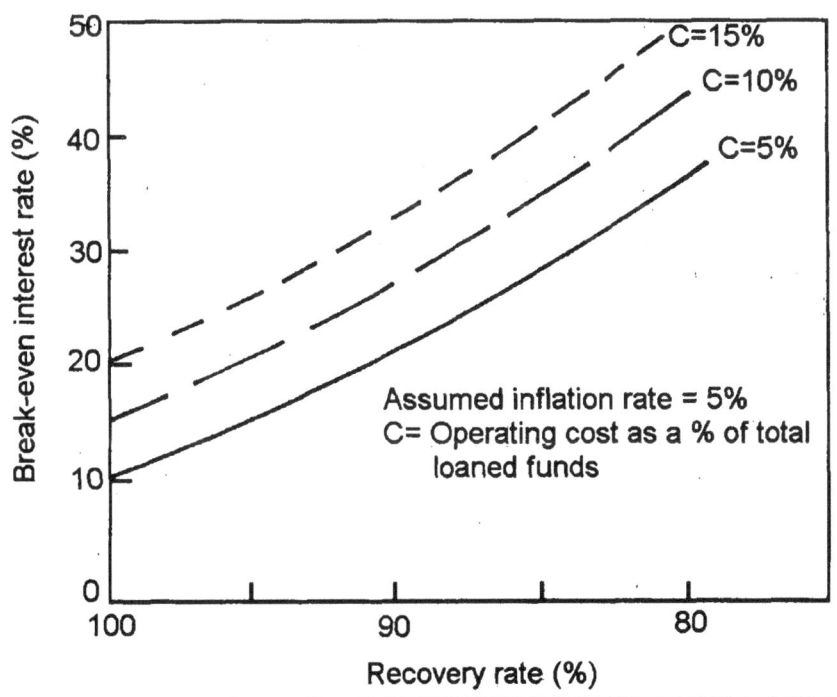

A plan for implementing action

I hope that this discussion of the problems of animal-based transport and potential solutions for them will provide a stimulative base for discussion to refine and redefine the constraints on animal-based transport and to define a possible plan of action.

The next stage is the most difficult – how to implement a plan of action? It seems clear that at present there is not a sufficient commercial basis to animal-based transport for action to occur through pressure of market demand. This suggests that considerable inputs will be required to promote the action proposed and that substantial resources and funding will be needed for this. I suggest that a coordinated approach across the region is likely to be most effective with a high level of networking which builds on the networks established by ATNESA and the International Forum for Rural Transport Development. There are clearly a number of common constraints across the region which can be tackled through a coordinated approach, but at the same time the programme must have sufficient flexibility to deal with local issues. It is possible that a coordinated approach may also be more effective in raising funding.

A coordinated programme might include the following components:

- Consolidation of the work started in the questionnaire into a database on animal-based transport in the region. There is probably a great deal of information available, particularly in project appraisal reports, which could be usefully collated and disseminated. Lessons learnt are especially important. For example, why has animal-based transport been widely adopted in some areas and not others?, what are the catalysts or conditions?

- Much more information is required on transport needs and priorities in rural areas, particularly for lower income households. This should give a clearer picture of where interventions are worthwhile and which transport options are likely to be appropriate and acceptable. Universities in the region could take a lead in this through student project and research work.

- There are some options which seem logical solutions to overcoming affordability constraints, for example single-animal carts, but it is not clear whether these will be

---

Box 2: Credit for animal carts in Zimbabwe

One of the central development objectives of newly-independent Zimbabwe was to encourage a substantial increase in smallholder agricultural output. Towards this end, a package of assistance was aimed at small-scale farmers including improved extension and crop marketing facilities and the provision of credit facilities for a range of agricultural equipment, including animal carts.

The principal source of credit for small farmers since independence has been the Agricultural Finance Corporation (AFC), a parastatal with offices in every district in the country. At first, loans were generally made to individuals, but in recent years, groups have become more important as AFC clients - today they account for around 40% of new loans. Group loans have the advantage of being cheaper to administer, as well as having higher recovery rates since peer pressure tends to support credit discipline.

Only farmers with no loans in arrears qualify for AFC credit. While no collateral is required, the borrower must contribute 25% of the supplier-quoted price as a deposit. The interest rate on a scotchcart is 21 per cent plus a 1.5% life insurance cover. The repayment period ranges between 3 and 5 years. While the AFC retains the right to reclaim the vehicles for which the credit was used in the event of default, this is seen as a last resort.

Today, there are impressive numbers of ox- and donkey-drawn carts to be found in all parts of Zimbabwe and they have contributed to the leap in smallholder output. Access to credit facilities has been essential in making this possible.

Source: IFRTD, 1994.

---

acceptable. I suggest that prototypes are developed and tested in areas where they are most likely to be accepted to gain experience in their operation and impact.

- Where there is clearly a demand for animal-based transport, but it is restricted by lack of production capacity or capability, the intervention needed may be fairly straightforward to identify and introduce. In these cases the action needed will be to obtain the funding required.

I suggest that the programme could be carried out through a network of support centres that would be responsible for implementing and coordinating the programme. These would be located in institutions which have the interest, resources and commitment to be involved in the programme. They would probably cover a fairly large area of a country and be linked to localised support centres.

Acknowledgements

The author wishes to thank the many members of ATNESA who completed and returned the questionnaire. The response was very good and it is clear that respondents have given considerable thought and time to filling in the questionnaire. The author acknowledges the significant contribution that this has made to the paper.

References

ATNESA, 1996. *The design, testing and production of animal-drawn carts*. Resource book of the Animal Traction Network for Eastern and Southern Africa (ATNESA) based on the workshop) held 25–29 January 1993, Harare, Zimbabwe. Intermediate Technology Publications, London WC1B 4HH, UK. 187p. ISBN 1 85339 338 X

Airey A, Barwell I J and Strandberg, 1993. *Local-level transport in Sub-Saharan Africa*. Consultancy report by IT Transport for World Bank Sub-Saharan Africa Transport Programme, International Labour Organisation, Geneva, Switzerland.

Anderson M and Dennis R A, 1994. Improving animal-based transport: options, approaches, issues and impact. Pp 378–404 in: Starkey P, Mwenya E and Stares J (eds), *Improving animal traction technology*. Proceedings of the first workshop of the Animal Traction Network for Eastern and Southern Africa (ATNESA) held 18–23 January, 1992, Lusaka, Zambia. Technical Centre for Agriculture and Rural Cooperation (CTA), Ede-Wageningen, Netherlands. 480p.

Devereux S, Pares H and Best J, 1987. *Credit and savings for development*. Oxfam, Oxford, UK. 71p.

IFRTD (International Forum for Rural Transport and Development), 1994. *Forum News* 2(2)

Oram C, 1995. An appropriate approach to intermediate technology? *Draught Animal News* 22 (June 1995): 21–23. Centre for Tropical Veterinary Medicine, University of Edinburgh, UK.

Relf C, 1995. *Operationalisation of Intermediate Means of Transport*. Guidelines for the World Bank Sub-Saharan Africa Transport Programme, Rural Travel and Transport Project. IT Transport Ltd., Ardington, UK.

Starkey P, 1995 (ed). *Animal traction in South Africa: empowering rural communities*. Development Bank of Southern Africa, Gauteng, South Africa, 159p.

World Bank, 1995a. *Transport Sector Policy Review*. World Bank Transportation, Water and Urban Development Department, Washington DC, USA.

World Bank, 1995b. *World Development Report, 1995*. Oxford University Press, Oxford, UK.

Appendix 1: questionnaire on animal-based transport

1. What do you estimate to be the level of household ownership of animals in your region? Please tick where appropriate.

Level of ownership	Donkeys	Oxen	Other
Less than 10%			
10 to 30%			
30 to 50%			
50 to 70%			
Greater than 70%			

Comments:

2. What do you think are the main constraints on wider ownership of animals? Please indicate by marking **L** - low constraint; **M** - medium constraint; **H** - high constraint.

Constraint	Donkeys	Oxen	Other
Animals not available			
Inadequate skills or knowledge in looking after the animals			
Disease			
Environmental conditions -lack of water -lack of grazing land or alternative feed			
Security/risk of theft			
Cost too high - cannot afford			
Cultural attitudes such as status of ownership			
Other			

Comments:

3. (i) What do you estimate to be the level of use of animals in transport? i.e. what percentage of the animals that are available do you estimate are used for transport activities. Please tick where appropriate.

(Ii) What is your estimate of the level of household ownership which could be achieved if constraints listed in 4 are removed or lessened? Please mark **X** where appropriate.

	Donkeys		Oxen		Other	
Percentage	Pack	Carts	Sledges	Carts	Pack	Carts
Less than 10%						
10 to 30%						
30 to 50%						
50 to 70%						
Greater than 70%						

4. What do you consider are the main constraints on the wider use of animals for transport in your region? Please indicate **L** - low constraint; **M** - medium constraint; **H** - high constraint.

Constraints		Donkeys		Oxen	
		Pack	Carts	Sledges	Carts
Limited availability					
Lack of awareness or knowledge of use of animals for transport					
Terrain too hilly					
Roads, tracks not suitable					
Limited supply of implements (eg carts)					
Technical problems in manufacture/production					
Lack of information on designs and methods of manufacture					
Materials not available					
Components not available	Wheels				
	Bearings				
	Tyres				
Affordability	Cost too high				
	Lack of access to credit				
Not acceptable due to:					
poor quality devices					
low status of animal-based transport					
cultural issues					
male/female allocation of duties					
Lack of funding for development and dissemination					
Other forms of transport more effective, appropriate or acceptable, such as:					
Other constraints; please list below					

5. Please list in order of priority the **three** (3) issues which you feel are most important in your region to promote greater use of animal-based transport.

Appendix 2: Brief country profiles

Although the patterns of animal-based transport vary considerably across a country due to different topographics, cultures, soil fertility etc. some general trends can be drawn from the responses. These are summarised briefly below for countries from which at least two responses were received.
(L - low constraint; M - medium constraint; H - high constraint)

Botswana (2 responses)
- donkeys common (owned by 30–50% of households) - little used for pack carrying, mainly for carts (about 30%)
- oxen less common, little used for transport
- constraints on oxen - disease, cost, lack of grazing
- limited supply of carts due to constraints of design and manufacture
- improved harnessing for donkeys is a strong need

Ethiopia (2 responses)
- donkeys widely owned (50% of households) - mainly used for pack loading
- oxen widely owned (50% of households) but not used for transport
- constraints - grazing for donkeys and oxen (2 H)

Kenya (11 responses)
- donkeys quite common (generally around 20% of households with local variations) - used for pack-loading in several areas, but not widely for carts
- oxen common, and high ownership in some areas (over 70% of households in two areas) -sledges not common, although over 50% in one location (SE from Nairobi). Cart usage generally less than 20%
- constraints on donkey mainly cultural but not strong
- constraints on oxen, disease (8 M, 2 H), lack of grazing (3 M, 4 H) and cost (3 M, 4 H)
- shortage of carts is a significant constraint (2 M, 6 H) for both donkey and ox carts. Causes are technical problems in production (2 M, 5 H) and lack of wheels and bearings (3 M, 3 H)

Mozambique (2 responses)
- great shortage of animals resulting from war
- negligible use of animal-based transport

Namibia (4 responses - all from N or NE)
- donkeys common in North (30% of households) but not in NE (<10%). Mainly used with carts
- oxen less common in North (<10% of households) than NR (50%). High use of sledges in NR (60%) but few carts
- lack of water and grazing significant constraints for oxen (M, 2H), less so for donkeys. No other serious constraints apart from high cost of oxen
- shortage of carts is not a serious constraint. Cost of carts and need for credit are lesser constraints than in other countries (probably because of significantly higher household incomes)

South Africa (4 responses)
- donkeys not widely owned (20% of households) - mainly used with carts (50–60%) - pack-loading not common. Main constraints on ownership are strong cultural resistance (M, 3 H) and lack of grazing (M, 2 H)

Tanzania (12 responses)
- donkey ownership is not high (<10% in 7 responses and 20% in 5). Mainly used for pack loading - carts not common (although 50% in Tanga region). Constraints are lack of skills (2 M, 4 H) and cultural restrictions (2 M, 5 H)
- oxen much more common (greater than 50% of households in some areas). Quite high use of sledges in most areas and also of carts. Constraints are disease (5 M, 3 H) and cost (2 M, 6 H)

- shortage of carts is not a serious constraint, although with manufacture and lack of information are mentioned by 7 respondents. Lack of components - wheels, bearings, tyres - is also a medium constraint
- more information on harnessing of donkeys is mentioned as a constraint by 3 respondents

Uganda (6 responses)

- limited ownership of donkeys and no significant use in transport. Main constraint is lack of availability (M, 5 H), lack of skills in husbandry (4 H) and high cost (M, 3 H). Cultural restrictions not mentioned as a constraint on ownership, but reasonably strong on use for transport (4 M)
- oxen more widely owned, but in some areas seen as an investment for security and not used for draft. Sledges common in two areas and carts in 3 areas. Lack of skills in handling oxen mentioned as a constraint (3 H). High cost is a significant constraint (3 M, 3 H)

- shortage of carts is a constraint (4 M), due to lack of information and capability for manufacture and also shortage of wheels, bearings and tyres.

Zimbabwe (7 responses)

- ownership of donkeys is generally low. Where they are available, use for pack-loading and carts are about the same. Main constraint is cultural (M, 5 H)
- ownership of oxen is still generally high (40–60% of households in 6 areas) despite losses from drought. Sledges are banned, but use with carts is quite high (40—60% in 3 areas). Constraints are lack of water (3 M, 2 H), grazing (3 M, H) and high cost (7 H - very strong due to losses resulting from drought)
- shortage of carts is a medium constraint (4 M) and high cost of carts a strong constraint (2 M, 5 H). Lack of access to credit is not such a constraint as in other countries.

The development of low-cost animal-drawn carts

by

C E Oram

Development Technology Unit, University of Warwick, Coventry CV4 7AL, UK

Abstract

Working in Nigeria and the UK, the Development Technology Unit of the University of Warwick has succeeded in generating designs for donkey- and ox-drawn carts and cart component designs which are substantially cheaper and easier to make than existing designs, but which have similar or better performance.

Cost reduction has been achieved by simplification of design. Most new DTU designs for carts avoid the use of drilling altogether in their construction, using only simple hand tools and welding equipment (which is very widely available). The number and difficulty of saw cuts have been severely reduced: all cuts (and there are not many) are normal to the length of the material. The result is that simple carts can be built in only a few hours compared to the days required for more conventional designs.

Most existing carts used in developing countries are made locally using wooden bodies and scrap automotive axles. Such axles are becoming scarce, their quality is falling and their price is rising. Alternative wheels, axles and bearings are required to reduce costs and provide farmers and transport operators, with very restricted capital, access to better transport. The DTU is developing designs for locally constructable roller bearing axles, again requiring only simple hand tools, welding equipment and widely available materials. A family of cast aluminium hubs and split-rim wheels with integral aluminium-race roller bearing provides another development which could allow complete axle sets to be made for only a few tens of dollars. These need only simple tooling, yet provide adequate or good performance at low cost.

With all cart developments low first cost and low repair costs, together with adequate performance, are of paramount importance to the potential purchaser and user. Quick and local repair is therefore important and is facilitated by appropriate designs, such as those mentioned above, which require zero special tooling and investment and make minimal demands on foreign exchange and distant and uncontrolled infrastructure, such as roads and dealer networks.

Introduction

Animal carts and other products intended for the rural and disadvantaged urban markets must be economical and efficient and they must expose both the manufacturer and user to minimal investment and risk. The importance of risk reduction in the uptake of new technology has been discussed by Ellis-Jones and Sims (1994) amongst others. Risk can be reduced if the initial price of the cart is modest and if the availability of the cart is high. However, the cart need not be totally reliable. Conventionally, engineers strive for long 'mean time between failures' for their products. But to achieve this high reliability, high quality materials and components are required which are commensurately more expensive and which must be manufactured in large numbers in only a few locations to achieve the consistency required. Starkey (1988) has discussed the problem of technically excellent products which are too expensive. The cart owner will probably then find difficulty in obtaining spares and difficulty in achieving early and cheap repair if spares and dealer facilities are available only at the end of a very long and muddy road. Long 'mean times to repair' result from this traditional 'mass production' approach, and overall availability suffers. Reynolds (1992) discusses some of these issues surrounding reliability and repairability in the context of rural water pumping.

Thus a more cost effective method of technology provision in low and very low income environments is probably to accept more breakdowns as long as they can be quickly and easily repaired. This opens the way for local production by artisans who then can not only repair the carts, but also manufacture them. The almost universal use of Japanese minibuses in Africa has been ascribed, not only to their low cost (vehicles are scrapped by Japanese owners because of extremely tough roadworthiness laws), but also because such vehicles are easily repaired.

An ATNESA resource book

If the economics of a technology are only marginal (as may be the case with tractors) then mean times to repair can become infinite and the technology is usually laid to rest in one of the machinery graveyards which border every developing country town. Rodriguez (1992) points out that taxies, minibuses and trucks will be driving past those same graveyards, maintained by the same societies which are apparently unable to keep tractors working. He suggests that one technology is cost effective whilst the other is not.

After many decades of development it might be imagined that designs of animal cart for emerging nations would be optimal or nearly optimal. Working in Nigeria and the UK however, members of the Development Technology Unit have developed cart components and cart designs which are significantly cheaper than existing designs. These cost reductions have been achieved by rigorous simplification of design, usually by providing only the minimum number of components. Such components as there are, are easy to make, of easily obtainable material and require only the minimum of skills and tooling to produce.

How is this lower cost possible? - the DTU approach to cart design

The requirements for low-cost, locally produced carts translate into profound differences in the approach needed in the design of such products. A number of design rules, features and manufacturing constraints and requirements become apparent:

- use of machine tools in manufacture should be reduced or avoided
- drilled holes should similarly be eliminated or reduced
- complicated joints and materials preparation (for example mitred and angled joints), should be avoided
- the number of operations involved in the production of the product should be severely reduced
- the number of components should be severely restricted
- requirements for high tolerances should be eliminated

- the number and complexity of moving parts should also be severely reduced
- threaded fixings should be avoided wherever possible
- symmetry and modular design is to be preferred so that lengths of components used in a product are similar to or multiples of each other. Asymmetry and handedness should be avoided
- material types and range of sizes used in designs should be reduced (makes stock control easier).

These restrictions may appear to render manufacture of products for the communities mentioned above difficult or impossible, but there are techniques which are technologically feasible but which are not preferred in industrialised countries and which are therefore unfashionable or unknown:

- holes can be produced by welding metal around them, rather than removing metal by drilling, and this can be quick and flexible
- punching holes is the normal method used by developing country blacksmiths
- components can be attached to each other by welding in such a way that the welds can be cut through when the product is required to be disassembled; this may be cheaper than bolting
- steel bar may readily be used for clenched nailing and riveting and other fixings. Its ductility may allow the fixing to undone and remade several times
- components can be bigger than their equivalents might be in industrialised countries and may still be workable at lower speeds and loads with inferior materials.

The section below shows a number of animal carts built from steel and timber and using clenched riveting and welding techniques. Particular features of each design are described.

Lightweight sideless donkey cart

The lightweight cart shown in Figure 1 is intended to be pulled by one donkey and is therefore as light as possible. Also it is intended for bulky loads such as firewood (which will be secured by tying) and hence it is not provided with sides. Only a very few material sizes are used in

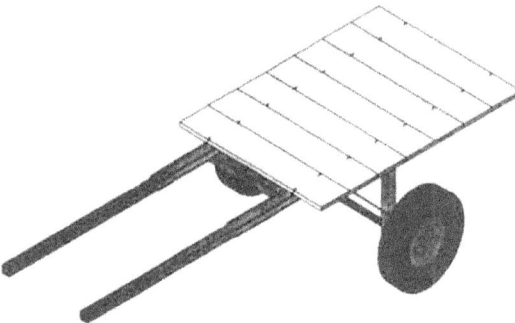

Figure 1: Lightweight sideless donkey cart

this design: 50 mm square box section tubing, 12 mm or similar reinforcing bar, 6 mm re-bar and 25 mm timber planking. All the components are cut off square from the material bars and there is no requirement to work closer than about 5 mm. No drilling is required and there are no moving parts within the design, other than the wheels. Only seven main components provide the chassis and only eight welds are required to join them. No jigs or fixtures are required for manufacture and the cart body can be made in only a few hours, which compares with as much as a week for conventional designs.

Donkey and ox carts with sides

The wooden planking used on the two designs shown in Figure 2 and Figure 3 provides the longitudinal bending strength of these carts. This is particularly helpful in reducing the mass and cost of production by using material more efficiently -

two functions are served rather than the usual one by these side planks. The designs use only a few material sizes and again no drilling and cutting at awkward angles are required.

Wooden carts

Figure 4 shows a wooden bodied cart. What is unusual is the method of jointing between components. Most wooden structures (for example furniture) are joined using methods in which large proportions of both meeting components are cut away to make, for example, a mortice and tenon joint. Such joints are time-consuming to make and require high levels of skill, together with a good tool base. The cart shown here uses joints which take only a few minutes to make.

Steel roller bearings – stub axles

Figure 5 shows one particular low-cost roller-bearing hub mounted in an angle iron central axle. The bearing does not give totally reliable performance, with some failures possible after only a thousand kilometers, but the axle is cheap and is readily repairable. With good lubrication some axles have lasted nearly 30 000 km. The method of mounting the wheelnut studs onto the hub tube removes the need for flame-cutting and turning a disc of steel plate and drilling the stud holes. The method also allows minor changes in the pitch circle diameter by simple bending of the struts and of more substantial changes in diameter by changing strut length. The hub can be made without machining and some versions of hub require no drilling either. This design is dictated by

Figure 2: U frame donkey cart

Figure 3: U frame ox cart

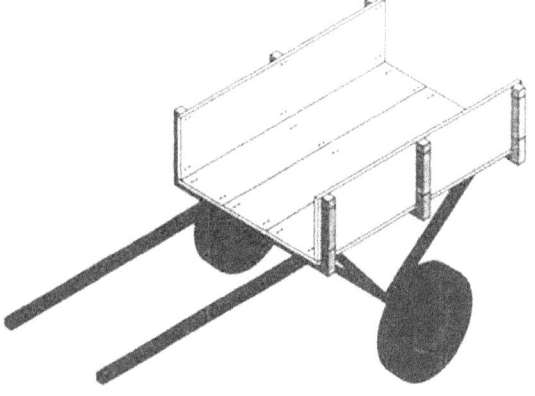

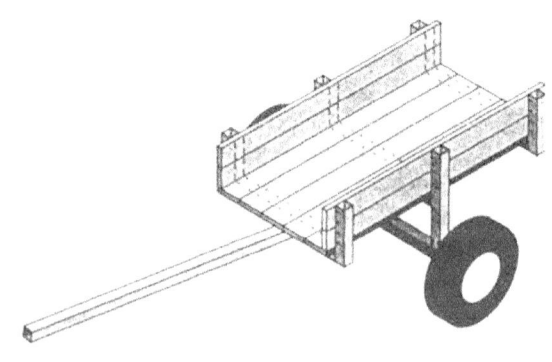

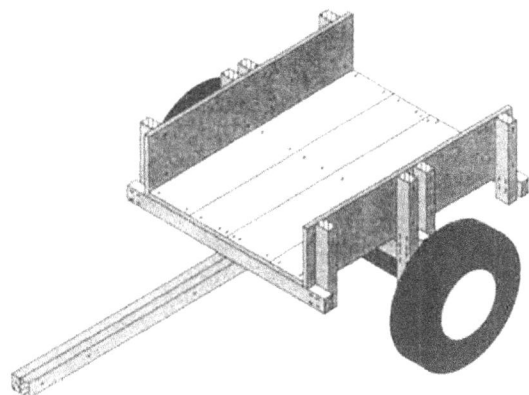

Figure 4: Wooden cart using U frames

the 63 mm hole in the middle of most automotive wheels.

Steel roller bearings - live axles

The separate stub axle system seen in Figure 5 can be difficult to make in some conditions and the performance of bearings with small rollers is not

ideal. Unfortunately the hole in the middle of most scrap car wheels usually prevents larger hub tubes than 2" BSP with an outer diameter of about 61 mm. The conventional live axle systems shown in the left of Figure 6 result in short axles and high loads on the bearings. The offset system shown on the right of Figure 6 allows long axles and avoids the need for a bearing mounting point in the middle of the cart. More importantly it allows large diameter bearings. Figure 7 shows how the low-cost pipe and roller bearings of Figure 5 could be modified for offset live axle systems. In these systems axial thrust loads are carried at the ends of the axle.

Aluminium wheel with integral roller bearing

Aluminium casting is quite widely practised in Africa, usually to make cooking utensils and the like. Aluminium wheels with integral roller bearings (Figure 8) could be made by these artisans and would provide a very low cost solution to the wheel and bearing problem. In the laboratory at Warwick and on a cart in a special

Figure 5: Pipe and roller donkey-cart axle

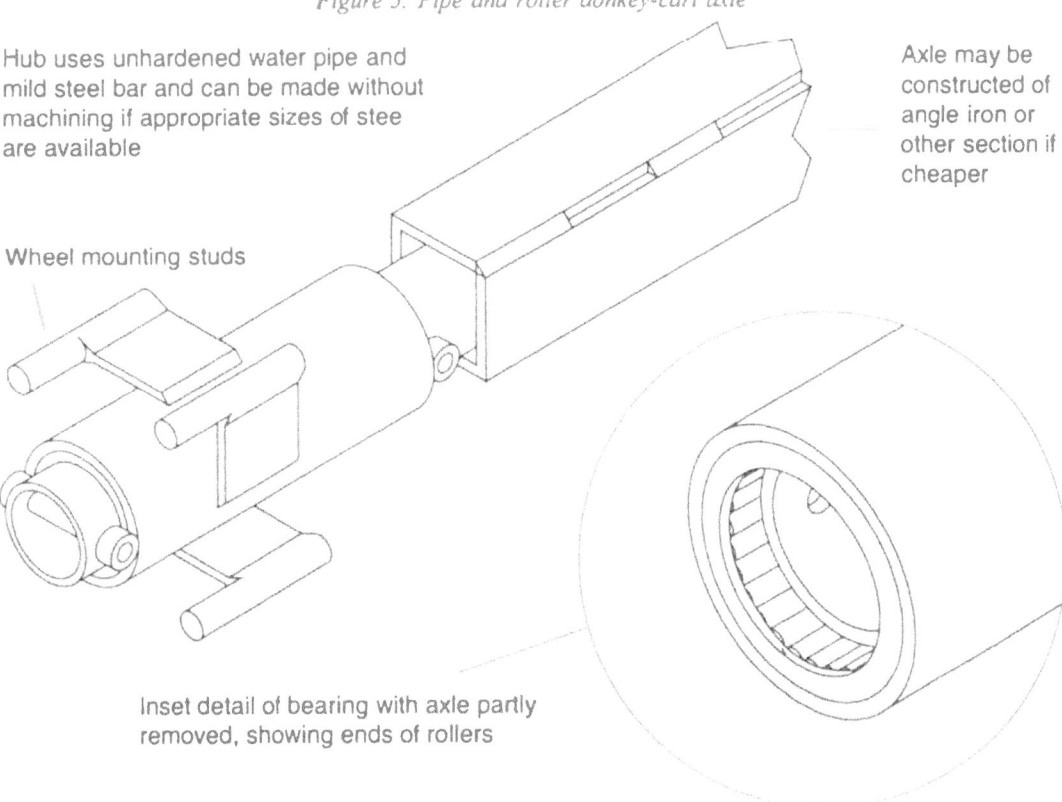

Hub uses unhardened water pipe and mild steel bar and can be made without machining if appropriate sizes of stee are available

Axle may be constructed of angle iron or other section if cheaper

Wheel mounting studs

Inset detail of bearing with axle partly removed, showing ends of rollers

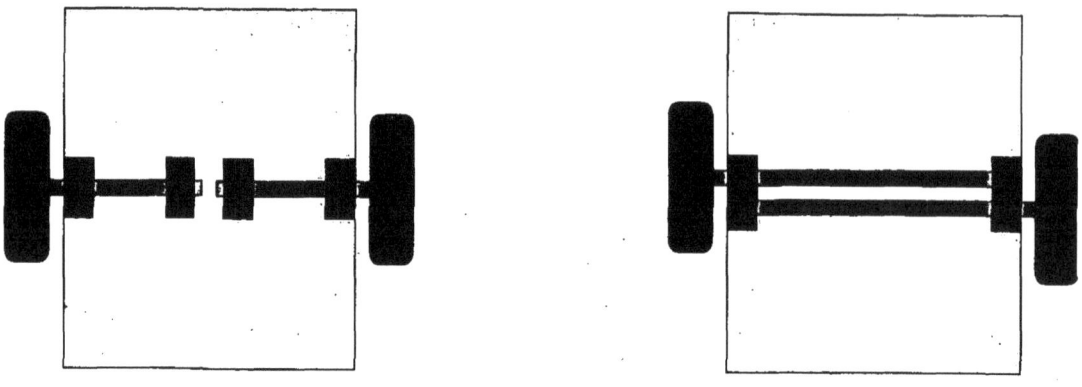

Figure 6: Conventional live axle system (left), offset live axle system (right)

test these bearings work very well – an endurance test was stopped after about 8000 km because of time pressure on the test machine, but inspection of the wheels and bearing surfaces showed little

Figure 7: Offset live axle system using pipe and roller bearings

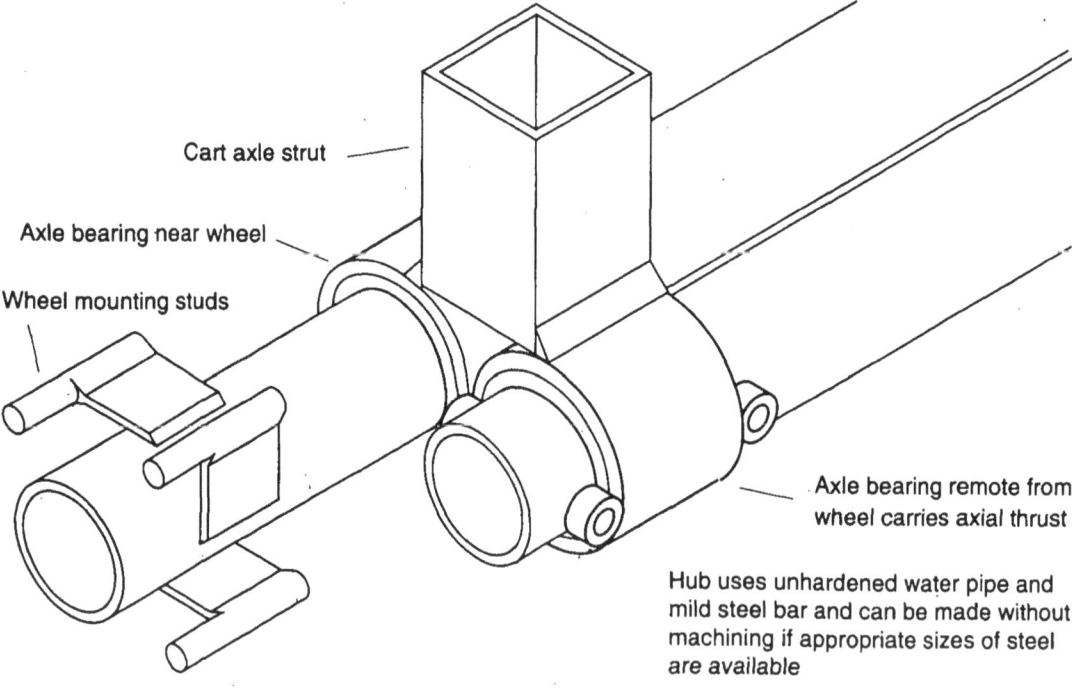

Cart axle strut

Axle bearing near wheel

Wheel mounting studs

Axle bearing remote from wheel carries axial thrust

Hub uses unhardened water pipe and mild steel bar and can be made without machining if appropriate sizes of steel are available

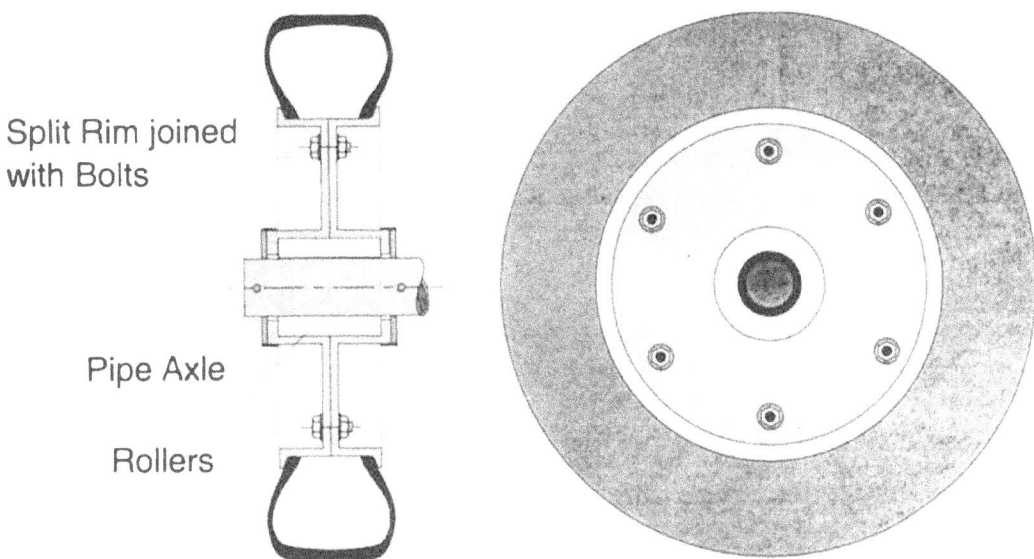

Split Rim joined
with Bolts

Pipe Axle

Rollers

Figure 8: Cast aluminium wheel with integral roller bearing

wear so that these bearings might be expected to last the full life of a cart. Further work will be performed on these wheels.

Artisanal training–examples of non-mass-produced products

A major problem, becoming more and more obvious to observers of the developing world, is the standard of technical education available to artisans in developing countries and moreover the effect that the shortfall in this area has on the ability of artisans to innovate. Conventional production engineering, with an emphasis on the low costs of sophisticated products through large-scale production in dedicated plants, fails frequently in developing countries. Local adaptation to local problems is seen by many of these observers as the only plausible solution to the problem of providing low-cost products to rural and low-income customers. Local designs locally generated stand a good chance of being locally owned, Poston (1990) has discussed many of these 'ownership' issues.

But how is this local adaptation to come about and thrive if the only group in a position to effect the work is struggling with the substantial handicap of inadequate design training – if the only examples of product design are those appropriate for mass production? Technical and design education must be made available to the artisans who can actually use it.

References

Ellis-Jones J and Sims B, 1994. *Increasing productivity on hillside farms: a farming systems approach.* Natural Resources Institute, Chatham, UK,

Poston D, 1990. *The development of manufacturing industry in central Africa, with reference to metalworking.* PhD Thesis, University of Warwick, Warwick, UK.

Reynolds J, 1992. Handpumps: towards a sustainable technology, UNDP-World Bank, Washington DC, USA.

Rodriguez M, 1992. Agricultural equipment: the maintenance myth. In: Carruthers I and Rodriguez M (eds), 1992. *Tools for Agriculture.* Intermediate Technology Publications, London, UK.

Starkey P, 1988. *Perfected yet rejected: animal-drawn wheeled toolcarriers.* Vieweg for German Appropriate Technology Exchange, GTZ, Eschborn, Germany. 161p. ISBN 3-528-02053-9

Ox carts in Kaoma, Zambia

by

Martin van Leeuwen [1] and Nawa Siyambango [2]

1) Advisor, 2) Agricultural Engineer
Western Province Animal Draught Power Programme, PO Box 940007, Kaoma, Zambia

Abstract

Eighty-three ox-cart owners were surveyed in Kaoma District, Zambia in 1994, to investigate the economic and technical aspects of ox-cart transport. About 13% of ox owners, or 1% of all households owned ox carts. Prices for new good-quality carts were US$600–750 and locally-made carts cost US$400–600. Cart owners had bigger households and cultivated more land than farmers without carts. Nearly half of all carts had been acquired in the last two years. About 60% of cart owners hired out their carts commercially. Carts could compete economically with trucks for distances of less than 20 km, especially for more remote areas with few roads and when it was difficult to arrange a sufficiently large load for a truck. There was a high incidence of breakdowns, especially for cheaper carts, but the rate was less severe than is often reported in the literature. Rural workshops become established and provide repair services without any support from government or donor programmes. Sales of carts made locally but using axles imported by the Western Province Animal Draught Power Programme were poor. Farmers preferred more expensive but durable ox carts from Lusaka.

Introduction

Kaoma District in the Western Province of Zambia is situated 400 km west of Lusaka on the road to Mongu, the provincial capital. Kaoma District is the most fertile area of the province, receiving 800–1000 mm of rain per year. During the average 113 days within the rainy season, there are usually two dry spells of more then 10 days (ARPT, 1993). The population of Kaoma District was 113,000 people in 1990 of which the majority live along streams and roads at densities between 10 and 100 people/km².

Agriculture in Kaoma is a mixed farming system with a higher proportion of market-orientated crop growers than in other districts of the province.Untill 1994 about 4000 farmers participated in registered maize marketing and use of seasonal credit, fertiliser and hybrids. Kaoma District produces about 90% of the marketable surplus production of maize and groundnuts of the Western Province.

The official production and yields are shown in Table 1, while Table 2 gives an overview of the production structure.

Animal draft power in Kaoma

Table 3 shows that only 10% of the rural households own oxen, while about 7% own a plow. However, it should be noted that the possession itself is less stable than the term 'ownership' suggests. Many owners sell, lose or slaughter their oxen in adverse times. The fact that relatives and neighbours can borrow and hire oxen for plowing their own fields implies that about 45% of all rural households have access to draft animal power (Kakwaba and van Leeuwen, 1999). Three percent (60) of the ox owners weeded with

Table 1: Crop production and yields in Kaoma

Crop	Area cultivated (ha)	Average yield (kg/ha)
Maize	9350	1530
Cassava	4800	810
Millet	1340	900
Groundnuts	890	450
Sorghum	590	630
Rice	370	1080
Mixed beans	200	360
Cotton	130	300
Soya bean	60	450
Sunflowers	30	600

Source: WP-ADPP, 1993

Table 2: Farms in Kaoma District

Area (ha)[1]	Number	% Female-headed
>20	25	8
>5–20	341	12
½–5	10,050	20

Notes:
1) The total number of agricultural households was estimated at 19,300 in 1990 by the Central Statistics Office. About 9000 rural households do not appear in the crop forecasting of the Department of Agriculture, since they cultivate less than 0.5 ha.
Source: WP-ADPP, 1993

oxen and ridgers in 1994/95 (WP-ADPP, 1995), while only about one out of every 12 ox owners possesses an oxcart.

Ox cart owners

A survey of 84 cart owners was carried out during 1994 in Kaoma District (van Leeuwen and Siyambango, 1995). All cart owners were interviewed in four distinct and different areas of the district.

Cart owners in Kaoma usually have bigger households than non-cart owners (average 10 members per household) and have an average age of 53. Almost all carts are owned by men, although one woman had two ox carts. Most cart owners have more than two oxen, while only 38% have two oxen or less. Figure 1 shows that almost half of all the carts (43%) have been acquired during the past 2 years. This fact, combined with the impression of local manufacturers and the substantial sales of new ox carts (about 20/year) during the past three years leads to the conclusion that investment in ox carts has increased during the past years.

Cart owners cultivate considerably more land (average 3.6 ha) than the average farmer (1.1 ha). Practically all cart owners used fertiliser and only 11% did not depend on seasonal loans from lending agencies. The average cart owner sold nine tonnes after the 1991/92 harvest and seven tonnes after the 1992/93 season. Cart owners received an average income of ZK 500,000 from surplus

Table 3: Quantities and distribution of draft animal implements in Kaoma District in 1993

Number of farm households	19,300
Total number of oxen	5,700
Cropped area (ha)	19,300
Households owning oxen	1,900

Item	Number of households owning	% of ox owners who own
Plow	1,314	69
Ox cart	243	13
Sledge	642	34
Harrow	40	2
Cultivator	17	1
Ridger	28	2
Planter	11	1

Source: Dibbits and Mwenya, 1993

agricultural production in 1994 (US$1≈ZK650 in 1994). The average income from transport services that year, estimated by the owners, amounted to ZK 137,000. Figure 2 shows the distribution of income from transport charges.

Typology of ox cart utilisation

Figure 2 reveals that a considerable portion (38%) of owners did not hire out their carts and consequently underutilised their transport resource. The reasons given by owners is 'that they want to save their cart or oxen'. Analysis of this group shows that it consists of all kind of cart owners. Amongst the 22% of owners who earn more with their cart than with their crop production, which one could label 'rural transporter', only very few (2%) did not hire out their cart. Among cart owners who sell over 4.5 t of maize/year, labelled as 'farmers', only 62% hired out their cart. The majority of cart owners with less agricultural production and low transport earnings do not hire their cart. Geographical analysis showed that cart owners in more isolated areas with lower population densities hire their ox cart less (and at lower prices), unless the cart is used for a

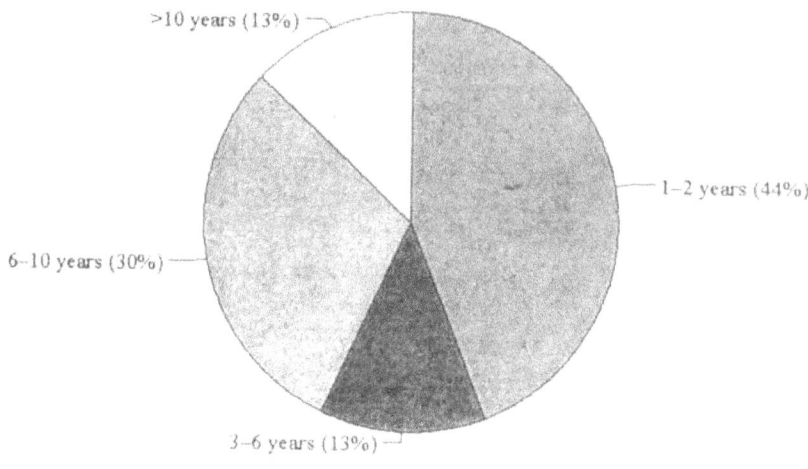

Figure 1: Duration of ownership of ox carts (n=83)

year-round business such as ferrying logs and planks to the main roads. Table 4 shows the main differences and findings concerning frequency of utilisation of carts.

Loads, distances and prices

Ox carts in Kaoma District transport a variety of products. Most products have a typical rural character such as maize, fertiliser, seed, firewood, planks, cassava and sweet potatoes. Table 5 provides an overview of different products that are carried, and the average distance and transport price in relation to the value of the load. The table shows that planks and fertiliser are the best items in terms of price per load. However, when time and distance are considered, bricks, shelled maize and fertiliser are more profitable to transport. Table 6 shows the *maximum* income from transport per day around town, as well as the produce and transport charge as a percentage of the value of the load (after being transported).

Figure 2: Income from transport in thousands of Zambian Kwacha (n=83); US$1≈ZK650 in 1994

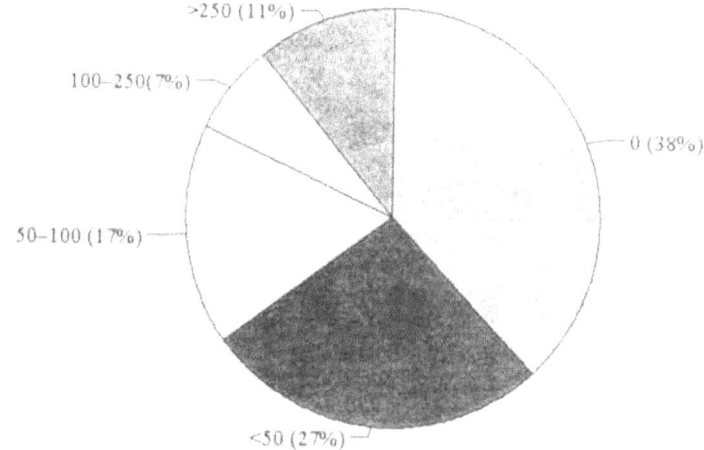

Table 4: Typology of owners of ox carts in Kaoma District

Area	Soil fertility	Population (people/km²)	Maize yields	Transport location	Cart utilisation	Cart owners
Town	–	High (>50)	Medium	Central	High	Transporter
West/South	Low	Low (<10)	Low	Central	High	Transporter
North/East	High–medium	High–medium	High–medium	Central	High	Farmer
				Isolated	Low	Farmer/ Transporter

Table 5: Products, carried by ox cart transport in Kaoma District, 1994

Produce	Load (kg)	1-way distance (km)	Price/load (ZK)	Price/load/km (ZK)	Price/t/km (ZK)
Maize (unshelled) n=46	490	1.8	2150	1190	2460
Maize (90 kg/bag) n=51	780	2.7	3320	1230	1570
Fertiliser (50 kg/bag) n=50	780	3.3	5690	1720	2230
Firewood n=8	870	3.4	3980	1170	1340
Planks (10 kg each) n=15	580	27.5	9120	330	1140
Charcoal (65 kg/bag) n=11	840	7.9	4640	590	700
Bricks (10 kg/brick) n=4	500	2.6	3950	1520	3000
Grass (23 bundles) n=10	350	8.4	4520	540	1540
Cassava/sweet potatoes (90 kg/bag) n=7	820	4.9	2550	520	640

Competitive transport

Ox carts in Kaoma face competition from trucks and sledges. Trucks ferry loads more cheaply for distances above 20 km, if there is a sufficient load (organised!), and if stumps, streams and loose sand do not block the truck. Ten-ton trucks on tarmac roads charged 50 ZK/t/km for grain and fertiliser transport during the period of the survey.

In isolated areas with loose sand, sledges often carry the same load (360–400 kg) as ox carts do in such areas, for the same price although more oxen are used and more time for transport is needed. In such inaccessible areas, sledges and ox carts may ferry loads over 50–70 km, charging 15–20% of the value of the grain which is transported. Cart owners in these areas are able to earn ZK 9000 per day during peak periods .

Apart from these exceptional inaccessible areas, the survey showed that, in general, transport prices of ox carts in Kaoma District are not high in comparison with prices cited in the literature (Mack, 1984; Starkey, 1991; Helsloot, Sichembe and Chelemu, 1991; Starkey, Dibbits and Mwenya, 1991).

Prices of ox carts

The survey did not confirm the idea cited by some authors (Mack, 1984; Helsloot, Sichembe and Chelemu, 1991) that the cost of ox carts can

Table 6: Maximum possible income per day from transporting various products by ox cart in Kaoma District

Product	Maximum income/day (ZK)	% of value of load
Maize (unshelled) 5.5 bags	11,000	8
Maize (shelled) (10 x 90kg bags)	10,000	7–13
Fertiliser (20 x 50kg bags)	20,000	6–11
Firewood (1 load)	4,000	50
Charcoal (20 x 65kg bags)	7,000	23
Grass (25 bundles)	10,000	50
Bricks (50 bricks)	10,000	17
Cassava (10 x 90kg bags)	9,000	5
Planks (40 planks)	7,000	28

Notes: ZK650≈US$1 in 1994

easily be repaid from transport charges. In Kaoma District it takes the average ox cart owner who earns ZK 137,000 per year, almost 4 years to earn the value of a good ox cart of around ZK 500,000. This price is equivalent to US$ 750, which is high compared with cart prices in West Africa (Wanders, 1992).

Locally-manufactured ox carts using Landrover, Datsun and Toyota axles cost around ZK 300,000, with the cart body (50%), wheels (30%), second-hand axle, hubs and bearings (20%) as the main costs. In the barter economy of Kaoma District such ox carts cost about 3,500 kg of maize or 3–4 young oxen. However, these cheaper ox carts do have some problems. In particular, the bearing-hub assemby and the rim-hub construction often develop problems within a short time.

Breakdowns and repair

According to cart owners the main problems with ox carts are bearings, inner-tubes and tyres, bolts/nuts, tube-valves and locknuts. A quarter of the 83 respondents bought spares like bearings, inner-tubes and tyres mostly from the capital, Lusaka.

In view of the attention in the animal power literature for breakdowns and absence of repair facilities (Malipaard, 1988; Starkey, 1989;

Starkey, 1991), the survey examined the frequency of cart breakdowns. Table 7 summarises the observations; they confirm the high incidence of breakdowns. However, it appeared that the period for which ox carts are out of order is relatively short. Moreover, it was found that the technical problems of ox carts which are broken-down for more than one year, did not differ from those which were repaired more quickly. Broken bearings, axles and hubs appeared to be the main causes of breakdowns for both categories.

Workshops in Kaoma District

There are at least seven workshops in the district where ox carts are assembled and repaired of which four are in Kaoma town. The repair services and manufacturing of carts are in all cases a minor part of the business run by these 'workshop' managers. The number of carts under repair or being constructed at each workshop ranged from 1–10 for 1994 (total 25), whilst in 1993 the number of carts produced and sold ranged from 0–5 (total 14). One of the most important qualifications for repair and assembling services seems to be the ability to scout around to find hubs, wheels etc at the most advantageous price. Although each manufacturer aims for at least 12–15% (≈ZK 50,000) profit (including labour) and costs for ox-cart bodies vary little, prices of ox

Table 7: The incidence of breakdowns according to period of ownership and type of purchase

Length of ownership	Carts bought new (n=38)			Carts bought second hand (n=45)		
	Running	Out of order	% down	Running	Out of order	% down
1–4 years	21	4	16	30	6	17
5–12 years	6	4	40	3	2	40
>12years	2	1	33	3	1	25
Total	29	9	24	36	9	20

carts differ considerably. Interviews with the workshop managers made it clear that other factors are of great influence on the price and quality of the ox cart. For instance, the degree to which the manufacturer is 'hard up with cash', the manufacturer's confidence in clients' ability to pay in installments and the extent to which the client is able to check the work critically.

The survey did not confirm the idea that extra workshop support such as training or supply of toolkits (Malipaard, 1988; Starkey, 1989; Helsloot, Sichembe and Chelemu, 1991), is necessary to increase the quality of manufacturing and repair work.

WP-ADPP services

The Western Province Animal Draught Power Programme, WP-ADPP II, concentrated on the demand for high quality ox carts in the district, by importing and selling new ox carts on a cash basis.

Over 2.5 years WP-ADPP has sold about 50 ox carts of a type that was previously sold during 1987–1991 by the former district cooperative. The same type was sold to contribute to standardisation and to stimulate hardware shops within the district to stock the correct spares.

Moreover, WP-ADPP imported new axles and tried for six months to involve four local workshops in cart assembly, by giving new hubs on consignment. A few carts were produced by the workshops, while WP-ADPP added a quality control logo and extension leaflets about local repair options and prices. However, sales of these carts were slow and the workshops were paid only after a lot of trouble. The workshops commented

favourably on WP-ADPP's activities but explained the higher sales of the imported carts as a result of farmers' desire 'to buy products which come far from Lusaka'. WP-ADPP also concluded that Kaoma has a market for more expensive but durable ox carts. It appeared that workshops were not interested in the programme of buying and stocking second-hand hubs, bearings and wheels.

Finally, WP-ADPP decided to buy old rims and tyres locally, and assembled about 30 carts, which have been sold on a rent-purchase basis to 15 regional farmers who organised the barter of fertiliser against maize in 1995. This activity did not involve the established workshops, but young people with a welding machine. This programme is part of a plow rehabilitation project, whereby traders of WP-ADPP barter broken-down plows against fertiliser and seed, while a newly settled workshop reconditions the plow, modifying the hitch and wheel attachment.

Conclusions

- Investment in high quality ox carts has increased during the past few years.

- Prices of new durable carts are around US$600–750, while local ox carts are built for US$400–600.

- Only a slight majority of cart owners (61%) provide transport on a commercial basis, earning an average income of US$200 per year from transport.

- Transport charges for produce like maize and fertiliser amount to less than 10% of the value of the load at distances of 1–4 km. Ox carts ferrying the same produce over 10 km or more earn less money per ton and per km, but still charge more than trucks or pickups.

- The incidence of breakdowns is less severe than suggested in the general literature. Only 10% of the carts are grounded for more than a year. This does not reflect scarcity in repair service but reduced interest/ability of the owner to organise and finance maintenance of his/her transport resource.

- Rural workshops become established and provide repair services without any support from government or donor programmes. Questionable repair work seems to be a result of price negotiations with the client, whereby the cart owners with a low budget and know how receive low quality repairs.

- Credit for ox carts cannot be justified with an argument that ox carts are too expensive. Farmers with surplus production or lumberjacks can considerably lower transport costs and increase their farm/forest gate price of their produce, if they invest in ox carts.

- It appears that ox-drawn transport is much more diverse than suggested in the literature. Agricultural research and extension should break away from conventional technical and financial focuses and utilise critical inquisitiveness into the commercial utilisation of transport in the rapidly developing sector of semi-subsistence agriculture.

References

ARPT, 1993. *State of the Art ARPT 1990-1993.* Adaptive Research Planning Team (ARPT), and Ministry of Agriculture, Food and Fisheries, Mongu, Zambia.

CSO, 1993. *Census of Population and Housing, 1992.* Central Statistics Office (CSO), Lusaka, Zambia.

Dibbits H and Mwenya E, 1993. *Animal traction survey in Zambia.* National ADP Coordination Project, Agricultural Engineering Section, Department of Agriculture PO Box 50291, Lusaka, Zambia and Insitute of Agricultural and Environmental Engineering, Wageningen, The Netherlands. 119p.

Helsloot J, Sichembe H and Chelemu K, 1991. *Animal powered rural transport in Zambia, prospects and constraints for development..* Ministry of Agriculture, Lusaka, Zambia in association with IMAG-DLO, Wageningen, The Netherlands.

Kakwaba K and van Leeuwen M, 1999. Hiring and lending of oxen for plowing in Kaoma, Zambia. pp 308–312 in: Starkey P and Kaumbutho P (eds), *Meeting the challenges of animal traction.* Animal Traction Network for Eastern and Southern Africa (ATNESA), Harare, Zimbabwe. 326p.

van Leeuwen M and Siyambango N, 1995. *Ox carts in Kaoma District, an economic and technical assesment of ox carts in a rural district of Zambia.* Western Province Animal Draft Power Programme (WP-ADPP), Department of Agriculture, Mongu, Zambia and RDP Livestock Services, Zeist, The Netherlands.

Mack R P, 1984. *The impact of the introduction of work oxen within the frame of the IRDP in the NW Province of Zambia.* GTZ, Eschborn, Germany.

Malipaard E J, 1988. *Possibilities for improvement of intermediate transport means in the Western Province of Zambia.* Faculty of Civil Engineering, University of Delft, The Netherlands.

Rauch T et al, 1988. *The sustainability of the impact of the Integrated Rural Development Programme (IRDP) Zambia/ NW Province.* Centre for Advanced Training in Agricultural Development, Technical University of Berlin, Berlin, Germany.

Starkey P, 1989. *Harnessing and implements for animal traction.* Vieweg for German Appropriate Technology Exchange, GTZ, Eschborn, Germany. 244p. ISBN 3-528-02034-2

Starkey P, 1991. *Animal traction: constraints and impact among African households.* pp. 77-90 in: Haswell M and Hunt D (eds), Rural households in emerging societies. Berg, Oxford, UK. 261p. ISBN 0-85496-730-3

Starkey P, Dibbits H and Mwenya E, 1991. *Animal traction in Zambia: status, progress and trends.* Ministry of Agriculture, Lusaka, Zambia in association with IMAG-DLO, Wageningen, The Netherlands. 105p.

Wanders A A, 1992. Supply and distribution of implements for animal traction: an overview with region-specific scenarios. pp 226–243 in: Starkey P, Mwenya E and Stares J (eds) *Improving animal traction technology.* Proceedings of a workshop of the Animal Traction Network for Eastern and Southern Africa (ATNESA) held 18–23 January 1992, Lusaka, Zambia. Technical Centre for Agricultural and Rural Cooperation (CTA), Postbus 380, 6700 AJ Wageningen, The Netherlands. 490p. ISBN 92-9081-127-7

WP-ADPP, 1993. *Crop forecasts for Kaoma District 1983–1993.* Western Province Animal Draft Power Programme (WP-ADPP), Department of Agriculture, Mongu, Zambia.

WP-ADPP, 1995. Annual Report 1994. Western Province Animal Draft Power Programme (WP-ADPP), Department of Agriculture, Mongu, Zambia.

A note on improving animal-drawn transport in Uganda

by

Victor Ogwang.

Department of Agriculture, PO Box 5, Apac, Uganda

Introduction

Access to animal-drawn transport can offer a simple, manageable and affordable improvement over existing methods of transport, particularly human head-loading. Ownership of a cart or the ability to hire one frees farmers from dependence on tractor services, which are in most cases unreliable and costly.

In villages remote from the road system lack of efficient transport for farm activities results in a considerable loss of potential cash income for farmers. Existing methods of transport are often arduous and time consuming and restrict the ability to produce and consequently limit the incentive to produce more.

Ox-drawn transport alleviates women's drudgery

Over 75% of women in Uganda carry out agricultural work. During harvesting of farm produce a pair of oxen is used to transport farm produce to the store instead of the women carrying baskets on their heads. Produce is also transported from the store to market using oxen. Fuelwood is sometimes transported using oxen, rather than by women, and as some is sold for cash, animal draft power is an income-generating activity. Building materials, such as grasses, which would normally be carried by women can be carried on ox carts, or ox-drawn sledges.

Reduced wear on agricultural implements

Ox plows can be carried to and from the field on sledges or carts, saving the wear that would occur if the plow was pulled along the ground.

Carts

Two-wheeled carts drawn by oxen and donkeys are used commonly for farm transport in the northern part of Uganda and generally in the rest of the country. They have a considerable potential for more widespread use since they:

- are capable of meeting rural movement requirements efficiently
- are affordable and socially acceptable
- are not restricted to use on motorable roads
- can be made widely available because of their low capital cost.

A well-designed cart greatly increases the load that can be moved by animals in comprison with that which can be carried on their backs or pulled on a sledge. Carts should be lightweight yet strong, have efficient wheel/axle systems that are simple to manufacture, and be equipped with an efficient means of harnessing the animals. Experience has shown that if such devices are to be produced and used sucessfully then they must be:

- adapted carefully to local operating conditions in terms of terrain, type of use and the characteristics of indigenous draft animals
- designed to take account of local production conditions in terms of the availability of components, materials and manufacturing skills
- manufactured in a way and on a scale that matches the capability of local industry
- marketed at a price people can afford with credit available if necessary.

Two-wheeled carts pulled by a pair of oxen are the most common type of animal-drawn cart in Uganda. Two-wheeled carts are relatively simple to make, cheap to purchase and easy to manouvre and control.

Double neck yokes can be left permanently attached to the shaft of a cart. For transport use it may be advantageous if the width of the yoke is equal to the wheel track of the cart. This means that the animals walk directly in front of the wheels and are therefore likely to avoid objects that may obstruct or puncture a tyre.

Use of donkeys for transport

Donkey utilisation is being introduced in Uganda, but less than 10% of farmers in the region have started using donkeys. Farmers that do use donkeys use them as pack animals for transporting fuelwood, water and farm produce.

The 'Golovan' one-ox cart

by

Bruce Joubert

Department of Agronomy (Animal Traction Centre), University of Fort Hare
Private Bag X 1314, Alice, 5700 South Africa

Abstract

A recent animal traction survey in South Africa identified a 'felt need' for low-cost, lightweight and modernised animal-drawn carts in many of the areas visited. The University of Fort Hare has obtained plans for an ox cart designed by an Eastern Cape farmer for use on farms and by rural communities. Called the 'Golovan one-ox cart', an example has been constructed, and it is currently being used on the university's research farm to do on-farm work. This paper discusses the technical aspects of the cart and the practicalities of using it for a range of on-farm and rural activities. The cart has a gravel-hauling capacity of about 1 t-km/hr and its use for road gravelling schemes will probably be economical. The technology is compatible with local culture and skills and may offer indirect benefits to rural communities.

Introduction

An animal traction survey conducted recently in South Africa found that, in the more remote areas of the country, between 40 and 80% of smallholder farming families currently make some use of animal power (Starkey, Jaiyesimi-Njobe and Hanekom, 1995). Draft animals are used for a wide range of agricultural activities, but the main uses are for transport and plowing.

The survey found that carts are used to carry goods and people and that they are usually made locally from scrap material. No standard designs exist and the cost varies depending on the availability and demand for scrap axles. Most carts are strongly built, but are heavy and consequently can carry only small payloads (Starkey, Jaiyesimi-Njobe and Hanekom, 1995).

Two-wheel carts were observed, pulled by two to four draft animals. The weight of the disselboom and any unbalanced cart weight was transmitted, in the case of cattle, via the yoke to the hump without causing serious discomfort or injury. However, for equines, the force was transmitted via neck straps to the neck, which often caused problems including injury (Starkey, Jaiyesimi-Njobe and Hanekom, 1995).

During the past three decades considerable effort has been put into developing low-cost, lightweight animal-drawn carts in Africa. Although there has been some success with alternative designs, farmers in Zambia prefer quality carts, which have roller bearings and pneumatic tyres. They are prepared to pay for such carts, even though they are susceptible to punctures (Helsloot, Sichembe and Chelemu, 1993).

Animal-drawn carts can improve rural transport, even though small-scale farmers may not be able to afford them, and in so doing can enhance the quality of life of rural communities.

There is a large unsatisfied demand for carts in many parts of Africa, resulting from production problems which stem mainly from the limited availability of materials and components, in particular good quality wheel–axle assemblies (Anderson and Dennis, 1994).

In South Africa availability, affordability and profitability of animal-based transport systems, and their accessibility to women, are key issues in the development of small farming communities.

The 'Golovan one-ox cart' was designed by an Eastern Cape farmer for use on farms and by small rural farming communities. It has a number of important attributes which make it suitable for use in such communities and for this reason the Animal Traction Centre at Fort Hare University has recently constructed a cart and is carrying out investigations to determine its suitability and acceptability in small rural farming communities.

The design of the Golovan one-ox cart

A plan of the Golovan one-ox cart is shown in Figure 1, and the main features of the design are outlined below.

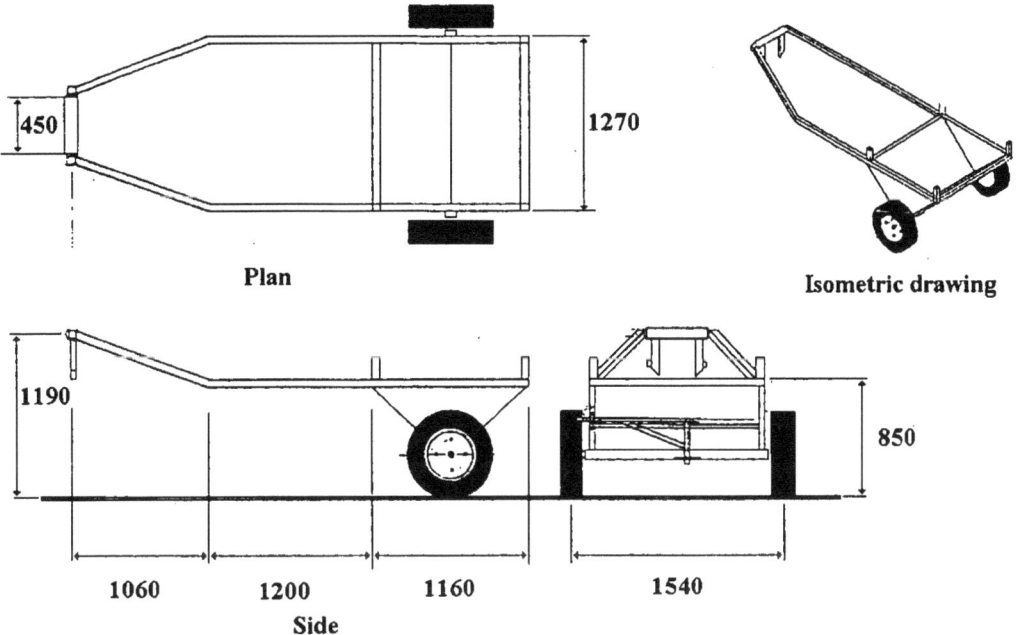

Plan

Isometric drawing

Side

Figure 1: The Golovan ox cart. All dimensions in millimetres.

- The cart is pulled by a single ox.

- The cart weighs about 320 kg. When pulled by an average sized ox, it therefore has a payload of some 500 kg.

- The volume of the cart-back is 0.33 m^3. Its shape is such that the centre of gravity of any load placed in it is directly over the rear axle, so there is no undue down or uplift force through the shafts onto the neck of the ox.

- The yoke 'skeis' are attached to a pipe, which forms the yoke and which rotates on an axle joining the two shafts. This 'swivel yoke' ensures that there is no chaffing of the animal's hump, that the action of the yoke on the neck is more gentle and that the ox is able to develop more power.

- The rear door for unloading is hinged at the top and opens at the bottom of the cart-back, which means that in most cases the load is discharged through the cart-back bottom onto the ground without the need for manual unloading.

- At each corner of the cart-back is a slot, which enables a seat to be fitted to accommodate two large or three small passengers. The seat is constructed such that the centre of gravity of the seat and passengers is positioned directly over the rear axle. The cart-back is also constructed so that loads are directly over the rear axle. Stanchions can also be fitted in these slots enabling low density loads such as baled hay to be carried.

- Pneumatic rubber tyres are fitted to 356 mm diameter rims, which are mounted on stub-axles with sealed, factory lubricated, tapered roller bearings.

- To reduce costs and weight, no springs or braking mechanism are included on the cart.

- Harnessing comprises the swivel yoke, a halter with long reins, and the breaching which enables the ox to provide the braking.

Practical use of the Golovan cart

The Golovan cart can be used for all of the carting activities typically carried out on farms or in rural communities. The technical features, notably the shape of the load-back and the bottom unloading rear door, enable various activities to be carried out with some convenience. For example:

The Golovan ox cart in use showing the rear door and breaching harness

- two 210-litre drums can be loaded, empty, into the load-back and, after filling, the drums can be discharged through the bottom unloading rear door onto the ground
- the cart is ideal for carrying road gravel or compost, which can be unloaded through the bottom unloading rear door and placed precisely where it is required. The cart is thus suitable for road maintenance, land fertilisation and harvesting
- the inclusion of stanchions in the four corner slots enables up to 20 average-sized bales of hay to be carried
- the seat is constructed and positioned in such a way that loading can take place without removing it.

Case study appraisal of the Golovan cart

The Eastern Cape Department of Agriculture has inspected the cart and is interested in its potential to carry gravel for the repair and maintenance of rural roads.

The Animal Traction Centre at Fort Hare University is presently evaluating the Golovan to assess the feasibility of using it for this purpose. The project will take about two years, and will involve a number of case studies, undertaken at different times of the year, which will aim to:

- establish the gravel haulage capacity of the cart
- determine the type and amount of supplementary feeding required for the ox
- determine the cost of operating the cart and one ox under rural conditions
- establish the acceptability of the cart to the inhabitants of rural communities in the neighbourhood of Fort Hare University.

Results from the first case study, carried out in September 1994, showed that over a six-hour average work day, the Golovan has a gravel haulage capacity of about 1 t-km per hour or 6 t-km per day (Ntili and Higa, 1995). In the case study, this was made up of 13 loads weighing an average of 630 kg each carried for journeys averaging 0.75 km (one-way).

Realistic maintenance rations and minimum supplementary feed requirements of the oxen during road gravelling work will be established during 1996, in collaboration with the Department of Animal Science at Fort Hare University.

The Eastern Cape Department of Agriculture has recently had two Golovan ox carts constructed. These will be stationed in two communities in the vicinity of Alice, where they will be used for various communal activities. In this phase of the project the performance of the carts will be monitored by the project supervisor, and their acceptability to the rural people will be assessed.

Evaluation on the research farm

The Golovan ox cart has now been doing work on the research farm for 16 months; during this time it has been used for collecting wood for farm staff, delivering water, putting down compost on lands, hauling soiled bedding to the compost heap, transporting lucerne hay for animals in the veld as well as in pens, and gravelling roads.

The road gravelling process is carried out by first putting down some 50 m of gravel dumps, spaced about 2 m (centre to centre) apart, with the golovan. Two oxen are then inspanned in a wide weeding yoke and hitched to a road smoother, which enables the oxen to walk either side of the dumps pulling the smoother which flattens the dumps. It would appear from the data collected to date that it will be possible to completely gravel and smooth 100m of road in approximately 10 working days using one golovan and a pair of oxen.

A few problems were enconterd in the 16 months of work with the cart:

- It took one whole month to train two oxen, both already trained to work as a pair, to accept the Golovan completely. Early in the investigation, the Golovan was extensively damaged when one of the oxen inadvertently broke away and crashed the cart into a tree. This was because the oxen had been trained to work as a pair and had to be retrained to work alone without a leader.

- The rear door opening mechanism has had to be modified to facilitate opening when the load back is full of gravel. This was due to the weight of the gravel on the door and latch making it difficult to open the door.

- Attempting to hitch the Golovan through a connection on the swivel yoke to a pick-up truck caused the side shafts to bend. This practice is not recommended with the existing design.

- Oxen do not perform well in extreme heat, so during hot weather it was often necessary to

start work early in the morning and to finish late in the evening, allowing the oxen to rest during the middle of the day. An outspan area and an adequate supply of feed and water are essential in the vicinity of any road gravelling operation.

- It is necessary to use good quality harnesses as the breaching, back straps and halters come under considerable strain during normal working, particularly when turning in confined spaces.

Discussion

Comparisons between tractor/trailer combinations operating at low gross utilisation rates and donkey-drawn cart systems in Botswana have indicated that using donkey-drawn carts is financially competitive for distances of up to about 2.5 km (McCutcheon, 1985). For practical reasons, haulage distances in the Alice area are generally less than 1 km (on average 0.747 km during the case study appraisal), so using the Golovan to cart gravel will probably also be financially competitive.

The introduction of the Golovan one-ox cart system could offer additional benefits, including the employment (job creation), the manufacture of carts and harnesses locally (small business) and the creation of entrepreneurial opportunities in rural communities (contract haulage). These would provide independence to the communities, keep local finances circulating within the communities and reduce the need to import fossil fuels into such areas.

It is important to note that animal-drawn cartage systems are compatible with local culture: the skills to manage them already exist and if something does go wrong it can be rectified by the people without the need to call in a suitably qualified expert.

Conclusion

Investigations carried out to date have led to the conclusion that the Golovan one-ox cart is suitable for consideration as a means of transport in rural small farming communities in the Eastern Cape. In certain respects it has advantages over other designs of ox cart.

It is necessary to fully establish the perceived needs of such communities in so far as animal drawn carts are concerned and to gain acceptance by the people for the Golovan.

If this can be successfully achieved, plans and specifications can be made available. This will enable the cart to be constructed locally, by small businesses, for supply throughout the region.

Acknowledgements

This project was made possible using funding provided by the Chairman's Fund of Anglo American and has been supported by members of staff from the research farm, the Faculty of Agriculture, the old Ciskei Government and the Animal Traction Centre. Special thanks are due to Messrs Beaumont, King, Dyan and Mene, from the research farm, to Professor Bester and Messrs Ntili and Higa from the Department of Agricultural Economics, to Mr Geof Meikle (recently of the Ciskei Department of Agriculture) to Mrs Ronell Grobler, to the late Willie Lawana, to Messrs Vula and Jacob, and to two casual labourers, Messrs Zixeshe and Prins, from the Animal Traction Centre.

References

Anderson M and Dennis R, 1994. Improved animal-based transport: options, approaches, issues and impact. pp 378–395 in: Starkey P, Mwenya E and Stares J (eds), *Improving animal traction technology*. Proceedings of the first workshop of the Animal Traction Network for Eastern and Southern Africa (ATNESA) held 18–23 January 1992, Lusaka, Zambia. Technical Centre for Agricultural and Rural Cooperation (CTA), Wageningen, The Netherlands. 490p. ISBN 92-9081-127-7

Helsloot H, Sichembe H and Chelemu K, 1993. *Animal powered rural transport in Zambia: prospects and constraints for development*. National ADP Coordination Programme, Department of Agriculture, Lusaka, Zambia.

McCutcheon R T, 1985. *The use of donkey-drawn carts in labour intensive road construction in Botswana*. International Labour Office, Geneva, Switzerland.

Ntili T P and Higa M W, 1995. *Appraisal of "Golovan one-ox cart" for cartage of road gravel*. Project report submitted in partial fulfilment of B Agric Economics (Honours), Faculty of Agriculture, University of Fort Hare, South Africa.

Starkey P, Jaiyesimi-Njobe F and Hanekom D, 1995. Animal traction in South Africa: overview of the key issues. pp. 17–30 in: Starkey P (ed), *Animal Traction in South Africa: empowering rural communities*. Development Bank of Southern Africa, Halfway House, Gauteng, South Africa. 159p. ISBN 1-874878-67-6

Photo (opposite): Calf suckling a yoked working cow in South Africa

Meeting the challenges of animal traction

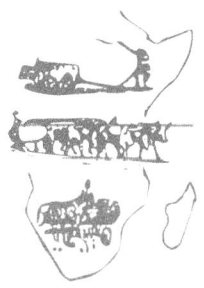

Animal issues:
donkey use, cow traction and feeding

The challenges in using donkeys for work in Africa

by

R A Pearson[1], E Nengomasha[2] and R Krecek[3]

[1]*Centre for Tropical Veterinary Medicine, Easter Bush*
Roslin, Midlothian, Scotland, EH25 9RG, UK
[2]*Matopos Research Station, P Bag K5137, Bulawayo, Zimbabwe*
[3]*Department of Veterinary Tropical Diseases, Faculty of Veterinary Science*
P Bag X04, 0110 Onderstepoort, Republic of South Africa

Abstract

A succession of dry years in sub-Saharan Africa has seen an increase in the use of the donkeys for transport and cultivation, particularly in areas where draft cattle numbers have declined. The challenges facing users of donkeys are to make the best use of the resources they have available. The challenges to livestock researchers and extension officers are to provide information that will help the user optimise resource use. The present state of knowledge and the main issues which would benefit from further understanding are discussed under the topics of management (nutrition and health), working practices (implements and harnesses) and the promotion of the donkey. Some issues need a better understanding in order to advance eg, strategic feeding and disease control, while other issues can largely be overcome by better communication, training and exchange of knowledge.

Introduction

Despite the increase in mechanisation throughout the world, donkeys are still well deserving of the name 'beasts of burden'. They have an important role to play in transport of people and goods in arid and semi-arid areas and where roads are poor or non-existent. This is shown by the widespread use of donkeys in rural and urban areas in Africa, as well as parts of central America and Asia (Table 1). In West, South and East Africa donkeys are used to power activities associated with crop production as well as transport. A series of dry years in sub-Saharan Africa has seen a notable increase in the use of donkeys for tillage, as draft cattle numbers on small farms have declined. This has resulted in changing perceptions of the value of the donkey in many rural communities that rely on animal power for crop production. A good example of this is in Zimbabwe. The drought in 1991/92 resulted in the death of many cattle and

therefore a shortage of oxen at plowing time. Other than family labour, donkeys are the only alternative to oxen on many smallholder farms in this country. The price of a donkey in communal areas in southern Zimbabwe increased from about 80 $Zim in 1990 to 600 $Zim in 1995, and donkey theft became a severe problem in rural areas, especially near plowing time. A similar situation has been reported in South Africa, where demand for donkeys exceeds supply.

For donor agencies and development organisations the donkey represents an attractive 'package' for promotion. It is the cheapest form of farm power other than human labour, and therefore within reach of the 'poorest of the poor'; it is 'available' to women in cultures where men usually manage the draft animals and can alleviate the drudgery of women's household activities such as water and firewood carrying.

These two situations have resulted in an extraordinary increase in interest in the donkey since 1990 by both farmers and aid agencies as well as a realisation that little is understood of donkeys' requirements, potential for improvement and contribution to rural livelihoods. The challenge facing farmers is to make the best use of the resources that they have available, while the challenge to livestock researchers and extension officers is to provide information that will help farmers do this. In this paper the issues that need to be addressed to assist farmers and urban donkeys owners make the best use of their donkeys are discussed.

The performance and capabilities of donkeys

The size of a donkey is a limitation to the amount of work that it can do. Most adult African

Table 1: Countries with the most donkeys in 1993, in the world and in Africa

	World			Africa	
Country	Donkeys (thousands)	1990s trend	Country	Donkeys (thousands)	1990s trend
China	10 983	-	Ethiopia	5 200	n
Ethiopia	5 200	n	Egypt	1 550	n
Pakistan	3 775	+	Nigeria	1 000	+
Mexico	3 190	+	Morocco	946	n
Iran	1 900	+	Sudan	670	-
India	1 550	n	Mali	610	+
Egypt	1 550	n	Niger	462	+
Brazil	1 364	+	Burkina Faso	436	+
Afghanistan	1 180	n	Senegal	364	+'
Nigeria	1 000	+	Somalia	356	n
Morocco	946	n	Algeria	340	n
Turkey	895	-	Chad	271	n

Key: + increasing, - decreasing, n no change
Source: FAO, 1994

donkeys fall in the weight range of 90–210 kg, which is less than the live weight of most cattle used for draft work (Pearson and Ouassat, 1996; Nengomasha, Jele and Pearson, 1995). However, if donkeys are well managed they can do many of the tasks undertaken by oxen, provided that they are teamed in sufficient numbers to provide the necessary draft force required to complete the task. Studies in Zimbabwe have shown that well-fed, well-trained donkeys teamed in fours are capable of sustaining a combined draft force of over 1 kN for a 4-hour working period. This power output is sufficient to plow relatively deep soil with a mouldboard plow, as well as complete most other agricultural tasks associated with crop production, in an acceptable time. However, animals are not always in such good condition, nor is it always possible to use a team of four animals. These problems can be alleviated by improving the management of the animals to improve the power supply or by reducing the demand for power by modifying the implements or tillage practices.

Management of donkeys

Although donkeys are known to survive with little management, their body condition may fluctuate during the year as feed supply fluctuates. For example in Zimbabwe body condition of donkeys falls in the late dry season when forage becomes scarce (Nengomasha et al, 1995). Seasonal patterns of disease incidence and morbidity have also been noted in some areas in Africa (e.g. parasitic diseases associated with rainfall patterns; Panday, Khallaayoune, Ouhelli, and Dakkak, 1994), suggesting that some form of seasonal disease management may also be beneficial.

Nutrition

Proper feeding of adult donkeys can enable them to resist disease challenge better, live longer, and have a higher rate of reproduction to provide replacement animals. In the growing donkey adequate feeding will allow it to reach its maximum growth potential, thus providing

Table 2: The energy costs of various activitites associated with work in donkeys

Observation	Location	Site	Energy cost	Source
Walking (J/m/kg liveweight)	Tunisia	sandy/gravel tracks	1.37	Dijkman, pers.comm
	Niger	laterite tracks	1.43	Pearson, 1994
	USA	gravel roads	0.98	Yousef, Dill and Freeland, 1972
	UK	treadmill	0.97	Dijkman, 1992
Carrying (J/m/kg carried)	Tunisia	sandy/gravel tracks	1.8–2.3	Dijkman, pers. comm.
	UK	treadmill	1.1	Dijkman, 1992
Pulling (J/m/kg pulled)	UK	treadmill	31.2	Pearson, 1994
	UK	treadmill	26.5	Dijkman, 1992

maximum working capacity. Some donkey users in Southern Africa, recognising the value of their animals, have been feeding supplements throughout the dry season; however, whether or not this expense returns value in increased work output or improved crop production has not been assessed. The challenge to scientists is to develop nutritionally sound recommendations on feeding strategies for donkeys that enable donkey users to make economically viable decisions regarding feed inputs and management. These decisions can be very different for different users. For example the urban donkey users working their animals daily are more likely to have to rely on purchased fodder, and have money to purchase it, than the rural users, who are more likely to have grazing land and time available to use it.

Energy requirements of donkeys for work have been determined under laboratory conditions (eg Dijkman, 1992) and under experimental conditions in the field (Yousef, Dill and Freeland, 1972). The data is summarised in Table 2. However, little information is available on donkeys' daily nutrient needs for work on farms in tropical conditions. Energy requirements can be estimated from the energy costs of the different activities associated with work (Lawrence and Stibbards, 1990) and a knowledge of the amount of work-done and distance travelled (eg Table 3). Until recently little

data has been available on the daily work done by donkeys in soil preparation and weeding, but data on distances covered and loads carried during transport operations are more readily available. For example Wilson (1991) reported that donkeys used in the salt trade in Ethiopia carried 50 kg salt loads distances of 160 km climbing 3000 m over four days. Loads of 60–80 kg are commonly carried for periods of 3-4 hours by donkeys carrying goods to markets in Mexico (A S de Aluja, pers. comm.).

Donkeys seem able to digest high fibre forage diets better than do horses, while maintaining similar or higher intakes of the feed. For example donkeys eating 15.3 g dry matter (DM)/kg live weight of an oat straw diet showed an apparent digestibility coefficient of organic matter (OM), neutral detergent fibre (NDF) and acid detergent fibre (ADF) of 0.52, 0.49 and 0.44 respectively. Horses eating 13.7 g DM/kg live weight of the same diet showed apparent digestibility coefficients of OM, NDF and ADF of 0.48, 0.41 and 0.37 (Cuddeford, Pearson, Archibald and Muirhead, 1995). Donkeys also seem able to compensate very accurately for the water deficit when drinking following a period of water deprivation (Yousef, Dill and Mayes, 1970). The mechanisms behind these two observations are not understood, but may account for the donkey's seemingly good body condition despite eating poor

Table 3: Three examples of the energy requirements for work for a donkey

Carrying a load over level ground on dirt tracks

Liveweight of donkey	120 kg
Distance travelled	15 km
Load carried	40 kg
Energy cost of walking (1.40 J/m/kg)[1]	2520 kJ
Energy cost of carrying (2.3 J/m/kg carried)[1]	1380 kJ
Total net energy of work	3900 kJ
Proportion of total energy cost of work used in walking	65%
Total net energy cost of work as a proportion of maintenance requirement	**0.31**

Plowing a field for 2.6 hours at an average draft force of 730N with a team of four[2]

Liveweight of donkey	120 kg
Distance travelled	5.5 km
Work done per donkey	1004 kJ
Efficiency of pulling	0.25
Energy cost of walking (2.0 J/m/kg)	1380 kJ
Energy cost of doing work pulling	4016 kJ
Total net energy cost of work	5336 kJ
Proportion of total energy cost of work used in walking	25%
Total net energy cost of work as a proportion of maintenance requirement	**0.42**

Carting a load over level ground at an average draft force of 140 N on laterite roads[3]

Live weight of donkey	120 kg
Distance travelled	15 km
Work done	2100 kJ
Efficiency of pulling	0.35
Energy cost of walking	2520 kJ
Energy cost of doing work pulling	6000 kJ
Total energy cost of work	8520 kJ
Proportion of total energy cost of work used in walking	30%
Total net energy cost of work as a proportion of maintenance requirement	**0.67**

[1] *Data from Dijkman (personal communication).* [2] *Data from Hagmann and Prasad (1994)*
[3] *Data from Slingerland (1989)*

quality feeds. Effects of frequency of watering on intake and digestibility of feeds and also on work output have not been determined under tropical conditions. Comparative studies of donkeys and ponies at Cornell University, USA showed that donkeys reduced food intake to a lesser degree that ponies when subjected to water deprivation for 36 hours (Mueller and Houpt, 1991). The suitability of different supplements as complements to forages and grazing, and the best time to feed in relation to work in tropical areas, have not been investigated. These are just some of the issues that need to be resolved so that recommendations on feeding systems for donkeys in the tropics can be made.

Health

The economic impact of disease on productivity of ruminants has been determined for some of the major diseases that affect livestock in the tropics. In 1984 in Africa for example, trypanosomosis caused an estimated annual loss of US$ five billion (excluding milk and hide losses) and theileriosis killed approximately three million cattle over the year (Murray and Gray, 1984). Chronic subclinical parasitism, usually accompanied by acute viral and bacterial diseases, may be the most important economic burden (FAO, 1991; Hansen, 1996). Virtually no information is available on the economic impact of any donkey diseases. This lack of information hinders development of policy decisions on disease control and prevention. The acquisition of this type of data provides a challenge to both economists and veterinarians.

Donkey disease agents

Donkeys harbour myriad infectious and parasitic agents, not all of which have been thoroughly investigated in this animal. Identification and treatment is often taken from knowledge of the disease in the horse. However, susceptibility or resistance to the effects of disease agents are not necessarily the same in the donkey as in the horse. The trypanosomes are some of the most important protozoal organisms affecting the well-being of livestock in the tropics. Horses do not survive long in the presence of infected tsetse flies, whereas donkeys are more tolerant of tsetse-transmitted trypanosomes and frequently appear to thrive in lightly infested habitats (Connor, 1994). Nevertheless Trypanosoma brucei causes acute disease in donkeys, as well as being a serious pathogen of horses. Donkeys, together with horses, mules and zebras, are susceptible to both the protozoan organisms Babesia equi and Babesia caballi (De Waal and van Heerden, 1994, Sahibi and Bakkoury, 1994). The South African vector the red-legged tick Rhipicephalus evertsi vertsi, is often observed on donkeys (De Waal and van Heerden, 1994).

Donkeys appear less susceptible to African horse sickness than are horses (Coetzer and Erasmus, 1994). Local indigenous animals usually undergo clinically inapparent or subacute infections, recover and are not normally vaccinated. However, introduced and imported exotic animals develop overt sign of the disease and if not vaccinated can die. Donkeys in the Middle East for example appear more susceptible to the disease than Southern African donkeys living in endemic areas (Alexander, 1948). Sometimes this is overlooked when donkeys are moved from one area to another and higher levels of mortality can result when animals encounter an unfamiliar serotype (Walton and Osburn, 1991). This should not be forgotten when development projects are planned involving the translocation of donkeys from one African country to another. Other diseases worthy of mention that donkeys are susceptible to are equine herpes virus, equine influenza, rabies, horse pox, mange and glanders.

Studies have shown that donkeys are host to a wide diversity and high prevalence of helminth parasites, which can lead to disease when the animals are underfed or overworked. Helminths reported from donkeys in Southern Africa include nematodes (roundworms), cestodes (tapeworms) and trematodes (flukes). Nematodes predominate and represent five taxonomic families and more than 40 different species (Theiler, 1923; Malan, Reinceke, and Scialdo-Krecek, 1982; Svendsen, 1989; Pandey and Eysker, 1990). Six species of the botfly also inhabit the gastro-intestinal tract of the donkey (I J Horak, pers comm).

The clinical effects of helminth disease on donkeys are less well known than in horses. This situation is exacerbated by the shortcomings of the diagnostic methods used to identify internal parasites. Worm egg count (eggs per gram, epg) is the main diagnostic tool used to identify internal parasites; this technique, however, does not necessarily provide an accurate assessment of the infection in the live animal. Despite this it is possible to make some comparisons between the donkey and horse. If a horse has a nematode egg count of 300–600 epg, treatment with an effective antiparasitic agent would be recommended (Krecek, Guthrie, van Nieuwenhuizen and Booth, 1994), but in African donkeys egg counts well above this (1600–2000 epg) are not uncommon

(Vercruysse, Harris, Kaboret, Pangui and Gibson, 1986; D Wells, pers. comm). The challenge to the veterinarians is to establish whether this difference reflects the host's response to the parasites, the high fecundity (egg-laying ability) of some worm species or simply an ability of the donkey to tolerate higher worm burdens without ill-effect than can the horse. The latter explanation would mean that the donkey resembles the zebra, which seems to have a high tolerance for internal parasites. Large numbers of blood-sucking worms are present in all zebra species in southern Africa and inhabit many of their organs including the liver. Whether this indicates a high tolerance to these worms or some symbiotic relationship that may exist in the zebra between the host and the parasites remains to be investigated (Scialdo-Krecek, 1983; Krecek, Malan, Reinecke and De Vos, 1987).

Further research to identify differences between donkeys and the other equids in their tolerance of common diseases would seem to be appropriate and would help in the development of low cost treatments.

Treatment and control of disease in donkeys

The production of disease control methods and disease treatments that are acceptable to the donkey owner presents a major challenge to the animal health specialist. Identification of the constraints to disease control or prevention in donkey owning communities can help in the development of acceptable methods. The role of good husbandry in disease control is frequently underestimated (Connor, 1994), although it may offer the donkey owner a financially more attractive alternative to expensive drug treatment at a later stage. Nutrition can play an important role and the provision of supplementary feed, especially during the dry season, can help in resisting or mitigating disease. In certain cases well-fed animals seem better able to resist the effects of trypanosome infections (Connor, 1994), and many animals attain an equilibrium with the parasites.

Trypanocides, one of which is toxic to donkeys, anthelmintic drugs and vaccines are often not available or are too costly to purchase, even when the need is recognised by the donkey owner. Simple local remedies warrant investigation, along with other simple management techniques to reduce contact between the animal and the vector or disease agent. A few examples are already available and the challenge to the veterinarian, livestock scientist and extension agent are to identify other practices acceptable to the farmers.

The resistance of worm eggs to many anthelmintics has developed at an alarming rate in the Southern Hemisphere (Waller, 1993), necessitating the investigation other methods of control. Removal of faeces from pasture to reduce contamination of grazing resulted in a decrease in worm egg counts, and less clinical disease in horses compared to those treated regularly with anthelmintics (Herd, 1986). Similar successes have occurred on Thoroughbred farms in South Africa, where faeces removal from yearling's pastures every 2–3 months helped to reduce worm egg counts (Krecek, unpublished data). This is a simple, low cost method of reducing worm egg counts which could be an affordable, effective and therefore acceptable practice to control parasite burdens in donkeys. Keeping animals in fly-proof accommodation at least part of the day when vector flies are prevalent can also reduce the incidence of disease.

Working practices

Implements and tillage practices

Most of the implements used in crop production in Africa have been designed for use with oxen. While these implements many prove satisfactory for a team of four donkeys in good condition, they are likely to be too heavy for smaller teams or single animals. Increasing the number of animals in a team is not always the answer. Teams of more than four donkeys are difficult to use, unless fields are large, and work output per animal drops as the number of animals in a team increases (eg Karim-Sesay, 1993).

A major challenge to the agricultural engineers is to identify, design or modify implements that can be used effectively by donkeys in primary cultivation. This would enable the farmer to reduce his reliance on oxen for these tasks. The implement has to be technically acceptable by and affordable to farmers in order to be adopted by them. Secondary cultivation, weeding and carting require lower draft forces than plowing and the low live weight of the donkey is less of a constraint. This is also the case on light sandy soils where conventional plowing is often unnecessary for crop establishment. The development of alternative tillage practices that require less power than conventional plowing in which the donkey can be used provide a further challenge to engineers and soil scientists.

Harnesses

Designs for suitable harnesses for donkeys are available, and consist of two types: collars and breastbands. The problem of harnessing is therefore not a technical one, but more one of acceptance, education and dissemination. This requires the expertise of the extension officer rather than the scientist. The challenge to the extension officer is to develop techniques to teach the farmers the benefits of using harnesses that are comfortable to the animal, and then encourage them to use these harnesses. The choice of harness depends on availability and cost in an area. In Africa this generally means a breastband harness, although there are a few areas where collar harnesses are available. Collar harnesses, while excellent for the donkey, are expensive for the farmers and require some training of artisans in their construction. Breastband harnesses are a cheaper alternative to collar harnesses, but these can cause bad sores if they are not made correctly or fitted correctly. The bands should be broad and of a material which will not readily rub the donkey, although padding of the harness with softer less abrasive material can overcome the latter problem. Fastenings should not be of a type that will rub the animal and adjustable straps over the back enable the size of the harness to be changed to fit different sizes of donkey. This is important if it is to be used on different animals.

Manufacturers of harnesses often make them too thin, to save on material and without adjustments or padding, in order to keep the price down. Purchase of such harnesses can prove a false economy to donkey users as they usually cause sores on the animals, thus reducing work performance. It is these messages that the extension officers need to get across. In areas where cattle deaths brought about by drought has meant that many farmers have to use donkeys rather than cattle for land preparation, farmers very often use the yokes they have for the cattle on the donkeys, despite such use being illegal in some areas. This can be the result of ignorance and/or

unwillingness to purchase donkeys harness. Again education, of donkey users as well as of artisans producing donkey harnessing, is required, rather than technical developments.

Load carrying and pack saddles

Most of the comments regarding harnesses apply equally well to the use of pack saddles. The best saddles are those that do not cause sores because sores lead to days off work. These are not necessarily expensive or complex and often make use of simple materials. The long term solution to saddle sores on pack animals lies in education and encouragement of farmers to adopt preventative measures rather than waiting for a problem to develop.

Donkey promotion

The donkey has a generally poor image in many cultures, partly due to the fact that work is often the only productive output of a donkey. Although pastoralists in northern Kenya make use of donkey blood and milk (Twerda, 1994), milk, meat and manure of donkeys are only used in a few cultures. Hence until recently donkeys in most areas had a low monetary value compared to other stock and therefore did not attract the attention of scientists, extension officers and farmers. Similarly they are not included in government policy documents on agriculture or rural development, although in many countries, particularly in Africa, they can make a substantial contribution to the economy of the country. Ethiopia is usually quoted as the classic example of this hidden economy. Wilson (1991) estimated that the annual return on a donkey used for carrying firewood to Mekele in Ethiopia, exclusive of the drovers labour was 1,388%.

People keeping donkeys tend to recognise their worth, even if ignorance, necessity or poverty mean that the animals are worked in poor condition and ill health. Those who do not own donkeys, however, whether they be farmers or government officials, often place considerable emphasis on the frequently unsubstantiated negative aspects of donkey ownership and use. Their low productivity compared with cattle, since meat and milk are seldom marketed, and their small size for draft work are often emphasised by agricultural officials. Some of the sayings attached to donkeys do not encourage wider use or better management of the animals. For example

Jones (1991) describes some of the negative sayings about donkeys that have arisen in Zimbabwe, partly as an indirect result of the introduction of legislation to control movement of animals in towns. Such sayings include the following: 'donkeys are dangerous, and kick people to death', 'donkeys destroy trees and the environment generally', 'donkeys are inedible except by lions'. Overcoming these prejudices offers a challenge to all those involved with donkeys.

Conclusions

Many challenges face those wishing to use donkeys for work. A better understanding of the issues surrounding working donkeys would help farmers to meet these challenges. For example the development of recommendations on feeding would benefit from a knowledge of what feeds are available to farmers, how farmers use these feedstuffs, the effects of work on the animal itself, and the effect of supplementary feeding on animal health, longevity and work output. Similarly, recommendations on working practices would benefit from a knowledge of the effects of donkey-driven tillage on timeliness of cultivation, weed cover and crop yield.

Other challenges in using donkeys for work include communication and transfer of knowledge to overcome ignorance and increase awareness of better practices. Acceptable messages are needed to prevent harness sores or lameness and to dispel the myths associated with donkey keeping.

Healthy animals, free from injuries associated with work, will live longer and be more productive. Providing information on health, management, and nutrition, and increasing the owners' awareness of the benefits of a healthy donkey, is a challenge to all. If the price of a donkey continues to rise in sub-Saharan Africa this task should become easier. The challenge may then become one of how to overcome the increasing incidence of donkey theft, or to produce larger donkeys rather than how to improve the management and use of those currently available. The challenges involved in donkey use are not fixed, but will change according to changes in attitude, prosperity and the needs of the donkey owning communities.

Acknowledgements

The authors are grateful to R J Connor, G R Scott and D Wells for their advice during preparation of the paper.

References

Alexander R A, 1948. The 1994 epizootic of horse sickness in the Middle East. *The Onderstepoort Journal of Veterinary Science and Animal Industry* 23, 77–92.

Coetzer J A W and Erasmus B J, 1994. African horse sickness. pp. 460–475 in: Coetzer J A W, Thomson G R, and Tustin R C (eds) *Infectious Diseases of Livestock with Special Reference to Southern Africa, Vol I.* Oxford University Press, Cape Town, South Africa.

Connor R J, 1994. African animal trypanosomiases. pp. 167–205 in: Coetzer J A W, Thomson G R, and Tustin R C (eds) *Infectious Diseases of Livestock with Special Reference to Southern Africa, Vol I.* Oxford University Press, Cape Town, South Africa.

Cuddeford D, Pearson R A, Archibald R F and Muirhead R H, 1995. Digestibility and gastro-intestinal transit time of diets containing different proportions of alfalfa and oat straw fed to Thoroughbreds, Shetland ponies, Highland ponies and donkeys. *Animal Science* 61, 407–418.

De Waal D T and van Heerden J, 1994. Equine babesiosis. pp. 293–304 in: Coetzer J A W, Thomson G R, and Tustin R C (eds) *Infectious Diseases of Livestock with Special Reference to Southern Africa, Vol I.* Oxford University Press, Cape Town, South Africa.

Dijkman J T, 1992. A note on the influence of negative gradients on the energy expenditure of donkeys walking, carrying and pulling loads. *Animal Production* 54, 153–156.

Food and Agriculture Organization, 1991. World production of animal protein and need for a new approach. Food and Agriculture Organization of the United Nations, Rome, AGA: AAP/75.

Food and Agriculture Organization, 1994. *Annual Production Yearbook* 1993, Vol 47. FAO, Rome, Italy.

Hagmann J and Prasad V L, 1994. The use of donkeys and their draught performance in smallholder farming in Zimbabwe. Project Research Report II. Conservation Tillage for Sustainable Crop Production Systems GTZ/AGRITEX/IAE/P.O. Box 415, Borrowdale, Harare, Zimbabwe.

Hansen J W, 1996. Tropical livestock industry: parasite control, food security and the environment. *Veterinary Parasitology* (in press)

Herd R P, 1986. Epidemiology and control of equine strongylosis at Newmarket. *Equine Veterinary Journal* 18, 447–452.

Jones P, 1991. Overcoming ignorance about donkeys in Zimbabwe a case study. pp. 311–318 in: Fielding D and Pearson R A (eds), *Donkeys, mules and horses in tropical agricultural development.* Proceedings of colloquium held 3-6 September 1990, Edinburgh, UK. Centre for Tropical Veterinary Medicine, University of Edinburgh, UK. 336p. ISBN 0907146066.

Karim-Sesay J A, 1993. The effects of harness and implement design on performance of draught animals: the case of the donkey in Botswana. pp. 123-124 in: O'Neill D H and Hendriksen G (eds), *Human and Draught Animal Power in Crop Production .*

Proceedings of the Silsoe Research Institute/CEC/FAO Workshop 18-22 January 1993, Harare. Printed by FAO, Rome.

Krecek R C, Malan F S, Reinecke R K and De Vos V, 1987. Studies on the nematode parasites from Burchell's Zebra in South Africa. *Journal of Wildlife Diseases* **23**, 404–411.

Krecek R C, Guthrie A J, van Nieuwenhuizen L C and Booth L M, 1994. A comparison of the effect of conventional and selective antiparasitic treatments of nematode parasites of horses from two management schemes. *Journal of the South African Veterinary Association* **65**, 97–100.

Lawrence P R and Stibbards R J, 1990. The energy costs of walking, carrying and pulling loads on flat surfaces by Brahman cattle and swamp buffalo. *Animal Production* **50**, 29–39.

Malan F S, Reinecke R K and Scialdo-Krecek R C, 1982. Anthelminthic efficacy of fenbendazole in donkeys assessed by the modified non-parametric method. *Journal of the South African Veterinary Association* **53**, 185–188.

Mueller P J and Houpt K A, 1991. A comparison of the response of donkeys (*Equuus asinus*) and ponies (*Equus caballus*) to 36 hours of water deprivation. pp. 86-95 in: Fielding D and Pearson R A (eds), *Donkeys, mules and horses in tropical agricultural development*. Proceedings of colloquium held 3-6 September 1990, Edinburgh, UK. Centre for Tropical Veterinary Medicine, University of Edinburgh, UK. 336p. ISBN 0907146066.

Murray M and Gray A R, 1984. The current situation on animal trypanosomiasis in Africa. *Preventive Veterinary Medicine* **2**, 23–30.

Nengomasha E , Jele N and Pearson R A, 1995. Phenotypic characteristics of working donkeys in Western Zimbabwe - present conference.

Pandey V S and Eysker M, 1990. Internal parasites of donkeys from the highveld of Zimbabwe. *Zimbabwe Veterinary Journal*, 21, 27-32.

Pandey V S, Khallaayoune K, Ouhelli H and Dakkak A, 1994. Parasites of donkeys in Africa. pp. 35–44 in: Bakkoury M and Prentis R A (eds), *Working equines*. Proceedings of second international colloquium held 20-22 April 1994, Rabat, Morocco. Actes Editions, Institut Agronomique et Vétérinaire Hassan ll, Rabat, Morocco. 412p. ISBN 9981-801-11-9

Pearson R A, 1994. Energy requirements of working equids. In: *Final Report 'Digestion et nutrition de l'ane*. Contract TSZ-0268F (EDB), INRA, Dijon.

Pearson R A and Ouassat M, 1996. Estimation of live weight and a body condition scoring system for working donkeys in Morocco. *Veterinary Record* **138**, 229–233.

Sahibi H and Bakkoury M, 1994. Equine babesiosis in Morocco: prevalence and equine ticks. pp. 65-74 in Bakkoury M and Prentis R A (eds), *Working equines*. Proceedings of second international colloquium held 20-22 April 1994, Rabat, Morocco. Actes Editions, Institut Agronomique et Vétérinaire Hassan ll, Rabat, Morocco. 412p. ISBN 9981-801-11-9

Scialdo-Krecke R C, 1983. Studies on the parasites of zebras. I. Nematodes of the Burchell's zebra in the Kruger National Park. *Onderstepoort Journal of Veterinary Research* **50**, 111–114.

Slingerland M A, 1989. Selection of animals for work in sub-Saharan Africa:- research at the ICRISAT Sahelian Centre. pp. 203–210 in Hoffman, D, Nari J and Petheram R J (eds), *Draught animals in rural development*. Proceedings of an international research symposium held at Cipanas, Indonesia 3-7 July 1989. ACIAR Proceedings Series No. 27. Australian Centre for International Agricultural Research, Canberra, Australia. 345p. ISBN 1 86320 003 7

Svendsen E D, 1989. The Professional Handbook of the Donkey. The Donkey Sanctuary; Sidmouth, Devon, UK. 284 pp.

Theile G, 1923. The strongylids and other nematodes parasitic in the intestinal tract of South African equines. *Report of Veterinary Research, Union of South Africa* **9-10**, 601–773.

Twerda M B C, 1994. The role of donkeys in Samburu and Turkana Society, MSc Thesis, Centre for Tropical Veterinary Medicine, University of Edinburgh, UK.

Vercruysse J, Harris E A, Kaboret Y Y, Pangui L J and Gibson D I, 1986. Gastro-intestinal helminths of donkeys in Burkina Faso. *Zeitschrift für Parasitenkunde*, 72, 821-825.

Waller P J, 1993. Workshop summary: sustainable production systems. *Veterinary Parasitology* **54**, 305–308.

Walton T E and Osburn B I (eds), 1991. Blue tongue, African Horse Sickness and related orbiviruses. *Proceedings of the Second International Symposium*. CRC Press, USA, 1042p.

Wilson T R, 1991. Equines in Ethiopia. pp. 33–47 in: Fielding D and Pearson R A (eds), *Donkeys, mules and horses in tropical agricultural development*. Proceedings of colloquium held 3-6 September 1990, Edinburgh, UK. Centre for Tropical Veterinary Medicine, University of Edinburgh, UK. 336p. ISBN 0907146066

Yousef M K, Dill D B and Mayers M G, 1970. Shifts in body fluids during dehydration in the burro *Equus asinus*. *Journal of Applied Physiology* **29**, 345–349.

Yousef M K, Dill D B and Freeland D V, 1972. Energetic cost of grade walking in man and burro, *Equus asinus*: desert and mountain. *Journal of Applied Physiology* **33**, 337–340.

The potential of cow traction in the East African highlands

E Zerbini[1], Alemu Gebre Wold[2] and B I Shapiro[1]

1) *International Livestock Research Institute (ILRI), PO Box 5689, Addis Ababa, Ethiopia*
2) *Institute of Agricultural Research (IAR), PO Box 2003, Addis Ababa, Ethiopia*

Abstract

To test the feasibility of the use of dairy cows for draft the International Livestock Research Institute (ILRI) and the Insititute of Agricultural Research (IAR) in Ethiopia have studied a herd of F1 Friesien x Boran crossbred cows on-station over three years together with on-farm experiments and economic analyses.

The total number of days in milk of working cows was similar to that of non-working cows. However, days in milk of working supplemented cows were 14 to 39% greater than non-working or working non-supplemented cows over two years. Milk yield of non-supplemented cows, whether working or not, was approximately half that of supplemented cows. Even though, over a period of two years, milk production of working cows was not significantly different from that of non-working cows, working non-supplemented cows had the lowest milk yield among all groups. This indicates that work with inadequate feeding would not be a feasible option for a production system involving the use of lactating cows for draft. On the other hand, total milk production of working supplemented cows over three years was only 10% lower than that of non-working supplemented cows.

There was a dramatic decrease in percentage of cows showing oestrus and in conception rate when work was applied to non-supplemented cows. Once pregnancy was established there was no effect of work on maintenance of pregnancy.

Over a period of two years the productivity index of supplemented cows was greater than that of non-supplemented cows (0.38 and 0.24, respectively), but it was similar between working and non-working cows (0.35 and 0.33, respectively). Work output more than compensated for the small decline in milk production and number of calvings and greater daily metabolic intake of working supplemented compared to non-working supplemented cows. The incremental benefit/cost ratio of having supplemented working cows over the traditional system of local cows and oxen is about 3.5 and the incremental internal rate of return is 78%.

Introduction

Due to increasing population and livestock pressure on the land, farmers in many developing countries may not be able to continue maintaining draft oxen for work purposes. The use of dairy cows for traction could benefit total farm output and incomes through increased milk production, while alleviating the need to feed draft oxen year-round and to maintain a follower herd to supply replacement oxen (Gryseels and Goe, 1984; Gryseels and Anderson, 1985; Matthewman, 1987; Barton, 1991). Besides contributing to a better utilisation of scarce feed resources, the use of dairy cows for draft would allow males to be fattened and sold younger, and could also lead to greater security of replacements. More productive animals on farm could result in a reduction of stocking rates and overgrazing, thus contributing to the establishment of a more productive and sustainable farming system.

The primary need of the working animal is to increase feed and metabolic energy intakes to meet energy requirements for work and avoid deleterious body weight losses. This becomes more critical in working cows requiring extra energy for lactation and reproduction, and where the main feed source is roughage.

A number of studies have reported no significant effect of work on feed intake in óxen (eg Soller, Reed and Butterworth, 1991; Pearson and Lawrence, 1992) and buffalo cows (Bamualim, Ffoulkes and Fletcher, 1987; Bakrie and Teleni, 1991). Other studies indicate an increased feed intake in working buffalo cows (Ffoulkes, 1986; Ffoulkes, Bamualim and Panggabean, 1987) and dairy cows (Gemeda et al, 1994). Furthermore, some authors have reported negative or no effect of work on digestion in buffalo and cattle, depending on the diet fed (Bamualim, Ffoulkes and Fletcher, 1987; Pearson, 1990; Soller et al, 1991; Pearson and Lawrence, 1992·), while others

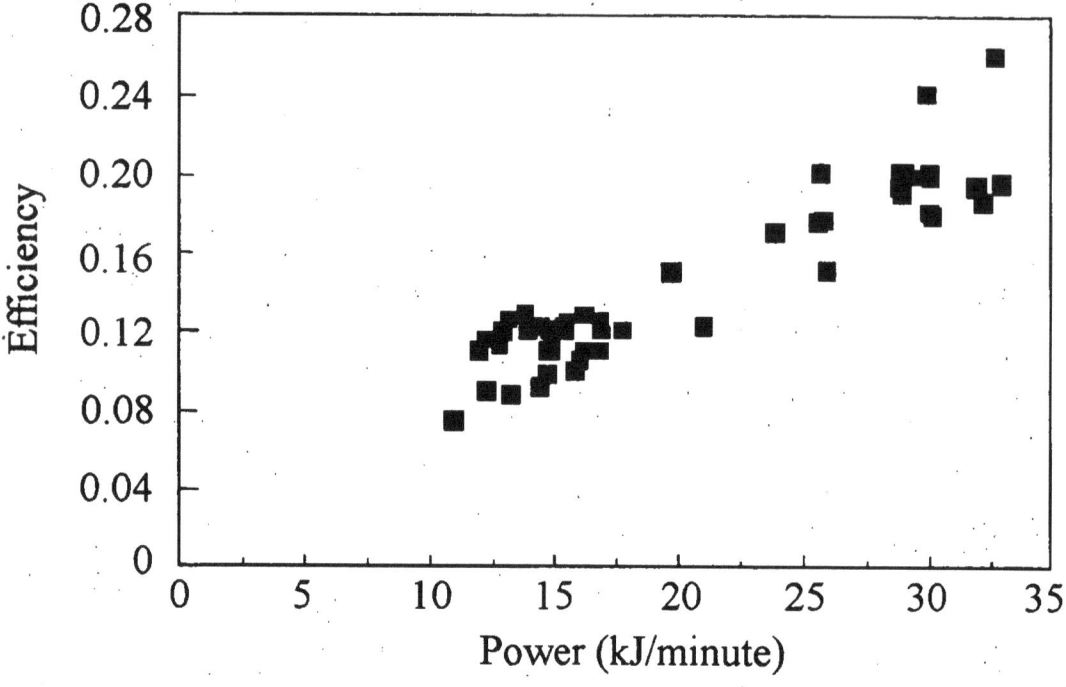

Figure 1: Work efficiency of the F1 Friesian x Boran cows used for draft

have shown a positive effect of work on digestibility (Ffoulkes, Bamualim and Panggabean, 1987; Pearson and Lawrence, 1992). How work could affect either rumen fermentation processes or digestion in the lower digestive tract, as well as other processes involved in intake regulation of roughage diets, is uncertain.

Production performance of cows is an important factor which determines whether cows are adopted for draft power. Working cows could perform at higher levels of efficiency than oxen, but only if nutrient inputs are adequate to meet their greater requirements and milk production and reproduction are kept at levels comparable to non-working cows (Mathers et al, 1985; Matthewman, Djikman and Zerbini, 1994). Energy deficits during the working season could result in body weight and body condition losses, thus affecting the production and reproduction efficiency of cows (Teleni and Hogan, 1989; Teleni et al, 1989; Zerbini et al, 1993a).

ILRI and IAR have researched different aspects of the use of dairy cows for draft work. This paper deals with work output, milk production and

reproduction of crossbred dairy cows used for draft. Information generated from ILRI and IAR research is used to elaborate the interplay of factors affecting work output, milk production, metabolism, physiological responses and the reproductive physiology and performance of dairy cows used for draft. The economic implications and the potential for adoption are also discussed. Highlights of an expert consultation on the transfer of technology for multipurpose cows in smallholder mixed farming systems organised by ILRI, FAO and ACIAR (Australian Centre for International Agricultural Research) in September 1995 in Addis Ababa are also reported.

Work output and efficiency of draft cows

Determination of the optimum work load that dairy cows can undertake is an essential component for successful adoption of cow traction technologies into smallholder mixed farming systems. Results of our investigation with crossbred cows in the Ethiopian Highlands (Zerbini et al, 1992) showed that dairy cows were able to work at a rate of about 500 W. This work rate, at a speed of 0.75 m/s, implies that the

Table 1: Cummulative work output (MJ) of the F1 crossbred cows used for draft over a period of two years

		Days	
Treatment	n	0-365	0-730
WNS	10	260	500
WS	10	270	508
se		6	12
F test		NS	NS

WNS - Working non-supplemented
WS - Working supplemented

Table 2: Average dry matter intake ($g/kg^{0.75}$) of the F1 crossbred cows over two years

	Days	
Treatment	0-365	0-730
NWNS	89	94
NWS	108	105
WNS	97	111
WS	121	117
se	4	4
Ftest		
Work	*	**
Supplement	**	***

NWNS - Non-working, non-supplemented
NWS - Non-working, supplemented
WNS - Working, non-supplemented
WS - Working, supplemented

sustainable horizontal draft force was roughly 670 N. This represented about 14% of mean body weight, in line with what would be expected (eg Barwell and Ayre, 1982). The work efficiency of cows increased from about 7% to 26% as the workload increased to its maximum (Figure 1). Cardio-respiratory measurements indicated that during work each additional heart beat transported approximately 72 ml of oxygen which is in turn equivalent to 1 kJ of additional mechanical work output (Zerbini et al, 1992). Over a period of two years, net work output of dairy cows averaged more than 200 MJ per cow per year, which was equivalent to or above that required by farmers for land cultivation (Table 1).

Lactation and reproduction performance
Dry matter intake

Dry matter intake was greater for working compared to non-working cows over a period of two years (Table 2). Working non-supplemented

cows increased roughage intake above that of non-working non-supplemented cows by 19% over two years. Similarly, working supplemented cows increased hay intake above that of non-working supplemented cows by 11%. Over a period of two years, dry matter intake of working cows increased by 15% compared to non working cows. Chemical composition of diet components is presented in Table 3.

Increased dry matter intake of working over non-working cows was also reported by Ffoulkes, Bamualim and Panggabean (1987) and is supported by findings of Zerbini et al (1994) who

Table 3: Mean chemical composition (g/kg) of diet components (n=14)

	Normal pasture hay	Concentrate
Dry matter	904	915
Organic matter	911	867
Nitrogen	9	45
Neutral-detergent fibre	696	364
Acid-detergent fibre	420	234

Table 4: Cumulative days in milk of the F1 crossbred cows over a period of two years

	Days	
Treatment	0-365	0-730
NWNS	291	501
NWS	302	579
WNS	280	476
WS	355	662
se	22	37
Ftest		
Work	NS	NS
Supplement	*	*

NWNS - Non-working, non-supplemented
NWS - Non-working, supplemented
WNS - Working, non-supplemented
WS - Working, supplemented

Table 5: Cumulative milk yield (kg) of the F1 crossbred cows over a period of two years

	Days	
Treatment	0-365	0-730
NWNS	849	1226
NWS	1792	3186
WNS	802	927
WS	1770	3044
se	152	219
Ftest		
Work	NS	NS
Supplement	***	***

NWNS - Non-working, non-supplemented
NWS - Non-working, supplemented
WNS - Working, non-supplemented
WS - Working, supplemented

reported that work increases the utilisation of feed energy. Even under conditions where adequate feed supplementation was not available to maintain body weight, such as for working non-supplemented cows, animals could still satisfactorily perform work by drawing on body reserves and increasing dry matter intake. However, Zerbini et al (1994) indicated that if such a situation exists for as long as one year, cows could loose more than 15% of their calving body weight and reduce milk production by more than 50% compared to working supplemented cows.

Days in milk, milk yield and completed lactations

The total number of days in milk of working cows was similar to that of non-working cows (Table 4). However, days in milk of working supplemented cows were 14 to 39% greater than the other treatment groups over two years, respectively. Days in milk were greater for supplemented compared to non-supplemented cows (Table 5). Milk yield of non-supplemented cows, whether working or not, was approximately half that of supplemented cows. In addition, in year two, milk production of non-supplemented

cows was only 30% that of year one, while in supplemented cows it was still 75% of that in year one.

In another study, Matthewman (1989) reported that cows using approximately 12 MJ metabolic energy per day for walking, reduced milk

Table 6: Number of completed lactations by the F1 crossbred cows

		Days	
Treatment	n	0-365	0-730
NWNS	10	6	13
NWS	10	7	16
WNS	10	6	11
WS	10	3	11
All	40	22	51

NWNS - Non-working, non-supplemented
NWS - Non-working, supplemented
WNS - Working, non-supplemented
WS - Working, supplemented

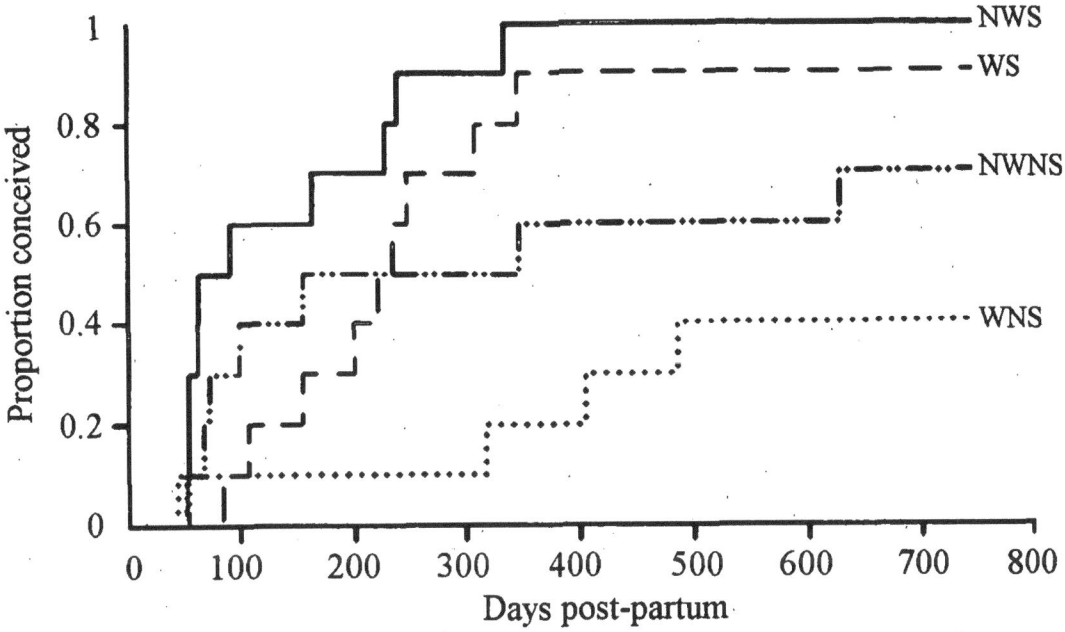

Figure 2: Distribution function of interval to conception after calving of the F1 crossbred cows.
NWS - Non-working supplemented; WS - Working supplemented;
NWNS - non-working non-supplemented; WNS - Working non-supplemented
Source: adapted from Zerbini et al, 1993a

production between 7 and 14% depending on diet fed. Also, Barton (1991) reported lower milk production and reproduction in draft cows in Bangladesh. On the other hand, on-farm trials in the Ethiopian Highlands, indicated that the effect of work on lactation of crossbred cows used for draft was minimal when feed supply was adequate and work requirements were modest (Gryseels and Anderson, 1985; Agyemang et al, 1991a).

In our study, over a period of two and three years, lactations completed by working supplemented cows in two years were 31 and 25% lower than those of non-working supplemented cows, consistent with greater days in milk of working supplemented cows (Table 6).

Even though, over a period of two years, milk production of working cows was not significantly different from that of non-working cows, working non-supplemented cows had the lowest milk yield among all groups. This indicates that work with inadequate feeding would not be a feasible option for a production system involving the use of lactating cows for draft. On the other hand, total milk production of working supplemented cows

over three years was only 10% lower than that of non-working supplemented cows. Differences could be attributed mainly to the lower number of parturitions and lactations completed among cows in this group.

Resumption of postpartum oestrus and conception

Bamualim, Ffoulkes and Fletcher (1987) indicated that in buffalo cows, work might reduce reproductive performance. On the other hand, the study by Winsugroho and Situmorang (1989) suggested that work, *per se*, was not a major factor influencing ovarian activity if energy reserves were adequate. Feed supplementation of thin working buffaloes induced a return to normal ovarian activity. Reh and Host (1985) reported that fertility was 6 to 7% lower in working than in non-working cows. However, research conducted in India, with working and non-working Red-Sindhi cows, over two lactations, showed no significant differences in milk production and length of lactation (Reh and Host, 1985). Agyemang et al (1991a) reported that the reproductive and productive performances of draft

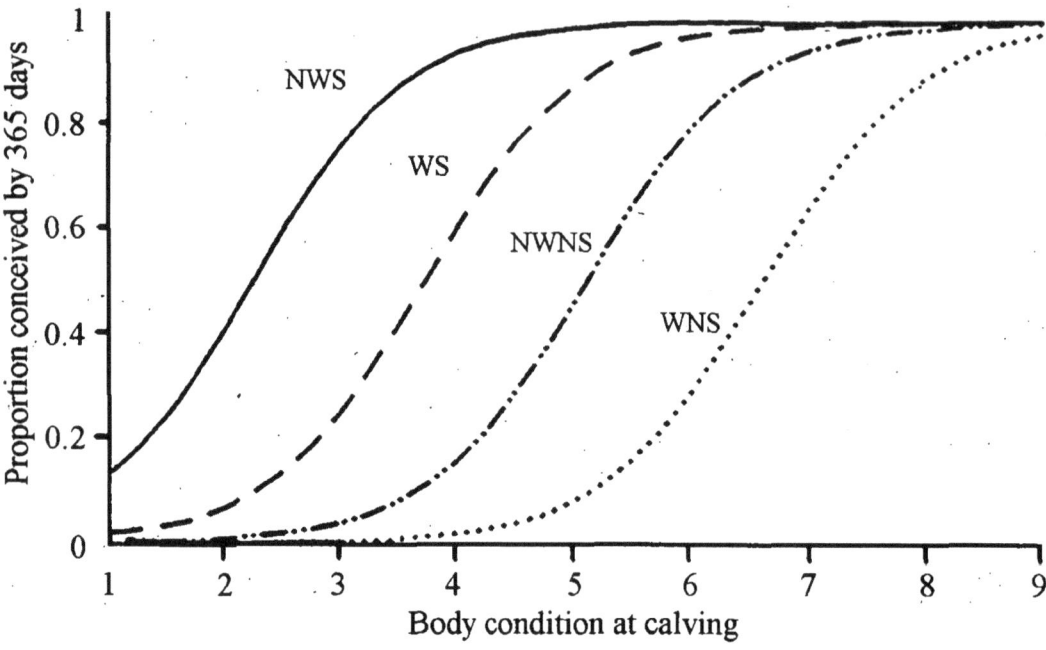

Figure 3: Predicted conception rate at 365 days vs initial body condition of the F1 crossbred cows.
NWS - Non-working supplemented; WS - Working supplemented;
NWNS - non-working non-supplemented; WNS - Working non-supplemented.
Source: adapted from Zerbini et al, 1993a

and non-draft cows were similar. However, the work done was lower than the amount required by a smallholder farmer.

Zerbini et al (1993a) reported that diet supplementation significantly decreased days to first oestrus and days to conception in non-working and working cows. When work treatment was superimposed on non-supplemented treatment, the effect on reproduction was deleterious. Differences in the first 200 days post-partum, in onset of oestrus and conception between treatment groups, seem to be related to work in the first post-partum period. However, if a 365-day period was considered, the differences are related to a greater extent to diet supplementation , suggesting a longer term effect of the supplementation than work (Figure 2). In supplemented cows, work significantly delayed days to conception. However, by 365 days post-partum, conception rate was similar for supplemented non working and supplemented working cows. For occurrence of first oestrus, the diet effect was considerably larger than the work effect (a probability factor of five versus a

probability factor of two). This was less pronounced for conception. This is contrary to results from other studies (Wells, Hopley and Holness, 1981) which indicated that supplementary feeding did not influence interval from calving to first ovulation, conception rate or interval from calving to conception. Body condition at calving significantly affected postpartum reproductive ability of non-working and working cows (Figure 3). This indicates that cows with greater body reserves at calving, and the ability to use these reserves during the post-partum period, can partly overcome the negative effect of dietary energy restrictions on oestrus onset and conception (Zerbini et al, 1993a).

The results of this study indicate a dramatic decrease in percentage of cows showing oestrus and in conception rate when work was applied to non-supplemented cows.

Postpartum anoestrous interval was extended in a larger proportion of working than in non-working cows. Work did not influence conception rate in supplemented cows, but had a substantial influence in non-supplemented cows. The significant delay

Table 7: Number of completed conceptions by the F1 crossbred cows

Treatment	n	Days	
		0-365	0-730
NWNS	10	6	8
NWS	10	10	17
WNS	10	2	4
WS	10	9	12
All	40	27	41

NWNS - Non-working, non-supplemented
NWS - Non-working, supplemented
WNS - Working, non-supplemented
WS - Working, supplemented

Table 8: Cumulative body weight changes (kg) of the F1 crossbred cows

Treatment	Days	
	0-365	0-730
NWNS	-60	-76
NWS	21	-10
WNS	-62	-86
WS	32	-23
se	13	10
Ftest		
Work	NS	NS
Supplement	***	***

NWNS - Non-working, non-supplemented
NWS - Non-working, supplemented
WNS - Working, non-supplemented
WS - Working, supplemented

of conception for supplemented working cows compared to supplemented non-working cows indicated that work output of cows might be associated with longer calving intervals and the economic trade-offs between the two factors should be examined in detail. Once pregnancy was

Table 9: Average condition scores of the F1 crossbred cows. Scoring was based on the ILCA system that uses values of 1–9, with higher numbers indicating better condition

Treatment	Days	
	0-365	0-730
NWNS	3.1	2.7
NWS	5.9	5.4
WNS	2.7	2.7
WS	5.8	4.9
se	0.3	0.3
Ftest		
Work	NS	NS
Supplement	***	***

NWNS - Non-working, non-supplemented
NWS - Non-working, supplemented
WNS - Working, non-supplemented
WS - Working, supplemented

established there was no effect of work on maintenance of pregnancy. A greater proportion of supplemented working cows cycled between 120 days and one year postpartum indicating that work applied soon after calving delayed, but did not suppress, oestrus and conception in subsequent resting or working periods (Zerbini et al 1993a, 1993b).

Conceptions over multiple lactations

Gemeda et al (1994) found that the number of working supplemented cows which conceived in year one was similar to that of non-working supplemented cows. However, over a period of two and three years, the number of working supplemented cows which conceived was 29 and 20% lower than those of non-working supplemented cows. In addition, over a period of one year, number of conceptions of non-supplemented cows could be reduced by 78% compared to those of adequately fed cows (Table 7).

Body weight losses have been reported to impair ovarian activity in female buffaloes and cows (Teleni et al, 1989; Agyemang et al, 1991b).

Further, over a period of two years, supplementary feeding reduced body weight loss of cows by 80% and was associated with a 59 and 63% increase in the number of conceptions and parturitions, respectively, compared to a non-supplemented diet (Table 8). In particular, supplementation of working cows reduced liveweight loss by 73% and doubled the number of conceptions and parturitions compared to working non-supplemented cows (Gemeda et al 1994). Body condition score followed a similar pattern to that of body weight change over the two-year period (Table 9). The probability of conception was not greater than 20% in cows with a body condition score lower than three (range 1–9) and with body weight losses greater than 15% from calving body weight (average of 412 kg).

In year one the number of calves born from working supplemented cows was 80% lower than that from non-working supplemented cows, despite the fact that the number of working supplemented cows which conceived was similar to that of non-working supplemented cows. This is consistent with the delay in conception after parturition reported for working-supplemented compared to non-working supplemented cows by Zerbini et al (1993a). Relatively fewer lactations and parturitions, and greater days in milk of working supplemented cows, over a period of three years, reflects the delayed conception in working supplemented compared to non-working supplemented cows. This is due to both a direct effect of work *per se* and to a deficit of energy yielding substrates, particularly during the working/lactating periods. Over a period of three years, diet was the main factor which affected reproduction of dairy cows used for draft work (Zerbini et al, 1994).

Recovery after work: long term effects

Even after extended periods of underfeeding, acyclic and anoestrus cows resumed ovarian cyclic activity in an average of 46 days and conceived in 75 days when fed about twice their maintenance energy requirements (Zerbini et al, 1994). The economic implications of long periods of low productivity or maintenance, in working and non-working cows, and the requirements for resuming reproductive activity need to be evaluated in detail especially for farming systems

Table 10: Average productivity index of the F1 crossbred cows over the two-year period

Treatment	Productivity index
NWNS	0.23
NWS	0.37
WNS	0.25
WS	0.39
se	0.03
Ftest	
Work	NS
Supplement	***

NWNS - Non-working, non-supplemented
NWS - Non-working, supplemented
WNS - Working, non-supplemented
WS - Working, supplemented

with large fluctuations in feed resources availability.

Productivity index (output/input)

Over a period of two years the productivity index of supplemented cows was greater than that of non-supplemented cows (0.38 and 0.24, respectively), but it was similar between working and non-working cows (0.35 and 0.33, respectively). While the productivity index of non-working supplemented and working supplemented cows remained relatively constant over three years, productivity index of non-working non-supplemented and working non-supplemented decreased by 21 and 34%, respectively (Table 10). This resulted mainly from reductions in milk yields and reproduction performance in non-supplemented cows.

The productivity index was used to describe the overall productivity of cows in each treatment group. Similar values of the productivity index for non-working supplemented and working supplemented cows over the total three-year period indicates that work output more than compensated for the small decline in milk production and number of calvings and greater daily metabolic intake of working supplemented compared to non-working supplemented cows. For on-farm situations working supplemented cows would

Table 11: On-farm values of lactation and reproductive parameters of crossbred cows worked an average of 26 days per year

| Status | First lactation | | First 2 years' milk (kg) | Calving interval (days) |
	Milk (kg)	Days		
No work	2252	410	2980	495
Work	1864	376	2620	525

provide the additional advantage of alleviating the need to maintain draft oxen year-round, could reduce stocking rates, and therefore could result in the allocation of more on-farm energy towards milk and meat production while maintaining draft power.

On-farm testing of cow traction in the Ethiopian highlands

The on-farm testing of cow traction technologies is designed to evaluate the effect of draft work and improved management on production and economic performance of crossbred dairy cows at the level of the smallholder farm. Pairs of crossbred cows (140 Friesien x Boran F_1) were purchased by selected farmers in 1993 and 1995 in the Holetta area. To ensure that even the poorest farmes can take advantage of these technologies, stratification of participating farmers into low, middle, and high income groups was carried out based on land and livestock holdings, livestock type, labour availability, total farm assets and location. Production and economic data of years 1993 and 1994 are presently being analysed and a whole-farm model based on the two years' data is being constructed.

During the first two years, milk production of working and non-working F1 crossbred cows on-farm was similar (2620 vs 2980 kg), ranging from 2010 to 3400 kg for working cows and from 2018 to 3907 kg for non-working cows. Calving intervals for working and non-working cows were 525 and 495 days, respectively. First lactation average milk yield and days in milk of working and non-working cows were 1864 and 2252 and 376 and 410 days, respectively. Average service per conception for working and non-working cows was 2.1 and 1.9, respectively. Over a period of two years cows worked an average of 26 days/year. These data are summarised in Table 11.

Economic implications and potential for adoption

Before going on-farm, the economic potential of the use of crossbred cows for milk production, meat, and traction was substantiated with the ILRI bio-economic herd model. Using the on-station results from three years, the production parameters and investment returns were simulated over a ten-year period (Shapiro, Zerbini and Geneda, 1994). The Incremental Internal Rate of Return (IIRR) of supplemented working cows over supplemented non-working cows was about 125%. This IIRR is very high because the incremental investment cost is very low while the benefits of work are large.

The simulation results show that the value of work more than compensated for the small reduction in milk production and longer calving interval found in working cows when supplementation took place to ensure adequate nutrition. The greater returns to investment in supplemented working crossbred cows were thus mainly a result of the higher value of the work output, in spite of the higher feed costs and lower offtake (milk, calves).

The effect over time of introducing crossbred dairy cows into a typical farm to replace the herd of local cattle for work and milk production were also simulated and compared to the traditional system of using the local cows for milk production and local oxen for traction. Again, the financial implications were investigated using incremental benefit/cost analysis. The incremental benefit/cost ratio of having supplemented working cows over the traditional system of local cows and oxen was about 3.5 and the IIRR 78%. The incremental benefit/cost ratio is high because of the very high productivity of the crossbred cows (5–7 times higher milk yield) relative to local cows.

Ethiopians are known as 'the people of the plow' because the rural culture has, for many centuries, been tied strongly to the use of oxen for draft work. Cultural attitudes could therefore also be obstacles to the adoption of crossbred cows for traction. Systematic, periodic anthropological surveys have been carried out to assess the evolution of attitudes of farmers participating in the on-farm trials, as well as neighbouring farmers who should be affected by the demonstration effect of having the on-farm trials being carried out in the vicinity of their farms. The first round of survey results showed that only 20% of those interviewed (on-farm trial participants, as well as non-participating neighbours) were unwilling to even consider using crossbred cows for traction. The other 80% said they would try using cows for traction and if it proved feasible they would use them regularly.

One conclusion of this first in-depth anthropological survey was that, in conjunction with the technical factors, careful study in the early part of the on-farm testing and adoption efforts needs to be made of the effects at the micro-level of socio-economic factors. Besides profitability, resource endowment has to be considered since it could hinder adoption by the poorer segments of the population, and access to supportive programmes such as credit and insurance could make a big impact on adoption. Such research would help policy makers to choose more effective policies and support programmes to promote widespread diffusion of new technologies.

The anthropological survey also concluded that while in the medium term the technical feasibility and the investment/cost ratio, as well as social factors will affect the acceptance of cow traction technologies, in the long run the diffusion of crossbred cows will depend on the effective extension of the results of the study. The environment for dairy development, including government policies and services, especially credit, veterinary and breeding services will also be critical.

Expert consultation

In sub-Saharan Africa, the greatest impact of using crossbred cows for milk and traction is expected in high potential highlands regions (ie Ethiopia, Kenya and the Great Lakes Region – Uganda, Malawi, Tanzania, Rwanda, Burundi). The highlands have the highest population densities, market accessibility, and stocking rates found in sub-Saharan Africa. Disease pressure is relatively low and low wages and small farm sizes make it unattractive to substitute tractors for animal traction. Now that the crossbred cow technologies have proven to have considerable potential both on-station and on-farm under Ethiopian Highland conditions, it was decided to start the process of adaptive research and on-farm testing in other East and Southern African countries with conducive agroecological and market conditions. An expert consultation on the transfer of technology for multipurpose cows including feed technologies for milk, meat and traction in smallholder mixed farming systems was therefore organised by ILRI, FAO and ACIAR in Addis Ababa in September 1995.

There is a clear need to devise better methods of ensuring the transfer and hence impact of promising new technologies. Technologies developed by international agricultural research centres need to be field-tested and adapted by national-level centres and then diffused by national extension organisations, as well as by NGOs. According to a recently completed ISNAR study of technology transfer a critical requirement for success is establishing better coordination of technology transfer efforts and formal linkage mechanisms involving international and national research centres and extension organisations. This process could be assisted and facilitated greatly by the support of international organisations such as FAO.

The expert consultation took the form of a workshop with participants drawn from East African research and extension institutions that have indicated a strong interest in the technology. Traction experts also came from Asia to share their considerable experience with cow traction technologies and discuss the possibility of extending the project to appropriate Asian countries. One objective of the workshop was to develop a means of coordination, as well as linkage and transfer mechanisms to ensure the transfer of the cow-traction technologies. Setting up institutional arrangements with development organisations such as FAO, could help to ensure successful technology transfer.

The principal objective of the workshop was, however, to develop a funding proposal to implement and study the transfer of the IAR/ILCA multipurpose crossbred cow technologies to East African smallholders. The proposal included a formal means of coordination as well as linkage and transfer mechanisms needed to ensure the transfer of the technologies.

The secondary purpose of the workshop was to learn from the experiences of experts from countries in Asia where female buffalo and cow traction already exists and to discuss the possibility of extending the project to Asia, if additional funding can be found. The consultation also benefitted greatly from participation of the ACIAR Draught Animal Power (DAP) project. This project included "multidisciplinary studies of draft animal power systems in Southeast Asia and feeding and management strategies for production and draft power in large ruminants."

Some initial efforts of information sharing between Asian and ILCA traction experts began during the DAP project, but these efforts were not continued, or formalised into institutionalised relations, and did not result in joint projects. The occasion of the expert consultation was then used to set up mechanisms for information exchange and collaboration between scientists and development organisations involved in research and transfer of multipurpose cattle and buffalo technologies in East Africa and Asia. This process represents a further development of DAP collaborative programs initiated by ACIAR and centres of the CGIAR system.

FAO played a principal role in the expert consultation and will play the lead role in the regional technology transfer project. Along with other international development agencies, FAO addresses livestock development across a broad spectrum. It provides technical advice and assistance to the agricultural community, to governments and funding agencies. It collects, analyses, and disseminates information as well as advises governments on policy and planning and

provides opportunities for government to meet and collectively discuss food and agricultural problems. Partnerships of this kind can provide essential critical mass, as well as state-of-the-art science, technology, and knowledge for the benefit of national research institutes.

The purpose of the proposed regional FAO/ILRI project is to promote proven IARC-developed technologies which are ready for diffusion to other countries. Included in the project will be national agricultural research centres and extension experts from countries in the region (more or less the same eco-region) where conditions conducive to the adoption of the technologies prevail. Each country project will be carried out with technical assistance from FAO experts, and with the ILRI scientists who developed the technologies acting as resource persons. An important innovation of these projects is the role of information sharing between regions. Although the projects are initially to be carried out in Eastern and Southern Africa, the possibilities of extending the project to Asia will be considered.

Conclusions

The results from this study indicate that draft work induced an increase in forage intake and digestibility in cows without decreasing solid phase transit time in the gut. Greater intake and digestibility could be related to increased retention time of the liquid phase and perhaps to increased gut volume. The attempt by working cows to increase intake to meet energy requirements even when fed relatively poor quality forage is important.

Over a period of three years, diet was the main factor which affected body weight and condition score, days in milk and milk production of crossbred cows, whether or not they were used for work. On the other hand, supplementation did not affect work output of cows. Work performed by supplemented cows had no adverse effects on lactation and reproduction. A similar productivity index for working and non-working supplemented cows over a period of three years indicates the potential of this technology to increase farm productivity and result in more efficient use of on-farm resources.

With appropriate feeding regimes dairy cows could be used for draft purposes without any detrimental effects on fertility, but calving intervals would be extended. Work, *per se,* does not influence postpartum ovarian activity when the energy reserve is adequate, but work does delay the interval from calving to conception in dairy cows.

Work increased the incidence of ovulation without oestrus and short luteal phases. However, these events did not influence pregnancy in subsequent normal oestrus periods.

Economic analysis based on on-station data showed that the greater returns to investment in supplemented working dairy cows compared to non-working cows or to the traditional system were mainly the result of the higher value of the work output, in spite of the higher feed costs and relatively lower offtake of milk and calves.

The results of the on-farm trials now being carried out are substantiating the potential of the cow-traction technologies to result in a more productive, more sustainable system than farmers' current practices. These results need to be carefully analysed within the whole farm context to ensure the fit of the cow-traction technologies into the farming systems, and to make sure that the technologies match farmer objectives and do not result in unsurmountable resource constraints.

There is a clear need to devise better methods of transferring promising new technologies to ensure impact. Technologies developed by international agricultural research centres and their national research colleagues need to be field-tested and adapted by other national research centres and then diffused by national extension organisations, as well as by NGOs. According to a recently completed ISNAR study of technology transfer, a critical requirement for successful transfer is establishing better coordination of technology transfer efforts and more formal linkage mechanisms involving international and national research centres, and extension organisations.

The technology transfer process could be greatly facilitated by the support of international organisations such as FAO. Substantial benefit could be derived from identifying proven technologies and encouraging adaptive testing and extension mechanisms and processes. Impact results would be improved, as well, if the process included an integrated study of the transfer process to provide feedback to the extension services and

guidance for future technology transfer projects. The proposed FAO/ILRI regional project aims at achieving these objectives.

In terms of understanding the production characteristics of working cows, there is a need to quantify the energy partition to different functions by working, lactating and breeding cows. The nutrient demand of the multipurpose cow is complex and the success of a nutritional management strategy will depend largely on the level of feed intake and, over short periods, on the level of body reserves (Egan and Dixon, 1993). The mechanism by which body reserves contribute to the energy expenditure of working cows is not clear. Future research priorities should include defining minimal nutrient requirements for pregnant and/or lactating working animals to allow for optimal reproductive performance.

To optimise the postpartum anoestrus period, draft dairy cows must regain weight during lactation and farmers must have management skills to integrate strategically physiological events such as pregnancy and lactation with draft work requirements.

On-farm comparison of working supplemented crossbred dairy cows used for draft with the traditional system is now underway.

References

Agyemang K, Astatke, A, Anderson F M, and Mariam W W, 1991a. Effects of work on reproductive and productive performance of crossbred dairy cows in the Ethiopian highlands. *Tropical Animal Health and Production* 23(4), 241–249.

Agyemang K, Little D A, Bath M L and Dwinger R H, 1991b. Effect of postpartum body weight changes on subsequent reproductive performance in N'Dama cattle maintained under traditional husbandry systems. *Animal Reproduction Science* 26:51–59.

Bakrie B and Teleni E, 1991. The effect of time of feeding in relation to work on draught animals 1: The effect of feeding high fibre diet either before or after work on feed utilization and physiology of working buffaloes. *Draught Animal Bulletin* 1: 39–45. Australian Centre for International Agricultural Research, James Cook University of North Queensland, Townsville, Australia.

Bamualim A, Ffoulkes D, and Fletcher I C, 1987. Preliminary observations on the effect of work on intake, digestibility, growth and ovarian activity of swamp buffalo cows. *Draught Animal Power Project Bulletin*, 3: 6–10. Australian Centre for International Agricultural Research, James Cook University of North Queensland, Townsville, Australia.

Barton D, 1991. The use of cows for draught in Bangladesh. *Draught Animal Bulletin* 1: 14–26. Australian Centre for

International Agricultural Research, James Cook University of North Queensland, Australia.

Barwell I and Ayre M, 1982. *The harnessing of draught animals*. Intermediate Technology Publications, London, UK.

Egan A R and Dixon R M, 1993. Feed resources and nutrition needs for draught power. Pp93 in: Pryor W J (ed), Draught Animal Power in the Asian-Australasian Region. *ACIAR Proceedings* 46. Australian Centre for International Agricultural Research, James Cook University of North Queensland, Townsville, Australia.

Ffoulkes D, 1986. Studies on working Buffalo - current research on nutritional aspects. Balai Penelitian Temakliawi, Bogor. *Draught Animal News* 6, Centre for Tropical Veterinary Medcine (CTVM), University of Edimburgh, UK.

Ffoulkes D, Bamualim A and Panggabean T, 1987. Utilization of fibrous feeds by working buffaloes. pp 161–169 in: Dixon R M (ed), *Ruminant feeding systems utilizing fibrous agricultural residues*. International Development Program of Australian Universities and Colleges, Camberra, USA.

Gemeda T, Zerbini E, Wold A G and Demissie D, 1994. Effect of draught work on performance and metabolism of crossbred cows 1: Effect of work and diet on body weight change, body condition, lactation and productivity. *Animal Science* 60:361–367.

Gryseels G and Anderson F M, 1985. Use of crossbred dairy cows as draught animals:experiences from the Ethiopian highlands. pp.237–258 in: *Research methodology for livestock on farm trials*. Proceedings of a workshop held March 25–28 at Aleppo, Syria. International Development Research Centre, Ottawa, Canada.

Gryseels G and Goe M R, 1984. Energy flows on smallholder farms in the Ethiopian highlands. International Livestock Centre for Africa (ILCA, now ILRI), Addis Ababa, Ethiopia. *ILCA bulletin* 17:2–9.

Mathers J C, Pearson R A, Sneddon C J, Matthewman R W and Smith A J, 1985. The use of draught cows in agricultural systems with particular reference to their nutritional needs. pp. 476–496 in Smith A J (ed) *Milk production in developing countries*. University of Edinburgh Press, Edinburgh, UK.

Matthewman R W, 1987. Role and potential of draught cows in tropical farming systems: a review. *Tropical Animal Health and Production* 19:215–222.

Matthewman R W, 1989. *Effect of sustained exercise on milk yield, milk composition and blood metabolite concentrations in Hereford x Friesian cattle*. PhD Dissertation. University of Edinburgh, Centre for Tropical Veterinary Medicine, Easter Bush, Roslin, Midlothian, EH25 9RG, UK.

Matthewman R W, Dijkman J T and Zerbini E, 1994. The management and husbandry of male and female draught animals: Research achievements and needs. pp 125–136 in: Lawrence P R, Lawrence K, Dijkman J T and Starkey P H (eds), 1993. *Research for development of animal traction in West Africa*. Proceedings of the fourth workshop of the West Africa Animal Traction Network held 9-13 July 1990, Kano, Nigeria. International Livestock Centre for Africa (ILCA), Addis Ababa, Ethiopia. 322p. ISBN 92-9053-276-9

Pearson R A, 1990. A note on live weight and intake and digestibility of food by draught cattle after supplementation of rice straw with the fodder tree *Ficus auriculata*. *Animal Production* 51:635–638.

Pearson R A and Lawrence P R, 1992. Intake, digestion, gastrointestinal transit time and nitrogen balance in working oxen: studies in Costa Rica and Nepal. *Animal Production* 55:361–370.

Reh I and Host P, 1985. Beef production from draught cows in small scale farming. *Quarterly Journal of International Agriculture* 24:38–47.

Shapiro B, Zerbini E and Gemeda T, 1994. *The returns to investment in dual use of crossbred cows for milk production and draught work in the Ethiopian Highlands*. Paper presented at the First Workshop of the Ethiopian Network for Animal Traction, held 27–28 January 1994, Addis Ababa, Ethiopia.

Soller H, Reed J D and Butterworth M H, 1991. Effect of work on utilization of cereal crop residues by oxen. *Animal Feed Science and Technology* 33: 297–311.

Teleni E and Hogan J P 1989. Nutrition of draught animals. pp 118–133 in: Draught Animals in Rural Development. *ACIAR Proceedings* 27, Australian Centre for International Agricultural Research, James Cook University of North Queensland, Townsville, Australia.

Teleni E, Boniface A N, Sutherland S and Entwistle K W, 1989. The effect of depletion of body reserve nutrients on reproduction in Bos indicus cattle. *Draught Animal Power Project Bulletin* 8: 1–10. James Cook University of North Queesnland, Townsville, Australia.

Wells P L, Hopley J D H and Holness D H, 1981. Fertility in the Afrikaner cow 1: The influence of concentrate supplementation during the post-partum period on ovarian activity and conception. *Zimbabwe Journal of Agricultural Research* 19: 13–21.

Winnugroho M and Situmorang P, 1989. Nutrient intake, workload and other factors affecting reproduction of draught animals. pp 186–189 in: Draught animal power for rural development. *ACIAR proceedings* 27. Australian Centre for International Agricultural Research, James Cook University of North Queensland, Townsville, Australia.

Zerbini E, Gemeda T, O'Neill D H, Howell P J and Schroter R C, 1992. Relationships between cardio-respiratory parameters and draught work output in F_1 crossbred dairy cows under field conditions. *Animal Production* 55:1–10.

Zerbini E, Gemeda T, Tegegne A, Gebre Wold A and Franceschini R, 1993a. Effects of work and diet on progesterone secretion, short luteal phases and ovulations without estrus in postpartum F_1 crossbred dairy cows. *Theriogenology* 43: 471–484.

Zerbini E, Gemeda T, Franceschini R, Sherington J and Wold A G, 1993b. Reproductive performance of F1 crossbred dairy cows. Effect of work and diet supplementation. *Animal Production* 57:361–369.

Zerbini E, Gemeda T, Wold A G and Demissie D, 1994. Effect of draught work on performance and metabolism of crossbred cows 2: Effect of work on roughage intake, digestion, digesta kinetics and plasma metabolites. *Animal Science* (in press).

Cow traction in Chokwe, Mozambique

by

O Faftine and A Mutsando

Instituto de Produção Animal, CP 1410, Maputo, Mozambique

Abstract

A survey was carried out among 60 farmers who were plowing with Nguni cows in Chokwe District in Gaza, Mozambique in July 1994. The histories of 110 cows raised primarily for breeding purposes but used also for traction were obtained from the owners. Calving intervals averaged 21 months and calving occurred throughout the year but peaked in the rains. Calf mortality from birth to weaning was not recorded but there was evidence of abortions and weakened calves in two villages.

These levels were equal to those obtained in the years with sufficient male animals for traction in the region and use of cows for this purpose was probably uncommon. They were also similar to those obtained in non-draft Nguni cows in better management situations on-station. Despite these positive results, using cow traction is apparently a temporary practice due to a shortage of cattle following drought and war. The main problem mentioned by farmers was the unavailability of cows for work for up to four months each year during calving.

Introduction

There is a general feeling among many farmers and agricultural scientists that the utilisation of cows for traction can adversely affect the reproductive performance of the animals. This study investigates an increase in use of cows for traction in an area of Mozambique. A preliminary survey in Chokwe in 1992 showed that more than 50% of animal traction pairs involved a female (Faftine and Massaete, 1994). A trend towards substitution of oxen by cows for animal traction had been found by Lexa (1985), Dionisio (1985) and by a Veterinary Faculty survey, referred to by Timberlake et al (1986) in Chokwe and by Rocha (1988) elsewhere. All referred to the scarcity of bulls/steers as the main reason for this change.

Background information

The present study was carried out in the largest irrigation scheme in Mozambique which is located near the city of Chokwe (260 km from Maputo) in Gaza Province. It was constructed in the early 1950s and the first Portuguese settlers arrived in 1954 (Myres, 1992). The smallholder agricultural sector traditionally cultivated rainfed crops but since 1985 a part of the irrigated area has been allocated to them (Woodhouse et al, 1986).

There are two main cropping seasons: the summer season (October to March), and the winter season (April to September). The crops grown in this area are maize, rice, cotton, wheat, cowpeas, beans, sweet potatoes, groundnut and vegetables. The soils are predominantly of heavy clay texture and the land topography is flat (Jimenez et al, 1990; 1991).

The population of draft animals in Chokwe has decreased from about 2000 pairs in 1985 (Lexa, 1985) to about 579 pairs of which 19% are cows (Uaila, 1992). The major causes of this decrease were drought (particularly during 1992), war and theft. Most (83%) of cattle owners use animal power for plowing their own land and to rent out (Uaila, 1992). Draft animals are used mainly for first and second plowing, harrowing and ridging while manual labour is used for first land cleaning, planting and weeding (Jimenez et al, 1991).

It has been estimated that there is one plow per 2 ha and one harrow per 6 farms. Cows and oxen are also used for transporting the harvest, fuelwood and water, with an average of one cart per 3 families (Uaila, 1992).

The major constraints faced by the communities in the irrigated area are quarrels over land (due to skewed distribution and overfragmentation), boundary disputes and illegal occupation, damage to canals or grazing of crops (Myres, 1992) and a shortage of animals. The farmers interviewed were members of the Development Association for Women supported by United Nations Development Fund for Women (UNIFEM); Matuba village, Farmers' Associations supported by a Lutheran

World Federal project (Nwachicoluane, Lionde and Muianga villages) and Farmers' Associations in Chokwe.

Methodology

Five villages between Chokwe and Muianga were visited. The oldest person working with the first spans encountered was interviewed. The survey was completed in 7 days. Sixty farmers were observed plowing with cows and each was asked about the circumstances that led them to begin using cows for traction. Other questions concerned how the cows responded to use for work and how farmers overcame the constraints to cow traction.

In general, the farmers had a good recollection of the number of calvings, and calves weaned from their cows. Thus, in this study the case histories were used as criteria of reproductive efficiency. Information on the management, uses, and reproductive and working performance of 110 cows was collected.

Results

The majority of the draft pairs found comprised two cows (57%) and the remaining were mixed pairs of cow/ox or cow/bull. Of the 110 female cattle surveyed, 11% were heifers and 89% cows. The age of animals ranged from 3 to 15 years.

Body scores of the cows indicated poor to reasonable condition. The body condition score was influenced by location and was worst in Muianga. Some areas (Lionde and Muianga) are enclosed by crop lands and the pasture situation was critical; this may have been a cause of the poor body condition.

The farmers referred to abortions as a frequent problem in the region. In their opinion it was caused by shortage of grazing time.

At the time of the survey 16% (18) of the cows being used were pregnant and 49% (54) suckling. Of the pregnant cows 40% were pregnant for the first time, at a mean age of 3.5 years.

The interviewees stated that intervals between parturitions were normally 1 rainy season since the cows usually calved during the rains, during the following rains they were still suckling and in the third rains they calved again. From this observation it can be deduced that the calving interval was approximately 18–24 months. As the

Table 1: Number of parturitions in Nguni draft cows before and after starting working. Data from a survey in Chokwe, Mozambique in July, 1994.

Variable	Number	%
Parturitions before starting traction		
0	1	2
1	40	72
2	9	16
>2	5	9
Total	**55**	**100**
Parturitions after starting traction		
0	7	7
pregnant	10	10
1	56	57
2	16	16
>2	9	9
Total	**98**	**100**
Parturitions of heifers		
0	1	8
pregnant	8	67
1	3	25
Total	**12**	**100**

bulls and the herds were not allowed to graze in the irrigated area for most of the year, mating opportunities were limited.

Some farmers believed that working infertile cows could solve their fertility problems. This was primarily undertaken in the case of overweight cows.

Calf mortality was not mentioned. Males born to draft cows were often paired with their mothers for work while females were at their first calving.

The periods during which farmers allowed their pregnant cows to rest before calving were variable, ranging from 1 to 5 months. The most common practice among farmers was not to return cows to work for one month (until umbilical cords drop) after calving. During this period the farmer was forced to borrow or hire animals from neighbours.

Few farmers were strongly against using cows for traction. However, they said that bulls were not attracted to cows used for work.

Discussion

Some farmers in the study area were not convinced of the advantages of using cows for traction and requested a reintroduction of oxen. They were being forced to use cows because of recent losses in their herds. The use of cows appears to be temporary and a way to help to ensure a fast rebuilding of the herds after drought and war.

Although their attitudes were in some ways a reflection of their culture and beliefs, farmers were also motivated by real practical constraints to cow traction. The farmers interviewed mentioned the main disadvantage of cows listed by Matthewman (1987) and Matthewman, Djikman and Zerbini (1993), namely the unavailability of females for work for approximately two to four months per annum around calving. To varying degrees, each of the farmers was detrimentally affected by this rest period during late pregnancy. This fact seemed to be the major influence on farmers' decisions regarding whether to use cow traction as a permanent practice.

The length of the rest periods before calving depended on how farmers managed their cows. Some farmers said that it was possible to use the cow during late pregnancy provided the animals were beaten only on the hind quarters. Other experienced farmers controlled their animals with voice commands, sounds or names. Occasionally, the cow worked until calving day because the owner did not expect her/his cow to calve so soon.

The possibility of brucellosis should be considered in triggering abortions. This disease is widely distributed in the neighbouring zones including the Chobela and Mazimuchopes Animal Experimental Stations. The animals of both stations came into contact with farmers' cattle during emergency evacuations to Chokwe during the war.

Conclusions

The reproductive traits observed were similar to those obtained in the same breed in much better on-station conditions (Rocha, 1985) where cows were not used for traction. However, this conclusion is preliminary because, although 8% of cows involved in this study had more than two parturitions after starting work (2–7 parturitions), most of the females sampled were in their first parturition. It is therefore not possible to predict the real effect of work in the long term.

Despite the apparent positive results in using cows for traction, during the survey it was common among farmers to avoid using cows for more than one purpose at the same time to prevent a possible decline in calf production. This negative effect of traction could be reduced with appropriate feeding management. This management might involve the cow being fattened before the working season and fed a good quality diet before calving and post partum, especially when at rest in the kraal at night. Greater contact with bulls during the rains, when fertility is heightened due to better diets, may also increase conception rates.

Acknowledgements

This work could not have been accomplished without the help and patience of many farmers. We are grateful to all of them and to the LWF and UNIFEM projects. In particular we would like to thank Mrs Deolinda, Elisa and Mr Agostinho, extension officers of these projects, for their collaboration.

References

Dionisio A C, 1985. Evolution of livestock production in Mozambique with special emphasis on beef cattle. pp 1–61 in: Jordão and Timberlake (eds), *Proceedings of the Livestock Production Seminar* held 2-7 December 1985, Maputo, Mozambique. Ministry of Agriculture, Maputo Mozambique and FAO, Rome, Italy. 299p.

Faftine O and Massaete E, 1994. *Suplementaçao dos animais de tracção: alguns aspectos para reflexão.* Presented at III workshop of the Veterinarian Association for Mozambique (AVETMO) held 3-4 March 1994, Maputo, Mozambique.

Jimenez H, Picciot G and Bata F, 1990. Sistemas de Produção Tradicional e Melhorado em Chokwe. I. Areas de solos leves, sistemas milho-milho. *INIA documento' de campo no 7.* Instituto Nacional de Investigação Agronomica (National Agronomic Research Institute), Maputo, Mozambique.

Jimenez H, Picciotto G, Chongo S and Bata F. 1991. Sistemas de produção tradicional e melhorado em Chokwe. Area de solos pesados- sistema arroz-milho. *INIA documento de campo no X.* Instituto Nacional de Investigação Agronomica (National Agronomic Research Institute), Maputo, Mozambique.

Lexa J P, 1985. Some aspects of the mixed farming systems in the family sector of Mozambique pp 218–239 in: Jordão and Timberlake (eds), *Proceedings of the Livestock Production Seminar* held 2-7 December 1985, Maputo, Mozambique. Ministry of Agriculture, Maputo Mozambique and FAO, Rome, Italy. 299p.

Matthewman R, 1987. Role and potential of draught cows in tropical farming systems: a review. *Tropical Animal Health and Production* 19: 215–222.

Matthewman R W, Djikman J T and Zerbini E, 1993. The management and husbandry of male and female draught animals: research achievements and needs. pp125–136 in: Lawrence P R, Lawrence K, Djikman J T and Starkey P H (eds) *Research for development of animal traction in West Africa.* Proceedings of the fourth workshop of the West African Animal Traction Network held 9-13 July, 1990, Kano, Nigeria. International Livestock Centre for Africa, Addis Ababa, Ethiopia. 306p. ISBN 92-9053-276-9

Myres G, 1992. *State farm divestiture in Mozambique property: disputes and issues affecting new land access policy - Chokwe region.* Report prepared for a collaborative project on land policy and state farm divestiture. Government of Mozambique and USAID/ University of Wisconsin, Madison Land Tenure Centre, Wisonsin, USA.

Rocha A, 1985. Review of investigations into animal selection and improvement in Mozambique. pp 130–155 in: Jordão and Timberlake (eds), *Proceedings of the Livestock Production Seminar* held 2–7 December 1985,

Maputo, Mozambique. Ministry of Agriculture, Maputo Mozambique and FAO, Rome, Italy. 299p.

Rocha A, 1988. The importance of animal power in a mixed farming system of the traditional sector of Southern Mozambique. In Namponya C R (ed), Proceedings of the SADCC workshop on Animal Traction, held July 1987, Maputo, Mozambique. *SACCAR Workshop series* 7: 38-45

Timberlake J, Jordão C and Serno G, 1986. Comunicão No 50 *Serie Terra e Aguas.* Instituto Nacional de Investigação Agronomica (National Agronomic Research Institute), Maputo, Mozambique.

Uaila R, 1992. Alguns dados sobre o sistema de produção pecuário do sector familiar no Siremo. O Agrário. *Revista do Instituto Agrario de Chimoio-Manica.*

Woodhouse P, Jimenez H, Heemskerk W, Spitttel M and Sloobbe W, 1986. Smallholder farming systems research in the Chokwe irrigation area. INIA. Project UNDP/ FAO/ Moz /81/014 *Field Document No 4.* Instituto Nacional de Investigação Agronomica (National Agronomic Research Institute), Maputo, Mozambique.

Comparative draft performance of oxen and heifers in northern Sierra Leone

by

G O Cole and J Steinbach

Tropical Sciences Centre, Justus-Liebig University, Ludwigstrasse 21, D-35390 Giessen, Germany

Abstract

Twelve 2½-year-old N'Dama cattle were used to investigate the draft performance of oxen and heifers under the agroclimatic conditions of northern Sierra Leone. The results indicated higher feed intake of heifers over oxen during the working and non-working period. Feed intake (dry matter) increased by 53% during the working period in both sexes. Heifers consumed significantly more (+16%) feed than oxen. The work performance as measured by speed (m/s), draft force (kN), power (kW) and area plowed and harrowed in 4 hours (ha), was significantly lower in heifers by 26, 22, 24 and 42% respectively. On wet land less draft force (-14%) and power (-21%) was required, whereas speed (+11%) and work completed (+28%) increased in comparison to dry land.

Although there are some disadvantages in using female animals for draft work, their draft performance suggests that heifers could be successfully utilised as dual-purpose animals in many smallholder mixed-farming systems provided that the draft requirement is not very high and adequate feed is available.

Introduction

In Sierra Leone, oxen are the only important draft animal species. The cost of supporting oxen can be considerable because on average they are only used for 41 working days a year for plowing and 3 days a year on harrowing, with other operations being of negligible importance (Corbel, 1988). Trained oxen may therefore spend about nine months of the year doing little or no work, yet consume valuable feed throughout the year. The possibility of using females for draft as well as milk and calf production offers a way of overcoming the disadvantages of using oxen. Besides contributing to a better utilisation of scarce food resources, the use of female animals for draft purposes would allow males to be fattened and sold at a younger age. Fewer but more productive animals on farms could reduce

stocking rates and overgrazing, thus contributing to the development of a more productive and sustainable farming system. It would seem appropiate therefore to consider seriously the use of female animals rather than oxen alone for draft purposes and to encourage this actively through extension programmes. Since there is a shortage of empirical data on the draft capacity of N'Dama heifers under the agroclimatic conditions of Sierra Leone, this study was undertaken to compare the draft performance of oxen and heifers.

Materials and methods

Twelve 2½ year old N'Dama cattle (six oxen and six heifers) were selected for this study. The animals were housed individually in a shed, had free access to water except when working and were trained for agricultural traction as described by Starkey (1982). The animals were fed local grass plus ground maize grain given at the rate of 0, 0.5 and 1 kg/head/day during the working period. Feed intake was measured during both the working and non-working period. Three groups of draft animals, each consisting of a pair of oxen and a pair of heifers were formed on the basis of body weight, physical fitness, size and compatibility. They were compared when they plowed and harrowed well-developed and partly-developed uplands and inland valley swamps for four hours a day during the dry season. The animals worked three days a week, at high (plowing) or low (harrowing) intensity following a change-over design. A different group of two teams worked every day beginning at 08:00h and finishing at 12:00h. Draft force, speed and area cultivated were measured. Power was calculated by multiplying force (kN) by speed (m/sec).

Least square means were obtained for the various work performance parameters, and differences between the means were tested for statistical significance using Student's t-test.

Table 1: Mean dry matter intake (grass) of oxen and heifers during the working and non-working period

Sex	Non-working period	Working period	Average
	Dry matter intake (g/kg liveweight $^{0.75}$)		
Oxen	91.86 ± 0.45^a	136.12 ± 1.12^a	113.99 ± 0.79^a
Heifers	102.68 ± 0.80^b	162.38 ± 1.45^b	132.53 ± 1.13^b
Average	97.27 ± 0.63^A	149.25 ± 1.29^B	123.26 ± 0.96

a,b) Within column means with different superscripts are significantly different (p<0.05)
A,B) Between column means with different superscripts are significantly different (p<0.05)

Results and discussion

Table 1 shows the mean dry matter intake of grass by oxen and heifers during the working and non-working periods.

Heifers consumed 16% more dry matter (g per kg metabolic body size) than oxen. Both sexes consume more during the working period. This difference was more pronounced in heifers (58 vs 48%). The difference observed in feed intake between the two sexes may be due to the fact that heifers have a lower ability to mobilise body reserves, especially during work. The increase in feed intake during the working period indicates that the animals were able to meet their high energy requirements by eating more, since they did not lose weight during the working period.

The speed, draft force, power and area cultivated by oxen and heifers on dry and wet land during plowing and harrowing are presented in tables 2, 3, 4 and 5, respectively.

During plowing and harrowing the animals were able to generate an average speed of 0.50 m/sec, a draft force of 0.41 kN and a power of 0.22 kW. They cultivated an area of 0.41 ha/day.

Oxen had a significantly (P<0.05) faster speed of working than heifers. The type of land and work done significantly (P<0.05) affected the working speed. An 11% higher speed was recorded on wet land compared to dry land. The speed of harrowing was 15% higher than the speed of plowing.

Table 2: Speed (m/sec) of oxen and heifers during dry and wet land plowing and harrowing

	Dryland			Wetland			Average		
	Oxen	Heifers	Mean	Oxen	Heifers	Mean	Oxen	Heifers	Mean
Plowing	0.50 ±0.18	0.35 ±0.22	0.43^{bB} ±0.20	0.55 ±0.33	0.40 ±0.27	0.48^{Ab} ±0.30	0.53 ±0.26	0.38 ±0.25	0.46^b ±0.26
Harrowing	0.57 ±0.18	0.42 ±0.13	0.50^{aB} ±0.16	0.62 ±0.25	0.48 ±0.21	0.55^{Aa} ±0.23	0.60 ±0.22	0.45 ±0.17	0.53^a ±0.20
Average	0.54 ±0.18	0.39 ±0.18	0.47^B ±0.18	0.59 ±0.29	0.44 ±0.24	0.52^A ±0.27	0.57^A ±0.27	0.42^B ±0.21	0.50 ±0.23

a,b) Within column means with different superscripts are significantly different (p<0.05)
A,B) Between column means with different superscripts are significantly different (p<0.05)

Table 3: Draft force (kN) of oxen and heifers during dry and wet plowing and harrowing

	Dry land			Wet land			Average		
	Oxen	Heifers	Mean	Oxen	Heifers	Mean	Oxen	Heifers	Mean
Plowing	0.56 ±0.18	0.45 ±0.15	0.51[aA] ±0.17	0.50 ±0.22	0.41 ±0.20	0.46[aB] ±0.21	0.53 ±0.20	0.43 ±0.18	0.48[a] ±0.19
Harrowing	0.40 ±0.21	0.31 ±0.23	0.36[bA] ±0.22	0.35 ±0.23	0.24 ±0.17	0.30[bB] ±0.20	0.38 ±0.22	0.28 ±0.20	0.33[b] 0.21
Average	0.48 ±0.20	0.38 ±0.19	0.44[A] ±0.20	0.43 ±0.23	0.33 ±0.19	0.38[B] ±0.21	0.46[A] ±0.21	0.36[B] ±0.19	0.41 ±0.20

a,b) Within column means with different superscripts are significantly different (p<0.05)
A,B) Between column means with different superscripts are significantly different (p<0.05)

Table 4: Power (kW) of oxen and heifers during dry and wet land plowing and harrowing

	Dry land			Wet land			Average		
	Oxen	Heifers	Mean	Oxen	Heifers	Mean	Oxen	Heifers	Mean
Plowing	0.30 ±0.20	0.22 ±0.18	0.26[aA] ±0.19	0.46 ±0.21	0.24 ±0.11	0.35[bA] ±0.16	0.42 ±0.22	0.22 ±0.12	0.32[b] ±0.17
Harrowing	0.24 ±0.16	0.20 ±0.13	0.22[bA] ±0.15	0.19 ±0.12	0.13 ±0.05	0.16[bB] ±0.09	0.22 ±0.14	0.17 ±0.09	0.20[b] ±0.12
Average	0.27 ±0.18	0.21 ±0.16	0.24[A] ±0.17	0.21 ±0.16	0.16 ±0.08	0.19[B] ±0.12	0.25[A] ±0.17	0.19[B] ±0.12	0.22 ±0.15

a,b) Within column means with different superscripts are significantly different (p<0.05)
A,B) Between column means with different superscripts are significantly different (p<0.05)

Table 5: Area cultivated (ha/d) by oxen and heifers

	Dry land			Wet land			Average		
	Oxen	Heifers	Mean	Oxen	Heifers	Mean	Oxen	Heifers	Mean
Plowing	0.38 ±0.22	0.20 ±0.13	0.29[bB] ±0.18	0.46 ±0.21	0.24 ±0.11	0.35[bA] ±0.16	0.42 ±0.22	0.22 ±0.12	0.32[b] ±0.17
Harrowing	0.55 ±0.20	0.31 ±0.12	0.42[aB] ±0.16	0.68 ±0.19	0.44 ±0.14	0.56[aA] ±0.17	0.62 ±0.20	0.38 ±0.13	0.50[a] ±0.17
Average	0.47 ±0.21	0.26 ±0.13	0.36[B] ±0.17	0.57 ±0.20	0.34 ±0.13	0.46[A] ±0.17	0.52[A] ±0.21	0.30[B] ±0.13	0.41 ±0.17

a,b) Within column means with different superscripts are significantly different (p<0.05)
A,B) Between column means with different superscripts are significantly different (p<0.05)

Oxen exerted a significantly higher average draft force than heifers (oxen: 0.46 kN, heifers: 0.36 kN). This is because they were able to sustain the effort required over a longer period. On dry land, the animals had to exert a 14% higher draft force than on wet land. Plowing had a (31%) shigher draft requirement (0.48 kN) than harrowing (0.33 kN). Oxen developed a significantly higher tractive power (0.25 kW) than heifers (0.19 kW). A significant difference ($P<0.05$) was observed in the tractive power developed on dry and wet land, and during plowing and harrowing. Both sexes had to develop higher tractive power when working on dry land (oxen: 0.27 kW, heifers: 0.21 kW). The tractive power developed during plowing was 17% higher than that developed during harrowing.

The draft capacity of oxen was significantly ($P<0.05$) higher than that of heifers due to a difference of 0.15 m/sec greater working speed on a 0.06 kW greater generation of power. As a result, oxen plowed and harrowed significantly more area per day (0.52 ha/d) than heifers (0.30 ha/d).

These results indicate that heifers can be effectively used for light cultivation which does not require high draft force. Sex differences amounted to 48% in plowing but only 39% in harrowing. This finding agrees with that of Monnier (1965). However, in order to generate enough force two heifers must be used in a team, whereas a single ox could most likely be used for light work such as harrowing. Comparing our results with those reported previously, differences could be observed. This may be explained by our experimental procedure in which the animals used were young (2 years old) and had a low body weight. In turn, this would result in the work performance parameters that we have reported showing lower values.

Conclusion

Draft animals are clearly a drain on farm resources and their numbers should be minimised within a given work programme. This objective can be achieved by using females instead of males for work purposes. In summarising the results of this study, it is evident that oxen generally exhibit a better draft performance than heifers. However, heifers with the appearance of poor muscle development could match their work output. Although heifers perform less work for more energy intake compared to oxen, their low draft performance is compensated by the production of other products in addition to draft work. Their favourable work output suggests that they could be used for heavy work (such as plowing) for short periods of time without adverse effects or can be effectively used for light work that does not require high draft force.

Acknowledgements

This study was implemented as part of a comparative study of the use of oxen and heifers as draft animals in northern Sierra Leone. The authors express their appreciation to the German Academic Exchange Service (DAAD), the Seed Multiplication Project (SMP) and the Sierra Leone Work Oxen Prgramme (SLWOP) for their much appreciated financial, moral and technical suppor in the execution of this work.

References

Corbel H, 1988. The economics of animal power in Koinadugu District, Sierra Leone: a case study of the work oxen introduction and credit programme. pp. 299–310 In: Starkey P and Ndiamé F (eds), *Animal power in farming systems*. Proceedings of the Second West African Animal Traction Networkshop held 17-26 September 1986 in Freetown, Sierra Leone. Vieweg for German Appropriate Technology Exchange. Deutsche Gesellschaft für Technische Zusammenarbeit (GTZ), Eschborn, Germany. 363p.

Monnier J, 1965. Contribution à l'étude de la traction bovine au Sénégal. I. *Machinisme Agricole Tropicale* 10: 3–25.

Starkey P H, 1982. N'Dama cattle as draft animals in Sierra Leone. *World Animal Review* 42: 19–26.

Dry season crop residue feeding for improved draft power in Zambia

by

J H Meinderts*, M Chibango and N Mwenda

Farm Power and Mechanisation Centre, Palabana Animal Draught Power Development Programme
P Bag 173, Lusaka, Zambia

Abstract

Farmers in Zambia indicated that the poor condition of draft animals at the onset of the rains is a constraint to early and timely planting. They suggested the inadequate grazing of the animals during the dry season is the main cause. Farmers explained that they did not conserve any roughage during the dry season. Together with farmers, the Palabana Animal Draught Power Development Programme scrutinised local feed resources for supplementation of dry season grazing and browsing. Leguminous crop residues received first priority. In an on-farm trial in six districts with 10 farmers each, groundnut stover was collected, stored and rationed to a pair of oxen for 3–4 months from August. During the cultivation period the trial was evaluated in a group discussion with the participating farmers. Farmers were enthusiastic about the performance of the animals. The main input was labour, which was not a constraint. The construction of a proper storage shed was the crucial input for a successful result. Extension materials based on these experiences have been developed to boost the message to more farms and increase the collection of residues within farms. During the harvesting season, field days were organised at farms of participating farmers to demonstrate to other farmers the construction of storage sheds and feeding practices. In the second year over 500 farmers in nine districts adopted the message and included collection of other crop residues such as cowpea and maize and sorghum stovers.

Introduction

Nobody will deny that an animal in proper condition will perform better than a hungry skeleton. Of course poor animals may be forced to plow a field, but these animals are slow, disobedient and may later refuse to work at all. If feeding is neglected, work oxen can lose 20% of

---

* *Subsequent address:* Johan Meinderts
Larenstein International Agricultural College
Brinkgeversweg 69, P O Box 7
7400 AA Deventer, The Netherlands

---

their body weight during the dry season (Bartholomew et al, 1993). At the onset of the rains when animals are needed for plowing, feed intake is often below maintenance level because time for grazing and availability of grass are limited. However, during the working period total estimated energy expenditures will often be 1.3–1.8 times maintenance (Lawrence, 1987). If an animal is in proper condition it will have fat reserves which may be mobilised during the working period. Research has been done by many people to identify an effective strategy to feed animals in an efficient way for a satisfactory work output. Many feeds have been reviewed and trials or programmes implemented to improve the utilisation of the feed resource, eg rangeland renovation, fodder production and conservation, agroforestry with fodder trees and preservation of crop residues.

In The Gambia, the feeding of crop residues, being a very under-utilised resource, was most successful. Farmers were encouraged to save and store cereal and groundnut stovers and feed restricted quantities to specific animals. In the same programme rangeland management with high inputs like fencing, seeds and herbicides, seemed unsustainable (Russo, 1990).

The period of supplementation has been examined by several researchers. Chikura (1994) supplemented draft oxen in a 90-day on-farm trial during the cultivation period with on average 3.9–4.2 hours plowing during 10–22 working days. The control group, as well as the supplemented animals, showed comparable weight gain, indicating that the animals did receive enough time for grazing the succulent young grass available during that time of the year. It was concluded that under these circumstances, supplementary feeding was not required. It was suggested, however, that in areas of lower potential, or in more severe years, it could be effective. Supplementation throughout the dry season was thought to be more appropriate. This strategy was examined by Prasad, Khombe and Nyathi (1994). They found less weight loss and a better performance (area plowed, rate of plowing and distance covered) for oxen supplemented with maize stover (1.5 kg per day) and groundnut halms (0.5 kg per day). They concluded that the feeding of a limited amount of crop residues throughout the dry season seems to be a sounder nutritional proposition than feeding a large amount in the late dry season.

Dry season supplementation was also studied by Francis and Ndlovu (1993). They demonstrated that cob-sheath and groundnut stover can offset dry season weight losses of draft oxen, especially if fed for a longer period during the dry season.

Bartholomew et al (1993) compared the work capacity of supplemented and non-supplemented oxen and found an increase of 28%, but indicated that this increase was obtained with a considerable cost in feed energy input. This increase was thought to be too costly if animals were used for less than three weeks in a year.

The Zambian situation

In Zambia, the use of draft animals has been intensified during the last decade. More farm operations (tillage, planting, weeding and harvesting) as well as transport are being done with animals. In Southern Province the number of draft cattle decreased becuase of high mortality caused by East Coast Fever and drought (Sinyangwe, 1995). In this important animal draft power province smaller numbers of animals have to do more work. For the near future it is foreseen that the use of draft animals in Zambia will intensify further.

The Palabana Animal Draught Power Development Programme found enough ground in the above experiences and opinions to discuss the issue with field extension staff and farmers. General complaints concerning poor condition and low work output were regularly brought forward by extension staff. In the past, extension messages on supplementary feeding (eg fodder banks, crop residues) were disseminated but seemed not to interest the farmers. Informal discussions were held with the parties concerned in Southern and

Eastern Provinces, which have high cattle populations and a long dry season of seven months. Farmers declared that they do feed crop residues to their animals. When probing the issues, however, they admitted that residues are only fed in the field after harvesting and that much is trampled and wasted or eaten by animals other than their own. Remaining crop residues and standing hay are widely burned as early as August to clear the fields. Between then and December very little feed is left for the cattle. Many animals from the villages go to the *dambos* often far away for longer periods. Some of the draft animals joined the herd and were not available for work. Others stayed behind for regular transport jobs without feed.

After discussing the special quality of groundnut stover farmers became interested, saying that they did not know its feeding value was high. Many farmers growing groundnuts regretted that they had never used the stover.

On-farm supplementation trial

Extension staff and farmers in six dsitricts decided to start an on-farm trial with 10 farmers in each dsitrict. Farmers would collect, store and ration groundnut stover as a supplement to one pair of oxen during August to December. In a formal one-week residential course at Palabana the extension staff concerned were invited to discuss the field programme. Practicals were conducted on collection and conservation of stovers, construction of storage sheds and rationing of the feed. A three-day intake experiment with different treatments was part of the practicals which clarified the increased intake from groundnut stover compared to standing hay. It was recommended to start feeding 2 kg of groundnut stover per day in August and increase this quantity gradually to 10 kg per day in December depending on the availability of grass and browse. For the total period of about 100 days the farmers would need about 600 kg of groundnut stover per animal. During the training it was learned by measurement that a reasonable stand of groundnuts would produce 2000 kg of stover per ha.

For the participating farmers a one-day field training exercise was organised by the extension staff. Timely collection and proper storage for protection of the stover against termites, sun and stray animals were discussed as prerequisites for successful preservation. The feeding of the stover to draft animals received only good support during the discussions. Many farmers suggested that it was not a problem to harvest the quantity of stover recommended for one pair of oxen.

First-year experiences

In each trial area the extension worker organised a group discussion among the farmers to exchange experiences and evaluate the results. A total of 60 farmers participated in eight discussions during the cultivation period. During the discussions a checklist was used to cover the issues systematically. The majority of farmers had never supplemented the animals before. The trial was implemented by the farmers as planned. In Eastern Province the groundnuts produced well. However, in Southern Province many fields did not produce any nuts because of drought. In the past these fields were left for grazing by the village herds. Collecting the groundnut plants for dry season feeding gave the farmers the good feeling that they were not planted in vain.

Storage sheds were constructed but some were of poor quality. Herds of stray animals broke these structures early in the season and finished the stover rapidly. Roofs were often absent but maize stover was used to cover the more valuable groundnut stover. Most oxen were fed after grazing in the late afternoon. Farmers did not ration the feed very well so the stover was finished by early November. Some farmers fed all remaining stover after the first rains as the stover became wet and they expected regrowth of the grass very soon. Unfortunately, the first rains were only a few showers and afterwards a dry spell of over six weeks was experienced. Draft cattle were separated during feeding and some farmers even constructed separate kraals for the draft cattle. As animals got used to supplementation in the evening they came home by themselves which was appreciated as an advantage by the farmers. Farmers who also supplemented early in the morning found the animals stayed around the farm the whole day which was an advantage if the farmers had regular transport jobs to do. It was the farmers' opinion that the condition as well as the performance of the animals was definitely better

compared to not-supplemented animals. Some of the comments were:

- animals moved faster
- animals were hard working
- animals worked longer without getting tired
- animals plowed more hectares in a week.

All farmers indicated that they would continue with this programme with some improvements on storage: the shed needs to be strong and constructed with a proper tin roof. The bin used for storing maize cobs was suggested as a proper alternative. Many farmers indicated that they would grow more groundnuts the following year to collect more stover.

Second-year experiences

In the second year (1995) it was expected that the programme would spread, with only little guidance, in areas where the trial was done the year before. The 'first-year' farmers were mobilised during field days to inform other farmers in the community. In general the majority of farmers participating in the first-year trial improved well on collection, storage and feeding. The group collected more stover, constructed more permanent sheds and improved on rationing.

Hay and other stovers (eg maize, cowpea) were collected in large quantities. Also new farmers picked up the message very well. In some villages all groundnut growers started to collect the stover even though it had never been practised before.

A number of new extension staff within the participating districts showed interest in starting the same trial in their area. Central Province and Copperbelt Province became involved as well. New extension staff were advised again to start with just 10 farmers. Starting in the first year with a total of about 80 farmers, in the second year over 500 farmers were collecting stover.

A setback in groundnut stover collection was experienced where over 70% of the groundnut fields were affected by the rosette virus. Groundnuts were not harvested at all and the stover was also destroyed in most fields. Farmers, however, collected maize and cowpea stover instead, indicating an adaptation of the message.

Extension staff showed great interest and enthusiasm by organising field days and developing extension material to educate farmers.

The preparation of a set of posters and a leaflet to be used in farmers' meetings and field days was initiated.

Discussion

On-farm research and extension programmes very often create enthusiasm within the target group when new concepts are tried with (often free) inputs like improved seed, 'modern' implements, chemicals etc. Frequently, enthusiasm changes gradually into realism with criticism from farmers on the risk and high investments involved. Identifying solutions to problems, risk and costs need to be considered first, espcially in small-scale farming.

In solving the problem of poor performance of draft animals during the times they are needed most, we concentrated on a low-input solution. Inputs should be available within or around the farm. It was assumed that labour would not be a constraint. With such an approach you have to realise that you have only ideas to offer. The implementation is completely in the hands of the farmers themselves investing their labour to collect and store the stover. We did not expect much enthusaism among the majority of farmers as long as the message had not been proven in the community. This was the main reason to start with just 10 farmers per extension worker to be able to encourage farmers regularly and individually. Once the message has been proven it should be possible to change the approach and disseminate the message to larger numbers of farmers. Also extensionists might have the feeling that they do not have much to offer farmers other than ideas. This was why extension staff had to come forward themselves before the Palabana Animal Draft Power Porgramme offered support by means of training and provision of extension materials. Using this strategy we experienced much enthusasm in the field with farmers as well as extension staff. The result of more than 500 farmers implementing the idea within two years is encouraging and supports continued dissemination of the message.

Conclusions

We may conclude that the approach of involving farmers and extension staff from the begining, discussing together the problem and its solutions, and the implementation and evaluation, provide a base for success. However, the best indicator of success will be an increasing adoption level during the next few years.

References

Bartholomew P W, Khibe T, Little D A and Ba S, 1993. Effect of change in body weight and condition during the dry season on capacity for owrk of draft oxen. *Tropical Animal health and Production* 25(1): 50–58.

Chikura S, 1994. Improving the management of feed resources for draft animals in Mangwende, Zimbabwe. pp 162–163 in: Starkey P, Mwenya E and Stares J (eds), 1994. *Improving animal traction technology*. Proceedings of the first workshop of the Animal Traction Network for Eastern and Southern Africa (ATNESA) held 18-23 January 1992, Lusaka, Zambia. Technical Centre for Agricultural and Rural Cooperation (CTA), Wageningen, The Netherlands. 490p .ISBN 92-9081-127-7

Francis J and Ndlovu L R, 1993. Improving work performance of Mashona oxen through strategic supplementation with locally produced feeds. *Draught Animal News* 19.

Lawrence P R, 1987. Nutrient requirements of working ruminants pp 61–79 in: Farley J L (ed) *An introduction to working animals*. MPW Australia, Melbourne, Australia, 198p.

Prasad V L, Khombe C T and Nyathi P, 1994. Feeding crop residue for improved draft power. pp 164–166 in: Starkey P, Mwenya E and Stares J (eds), 1994. *Improving animal traction technology*. Proceedings of the first workshop of the Animal Traction Network for Eastern and Southern Africa (ATNESA) held 18-23 January 1992, Lusaka, Zambia. Technical Centre for Agricultural and Rural Cooperation (CTA), Wageningen, The Netherlands. 490p. ISBN 92-9081-127-7

Russo S L, 1990. The use of crop residues for livestock feed by small farmers in The Gambia. pp 165–185 in: Dzowela B H, Asrat Wendem-Agenehu and Ketegile J A (eds). *Utilisation of research results on forage and agricultural by-product materials as animal feed resources in Africa*. Proceedings of the first joint workshop, held 5–9 December 1988, Lilongwe, Malawi. PANESA/ARNAB, International Livestock Centre for Africa, Addis Ababa, Ethiopia. 824p.

Sinyangwe P, 1995. Cattle distribution and supply for draft purposes. pp 43–45 in: Meinderts J H, Bwalya M and Chibango M (eds). *Livestock concerns in animal traction*. Proceedings of the fourth annual animal traction workshop held 31 August–2 September 1994, Lusaka, Zambia. Palabana Animal Draft Power Development Programme, Lusaka, Zambia. 76p.

Photo (opposite): N'Dama oxen pulling a cart in Sierra Leone

Meeting the challenges of animal traction

Social and economic challenges in West Africa

Ox traction in a long-term perspective: policy implications of a socio-economic study in Ghana

by

J H Hesse [1] and A Runge-Metzger [2]

[1]*Institute of Rural Development, University of Göttingen
Waldweg 26, 37073 Göttingen, Germany*
[2]*Delegation of the European Commission in Zimbabwe, POB 4252, Harare, Zimbabwe*

Abstract

Changes in the pattern of use of ox-traction technology in farming systems in northern Ghana are analysed. During the 1993/94 and 1994/95 cropping seasons data were collected in 42 households in three villages. The same households were visited during a study conducted in 1982/83; the same methods were used to ensure comparability.

Since the first study, patterns of adoption have changed substantially. The average proportion of farm area plowed by oxen has increased. However, the number of farmers actually owning oxen has decreased. This indicates that the rental market for animal traction has gained importance. These changes in adoption patterns can be attributed to changes in profitability. Recent findings show that utilisation of ox traction still results in a clear reduction in labour input requirements for crop production, especially in the amount of time spent on soil cultivation. However, a comparison of average physical and monetary productivities of labour and land did not show any statistically significant differences; this contradicts the earlier study. Other factors influencing the adoption pattern include changes in relative prices of farm inputs and outputs, increasing land scarcity, and obstacles to sustaining cattle herds including traditional inheritance laws.

The results of the survey indicate that rental markets are an important, but hitherto neglected, facet of animal traction adoption. Where farmers cannot expand their cultivated area rental markets are likely to be important in ensuring the profitability of animal traction.

Unlike the earlier investigation, this study did not find that animal traction increased yields per unit area. The impact of ox traction on sustainable land use needs to be monitored more closely to establish the cause-effect relationships, especially whether early yield advantages of ox traction reflect more rapid mining of soil nutrients.

Introduction: ox traction and the development of farming systems

The role of ox traction in the development process has been recognised widely. Adoption of the technology has been found to be closely related to its profitability. The profitability of the use of ox traction depends on the net cash flow derived from benefits realised by farmers due to changes in farm size, crop yields, labour costs, and income from hiring ox services, and costs including capital costs, cost of implements, costs of maintaining oxen and equipment, and hiring costs.

A number of authors have analysed the general economic impact of ox traction (eg Sargent et al, 1981; Kirk, 1984; Starkey, 1991; Strubenhoff, 1988; Pingali, Bigot and Binswanger, 1985; in this area Panin, 1988; Hailu, 1990; Runge-Metzger, 1993). Some general results include:

- Ox traction allows adopters to save labour when carrying out tillage operations, weeding, and transport (ie increasing labour productivity and substituting labour with capital).

- Where land is available, ox traction enables the expansion of farms, sometimes combined with increased cultivation of cash crops. This may lead to increases in farm income.

- ox traction permits more timely plowing, sowing, and weeding; it can also improve ridge formation and seedbed preparation. As a result yields per hectare can be increased.

Within the same agroecological zone, the magnitude of these immediate determinants of profitability are influenced by:

- population density
- relative factor and product prices
- institutional factors.

Population density

Boserup (1965) emphasised that agricultural mechanisation is affected by population density. She sees the shift from hand hoe to animal traction as the consequence of reduced fallow periods indicating increased land scarcity. Reduced fallow periods lead to a decline in soil fertility which can be offset by a higher labour input. Under such circumstances animal traction is adopted to increase labour productivity.

The effect of population density on the profitability of ox traction may not necessarily be linear. Increased population density is often associated with increased competition between land for cropping and land for animal feed. The consequences could be threefold. First, as Ruthenberg (1983) describes, the subsequent dangers of overgrazing and land degradation could reduce the potential yield advantages of ox traction. Second, the reduced availability of land may prevent farm expansion and so reduce the potential workload of oxen for traction. Consequently, there would be fewer economies of scale (McIntire, Bouzat and Pingali, 1992) and the costs of ox traction could only be reduced by hiring out (Runge-Metzger, 1991). Third, reduced availability of fodder tends to increase herd maintenance costs, since good pasture areas are increasingly scarce.

Prices and marketing

Hayami & Ruttan (1985) elaborated on the implications of relative price changes on the introduction of agricultural innovations that are closely interrelated with population growth and the growing intensity of land use. They found that the principal effect of population growth is to inflate the price of land relative to that of labour. In the absence of a growing non-farm labour market this would undermine the profitability of mechanisation.

The relative costs of farm implements to product prices have a direct impact on the profitability of ox traction. During the course of structural adjustment programmes input prices generally increase because in the past many African governments subsidised agricultural inputs heavily. However, it is also assumed that output prices increase because African governments tended to favour urban consumers. If trade is liberalised the overall effect therefore depends very much on the magnitude of subsidies at the various levels as well as the level of world market prices. Because of the potential impact of economic policies on relative prices, Lele & Stone (1989) argued that Boserup's hypothesis (Boserup, 1965) should be extended, and called for an inclusion of policies and incentives for explaining technological change.

Pingali, Bigot and Binswanger (1985) stressed the importance of market access and higher product prices. Usually it is assumed that with increased population growth road infrastructure improves, demand for transport increases, and so transport costs decrease. In addition, access to credit could be an important determinant for the introduction of ox-traction technology as the initial capital requirements for ox-traction owership are substantial.

Rental markets

An important feature of the adoption of mechanisation in smallholder farming systems are rental markets. Rental charges are likely to be high if farms are large (land is plentiful) and ox owners utilise the full capacity of the oxen on their own fields. If the net benefit of animal ownership decreases relative to renting costs, there is likely to be a shift from ownership of oxen to rental of traction services. This could occur if, for example, costs of herd maintenance increase due to disease or increasing fodder costs, or if crop production revenues decrease because of increasing land scarcity, decreasing yields per unit area or a decreasing ratio of output to input prices, or simply if actual rental charges decline or market transactions become cheaper.

At the level of the individual household, changes in adoption are related to the life-cycle of the households. One important institutional feature after the death of the head of a household is the prevailing customary law of inheritance of oxen, related assets and land.

In general, the theoretical remarks show that it is difficult to predict precisely the impacts of a changing socioeconomic environment on the adoption of ox traction. Therefore, empirical field studies, like the present one, are necessary to explore the relationships more closely. The objectives of this study are:

- to identify the major changes in the pattern of adoption of ox traction over the past twelve years in Northern Ghana

- to determine possible factors for changes in the pattern of adoption

- to identify the implications of the empirical findings for the formulation of policies which facilitate the introduction of ox traction into the farming systems.

Background to the study area

The Guinean savanna zone of West Africa is predominantly grassland savanna with scattered trees, and is believed to have the highest agricultural production potential in West Africa. There is one rainy season that lasts from April to October with an average rainfall of about 1000 mm. The general infrastructure of the area is not highly developed. The introduction of modern farm practices has been regarded as a promising way of exploiting the agricultural potential of the area. However, since the introduction of such practices into the area during the 1950s, 1960s, and 1970s the socioeconomic environment has changed drastically. For instance, population growth has been above 3% in the study area (Government of Ghana, 1984a) so competition for land has increased gradually. In addition, Ghana's economy has undergone major changes since 1983, when the national economic policy was harmonised in accordance with the concepts of the World Bank and the International Monetary Fund (IMF). As structural adjustment measures were implemented, the exchange rate of the Ghanaian Cedi was liberalised, state-owned assets were privatised, and a substantial part of the labour force in the public sector was retrenched (Herbst, 1993). The availability of consumer goods and farm inputs in particular of all imported goods, has improved. Compared to non-food items the prices of food items in the country have declined (Commander, Howell and Seini, 1989; Government of Ghana, 1994b).

Such dramatic changes affect the adoption of modern farm practices; this paper discusses changes in the utilisation of ox traction in northern Ghana. Animal traction was introduced to farmers in northern Ghana in the early1930s by colonial authorities (Lynn, 1937) but was neglected immediately after independence because it was considered to be outdated and inappropriate. After tractorisation attempts failed in the 1960s and early 1970s, ox traction received new attention. Subsequently, new government and NGO projects were launched to support further diffusion of ox traction (Famiyeh, 1993; Herbst, 1993). Some of these projects ended in the early 1990s.

In Nakpanduri, the main village in the study area of Northern Ghana, the Ghanaian German Agricultural Development Programme (GGADP) worked from the late 1970s to the mid-1980s to improve the agricultural extension station in cooperation with the Ministry of Food and Agriculture (MoFA). Today, the structures still exist, but no practical work is carried out on the station. As part of the structural adjustment programme MoFA adopted new extension strategies (training and visit), and extension activities are now concentrating on farmers' fields. MoFA also stopped the subsidy and distribution of farm inputs. Ox-traction implements which used to be available at the station are no longer supplied by the agricultural extension services. Private traders have taken over the supply instead and ship the implements from a factory in Tamale (200 km south west) or private blacksmiths in Bawku (60 km north). The availability of farm inputs including ox-traction implements is thus ensured, but prices have become higher.

Panin (1988) compared households owning oxen to those not owning oxen. His findings from 1982/83, which serve as a reference situation for the present study, can be summarised as follows:

- ox owning households were better equipped with land, labour and capital

- cultivated area per active worker did not differ between ox owning and non-owning households

- the area cultivated per active worker declined with the number of years of experience with the technology

- farmers with more experience cultivated more cash crops

- ox-owning households achieved higher yields per hectare. Ox traction was found to have a positive impact on the yields per hectare

- labour requirements of ox-owning households and ox-cultivated fields were reduced

- ox-owning households achieved higher net farm incomes and cash incomes

- the investment in ox traction technology was highly profitable

- the provision of credit facilities was recommended, because the heavy cash outlays caused by the initial investment in purchasing oxen were identified as a major constraint of adoption.

Methods

In principle, long-term changes in adoption can only be investigated by a longitudinal study in the same communities and households. This paper presents preliminary results of such a panel study. Fieldwork was carried out in the Northern Region of Ghana from March 1993 to April 1995. The study covers 42 households in the three villages of Nakpanduri, Sakogu, and Gbinbalanchet in Gambaga District in the Northern Region of Ghana. Data were collected at village, household, and field level. The villages and households are identical to the ones studied in 1982/83 (Panin, 1987, 1988, 1989,1990). In 1982/83, the household sample was a stratified random sample consisting

of 30 ox-owning households and 12 non-ox-owning households.

Direct measurement (record keeping), observation, formal and informal interviews were used for data collection. To achieve a maximum degree of comparability, it was attempted to employ the same methods as Panin (1987,1988,1989,1990) and Panin and de Haen (1989).

The dynamics of adoption patterns

A declining trend of ox ownership was confirmed by village census and informal interviews. In 1982, Panin (1988) found 20% of the households possessing oxen but ox ownership in all three villages declined to 16% within the last decade. However, this trend was not equal across the three study villages. In Gbinbalanchet an increase of 3% in the number of households possessing oxen was observed in 1994 as compared to 1982. On the contrary, the relative number of ox owners declined by 4% and 5% in Sakogu and Nakpanduri, respectively. Furthermore, the surveys show that within the original household sample of 42 households the actual

Table 1:Changes in proportion of farm area plowed with ox traction

	Number of housholds	% of total population	Mean farm area (ha)	Area plowed with oxen	
				ha	% of farm area
Non-ox-owning households					
1982/83	12	80	3.6	0.6	16
1993/94	28	84	2.7	1.0	36
Ox-owning households					
1982/83	30	20	5.6	4.2	75
1993/94	14	16	4.5	3.0	68
Estimate for total population[1]					
1982/83	–	–	4.0	1.3	32
1993/94	–	–	3.0	1.3	44

Notes:
1) Calculated on basis of total number of households (1982: 614; 1994: 1096 households) in the three villages studied.

Table 2: The impact of tillage technology on labour demand for various farming operations (times in work days per hectare)

	A) Hoe n=49		B) Hired oxen n=77		C) Own oxen n=77		Total n=203		
	Mean	CV (%)	Mean	CV (%)	Mean	CV (%)	Mean	CV (%)	Significance[1]
Clearing	6.1	91	5.5	106	8.8	128	6.9	121	AC, BC
Tillage	16.3	101	10.3	116	6.6	127	10.3	122	AB, AC, BC
Planting	14.1	112	9.0	96	10.9	122	11.0	115	AB
Weeding	26.4	136	16.1	92	20.9	132	20.4	128	AB
Harvesting	61.4	96	31.4	86	52.8	123	46.7	114	AB, BC
Applying fertiliser	0.1	509	0.1	708	0.1	724	0.1	659	
Other	–	–	0.8	105	0.6	160	0.5	130	AB
Total	124.4	68	73.2	60	100.7	102	96.0	86	AB, BC

Notes:
1) Letters represent columns which are significantly different at alpha=0.1 level using a Duncan test

number of ox-owning households dropped from 30 to 14 (-53%).

However, in 1993/94 the study revealed that a considerable number of farmers who did not possess oxen rented them. It was also observed that in some cases households that own oxen also plow fields by hoe. Ox ownership is only a simple index of technology adoption and the degree the technology is utilised at field level may be more revealing. The proportion of the total field area plowed by oxen is shown in Table 1.

The results of this comparison are surprising: while the non-ox-owning households increased the proportion of ox-plowed fields from 16% to 36% of their total farm area, ox-owning households reduced the ox plowed area from 75% to 68% of their total farm area. At the overall village level, the percentage of fields plowed with oxen increased, since non-ox-owning households represent the majority of households in the study area. It is estimated that the utilisation of ox traction increased from 32% to 44% of the total cultivated area, which is a relative increase of 35%.

Overall, the following non-uniform changes in the pattern of adoption were observed:

- in two out of three villages ox ownership declined in relative terms, while in one it has increased
- ox owners plowed a smaller proportion of their total farm area with oxen than they did in 1982/83
- non-ox owners hired ox-plowing services much more than in 1982
- overall, utilisation of ox traction has increased over the past decade.

Determinants of the changing pattern of adoption

As the data analysis is still in its initial stages only preliminary indications are presented which have to be interpreted with caution until a full model can be tested.

Impact on labour input

A total of 203 separate plots were analysed in order to identify the advantages of ox traction compared with those found by Panin (1988). Three groups of plots were compared: those tilled by hand hoes, those tilled by hired oxen and those

Table 3: The impact of tillage technology on soil and labour productivity

	A) Hoe n=49		B) Hired oxen n=77		C) Own oxen n=77		Total (A+B+C) n=203	
	Mean	CV (%)	Mean	CV (%)	Mean	CV (%)	Mean	CV (%)
Yield (kg/ha)	3,706	132	3,526	172	4,549	158	3,957	158
Returns (Cedis/ha)[1]	150,567	107	146,241	140	195,618	139	166,243	135
Variable costs (Cedis/ha)								
Seed	6,150	110	4,993	110	6,291	134	5,765	122
Fertiliser	2,054	438	509	809	625	877	926	660
Hired labour	7,065	215	7,084	176	4,312	218	6,028	202
Hired bullocks	–	–	3,453	181	–	–	1,310	320
Interest[2]	1,527	117	1,604	117	1,123	140	1,403	131
Gross margin (Cedis/ha)	133,771	123	132,051	123	183,268	147	152,122	149
Gross margin (Cedis/workday)	1,367	126	1,595	126	2,281	164	1,803	155

Notes:
None of the columns differed significantly (Duncan test at alpha=0.1 level)
1) In 1994 US$1=1100 Cedis
2) assumed to be 10% for 6 months

tilled with the farmer's own oxen. One of the striking characteristics of ox traction was its ability to reduce the workload for the farmer at times of labour peaks, for example during tillage. Table 2 shows a comparison of the amount of labour spent on different farming operations in 1993/4.

The saving in labour occured with hired as well as with owned oxen. Plots which were ridged by hoe also had a higher input of labour for most of the other operations. This might be due to economies of scale because the average size of hoe plowed plots is significantly smaller (0.38 ha) than for those plowed by hired or own oxen, which are 0.43 ha and 0.6 ha, respectively. The coefficients of variation of the figures are rather high; this is probably due to pooling of plots across different crop enterprises and land-use intensities.

Economics of ox traction

In a first, rather simple, approach the economic performance is compared by calculating average factor productivities in physical and monetary terms for the three groups of plots plowed with different tillage technologies. Table 3 shows that neither soil productivity nor labour productivity show any statistically significant difference depending on the tillage technology.

This result is substantially different from Panin's observations in 1982/83, which indicated a significant increase in yields with bullock traction. If consistent, it will certainly have a major negative impact on the profitability of investment in ox traction and hence discourage its adoption. In addition, this result means that it cannot be assumed that ox owners are able to maintain a higher degree of soil fertility. However, results have still to be interpreted with caution as yields and other factor productivities depend also on factors others than the tillage technique. These relationships will be analysed in more detail by production function analysis.

Population growth

One major influence on patterns of ox ownership is population growth, which affects the profitability of keeping oxen for traction. In particular, limitations on land availability directly affect the benefits of ox-traction ownership. As shown in Table 1 the mean farm size of ox owners declined by 20% between 1982 and 1993. This could mean that oxen cannot be employed fully by their owners on their own fields. Limited land availability has apparently facilitated the rapid expansion of rental markets for ox services.

However, population growth was not equal in the three study villages. The growth rate of 2.8% per annum in Gbinbalanchet was low compared to an average of 5.3% per annum for all the villages. As a result land scarcity is likely be less of a problem compared to the other villages. For these reasons, ox owners in Gbinbalanchet engaged their animals primarily on their own fields and were unwilling to hire them out. However, the situation in this village is exceptional as the rate of population growth was low due to violent ethnic conflicts at the end of the 1980s.

Changes in relative prices

Relative prices of inputs and outputs have changed dramatically since 1983. The ratio of output prices to agricultural input prices has declined (Runge-Metzger, 1993), ie farming is less profitable. The people in the study area also reported an increasing number of incidents of cattle theft. Farmers perceive cattle ownership to be more risky than a decade ago; this is negatively affecting the benefits expected from ox traction. Both of these factors tend to discourage farmers from investing in ox traction and may explain part of the stagnation in ox ownership that was observed in the two villages.

Cultural factors

In some cases, traditional inheritance rules clearly jeopardised the sustainability of cattle ownership within a household. According to observations, if a head of household passes away his closest senior male relative becomes the trustee of all the household's cattle. He is supposed to keep the cattle for the benefit of the family clan.

This is a much larger social unit than the household. Traditionally, cattle have an important function in the social security of the extended family. In emergency an animal of the family herd is supposed to be made available for sale to generate funds, for example for paying hospital bills. All clan members are entitled to this kind of traditional social support. The magnitude of the support is often subject to difficult negotiations. Remarkably, trained oxen suitable for traction are treated in the same way as ordinary cattle. In other cases, relatives of a deceased person were not in agreement on the mode of redistribution of cattle among the members of the family clan. In particular, when the number of disposable oxen is rather small, the animals are likely to be sacrificed and consumed at the event of the funeral. As can be seen from these examples, inheritance rules can have a major impact on the pattern of adoption of ox traction at the level of individual households.

Implications for agricultural policy

The direct benefits, population growth, relative prices and cultural factors discussed in this paper all contributed to some extent to the changes in the pattern of adoption of ox traction that was observed in northern Ghana. However, many other factors are likely to be imprtant. Although the situation has not yet been analysed in detail, some preliminary conclusions will be drawn with respect to the formulation of policies facilitating the introduction of ox traction into the farming systems in northern Ghana.

Farmers in the study area generally perceived ox traction as a potential method of improving crop production. For example, ox traction enabled farmers to reduce the labour burden of farm work. However, adoption of ox traction technology does not necessitate ox ownership by all farmers. Whereas the number of farmers owning oxen dropped on the average, the area plowed with oxen increased significantly. This finding generally helps to arrest fears that during the implementation of structural adjustment programmes, or due to rapid population growth resulting in smaller farm sizes, rates of adoption of agricultural technologies might be declining. On the contrary, scarce resources seem to be used more wisely.

So far, the potential of rental markets for further development of animal traction has been neglected by agricultural research and extension in Ghana. Very little is known about the advantages of this kind of market transaction. The renting of animal traction services has advantages compared to ownership: there are no learning costs for farmers who rent, there is less pressure on limited cash resources, and rental payments might be organised in accordance with local arrangements, for example, the provision of informal credits in kind. Moreover, if farmers cannot expand their farms it would probably not be economical to adopt ox traction. Under these circumstances, calls for formal credit schemes and programmes for purchasing oxen and implements seem to be justified only on a limited scale and when rental markets evolve at the same time. In addition, local potential for improved utilisation of oxen should be explored more intensely.

The yield advantages of ox traction reported by a previous study can no longer be observed in the study area. If these preliminary findings are confirmed by a more detailed analysis, the impact of ox traction on sustainable land use needs to be monitored more closely to establish the cause-effect relationships. In particular, investigations should study whether early yield advantages of ox traction reflect a more rapid mining of soil nutrients.

The effects of cultural rules for inheritance of cattle on the adoption of animal traction have been neglected. More detailed analysis of this aspect is necessary before definite policy conclusions can be drawn.

Acknowledgements

The authors are grateful for the close collaboration with the Savanna Agricultural Research Institute (SARI), Nyankpala, Ghana, which is supported by the Deutsche Gesellschaft für Technische Zusammenarbeit (GTZ), Eschborn, Germany. We owe thanks to the GTZ project Flanking Programme for Tropical Ecology (TÖB), Eschborn, Germany, for funding this particular study. The authors are also indebted to Dr Anthony Panin, Botswana College of Agriculture, Gaborone, Botswana, for preserving the only copy of the 1982/83 data set and his assistance in accessing these data.

References

Boserup E, 1965. *The conditions of agricultural growth - the economics of agrarian change under population pressure.* Allen and Unwin, London, UK.

Commander S, Howell J and Seini W, 1989. Ghana 1983-7. pp 107–126 in: *Structural adjustment and agriculture, theory and practice in Africa and Latin America.* Overseas Development Institute, London, UK.

Famiyeh J A , 1993. *A survey of agricultural policies in Ghana.* Paper presented at SADACC Workshop Oct. 12–15 1993, Ouagadougou, Burkina Faso.

Government of Ghana, 1984a. *Population census of Ghana.* Government of Ghana, Statistical Service, Accra, Ghana

Government of Ghana, 1984b. *Statistical newsletter.* Government of Ghana, Statistical Service, Accra, Ghana

Hailu Z, 1990. *The adoption of modern farm practices in African agriculture, empirical evidence about the impacts of household characteristics and input supply systems in the Northern Region of Ghana.* Verlag Josef Margraf, Weikersheim, Germany.

Hayami Y and Ruttan V,1985. *Agricultural development: an international perspective.* John Hopkins University Press, London, UK.

Herbst J, 1993. *The politics of reform in Ghana, 1982–1991.* University of California Press, Berkeley, California, USA.

Kirk M, 1984. *Ochsenanspannung in Westafrika: Probleme der entwicklungkonformen Technologiewahl für Kleinbäuerliche Betriebssysteme.* Herodot, Göttingen, Germany.

Lele U J and Stone S W, 1989. *Population pressure, the environment and agricultural intensification- variations of the Boserup hypothesis.* The World Bank, MADIA Discussion Paper 4, Washington DC, USA.

Lynn C W, 1937. *Agriculture in North Mamprusi.* Department of Agriculture Bulletin No 34, Ministry of Food and Agriculture, Accra, Ghana.

McIntire J, Bourzat D and Pingali P, 1992. *Crop-livestock interaction in Sub-Saharan Africa.* The World Bank, Washington DC, USA.

Panin A, 1987. The use of ox traction technology for crop cultivation in northern Ghana: an empirical economic analysis. *ILCA Bulletin* 29: 2–8.

Panin A, 1988. *Hoe and ox farming systems in northern Ghana - a comperative socio-economic analysis.* Triops Verlag, Langen, Germany.

Panin A and de Haen H, 1989. Economic analysis of ox traction - a comparative analysis of hoe and ox farming systems in northern Ghana. *Quarterly Journal of International Agriculture* 28(1): 6–20.

Panin A, 1989. Profitability assessment of animal traction investment: the case of northern Ghana. *Agricultural Systems* 30(2): 173–186.

Panin A, 1990. Profitability assessment of animal traction investment: the case of northeastern Ghana. pp 201–207 in: Starkey P and Faye A (eds), 1990. *Animal traction for agricultural development.* Proceedings of the Third Regional Workshop of the West Africa Animal Traction Network, held 7-12 July 1988, Saly, Senegal. Technical Centre for Agricultural and Rural Cooperation (CTA), Ede-Wageningen, Netherlands. 475p. ISBN 92-9081-046-7

Pingali P L, Bigot Y, and Binswanger H P, 1985. *Agricultural mechanisation and the evolution of farming systems in sub-Saharan Africa.* World Bank, Washington DC, USA.

Runge-Metzger A, 1991. *Entscheidungskalküle kleinbäuerlicher Betriebs-Haushalte in bezug auf Wirtschaftlichkeit und Akzeptanz ausgewählter landwirtschaftlicher Innovationen.* Vauk, Kiel, Germany.

Runge-Metzger A, 1993. Farm household systems in Northern Ghana. pp 31–170in: Runge-Metzger A and Diehl L (eds) *Farm household systems in Northern Ghana - a case study in farming systems oriented research for the development of improved crop production systems*. Verlag Josef Margraf, Weikersheim, Germany.

Ruthenberg H, 1983. *Farming systems in the tropics*. Oxford University Press, Oxford, UK.

Sargent M W, Lichte J A, Matlon P, and Bloom R, 1981. An assessment of animal traction in francophone West Africa. *African Rural Economy Working Paper* No 34. Michigan State University, East Lansing, Michigan, USA.

Starkey P, 1991. Animal traction: constraints and impact among African households. pp. 77–90 in: Haswell M and Hunt D (eds), *Rural households in emerging societies*. Berg, Oxford, UK. 261p.

Strubenhoff H W, 1988. *Probleme des Übergangs von der Handhacke zum Pflug, eine ökonomische Analyse der Einführung der tierischen Anspannung in Ackerbausysteme Togos*. Vauk, Kiel, Germany.

The potential for animal traction in south-western Nigeria

by

Adebiyi Gregory Daramola

Department of Agricultural Economics & Extension, Federal University of Technology
PMB 704, Akure, Nigeria

Abstract

On average, agricultural productivity in sub-Saharan Africa is poor. To increase the productivity of agriculture in sub-Saharan Africa cultivation must be both intensified and extended. Seasonal labour shortages are a major constraint to this. Tractors have not proved a viable technological solution for smallholder farmers. Animal traction may be a feasible intermediate solution that could result in increased overall productivity. In the south-west of Nigeria there is little livestock keeping because of the problems of trypanosomiasis: as a result there has been no history of animal traction. Farmer awareness of animal traction is very low and there are many barriers to the uptake of the technology including the supply of animals, poor infrastructure for veterinary care and implement distribution and farmers' lack of capital. However these problems are not insurmountable and animal traction could still have an important impact in the south-west of Nigeria.

Introduction

On average, agricultural productivity in sub-Saharan Africa is poor. To increase the productivity of agriculture in sub-Saharan Africa cultivation must be both intensified and extended. In other words both the total area cultivated and the yield per unit area cultivated must be increased. Manual cultivation is used commonly by farmers but it is technically inefficient and is labour intensive, so limiting the scope for expanding the area cultivated. It has been reported that in many farming systems of West Africa, seasonal labour shortages are one of the primary production constraints. To increase productivity there is therefore a need for improved, appropriate and sustainable technology. The experience in many developing countries, including Nigeria, is that tractors have not been an appropriate solution (see Daramola, 1987 unpublished;). Animal traction may be a viable intermediate option in the humid parts of West Africa. Although draft animal traction has had a fairly long history of success in

the northern part of West Africa, it is relatively uncommon in the southern parts because of ecological, socio-economic and cultural dissimilarities.

As is common in West Africa, the majority of Nigerian farms are small. However, small farms, usually less than 2 ha each, account for over 90% of the total agricultural production. In the light of this there is a need to direct attention to the small-farm sector. The small-scale farming sector offers good potential for increasing the output and income of the farming community, thus helping to reduce poverty.

Technological developments can help to increase agricultural productivity and efficiency and maximise the exploitation, utilisation and processing of natural resources (Naiz, 1978). However, most farmers in sub-Saharan Africa have benefited only marginally from recent developments in production technology. Some of the reasons include the inappropriateness of some of the developed technologies for sub-Saharan African conditions. For example, farmers have not accepted monocropping in place of the traditional mixed cropping. The development and adoption of technology appropriate to farmers' conditions along with essential services would considerably increase their productivity and income. In addition, a small increase in productivity on individual farms would add up to a substantial increase in total production. Attention should therefore be focused on 'appropriate technology' suitable to small farms.

Tractors have been shown to be inappropriate for small farms for reasons including the lack of spare parts and the high cost of fossil fuels. Animal traction appears to be an appropriate alternative technology. However, if animal traction is to realise fully its yield-increasing potential through mass adoption in south-western Nigeria, it is necessary to reinforce the elements that minimise variance because farmers have very low risk-bearing abilities (Norman, Simmons & Hays 1982). The requirements for successful introduction of animal traction into an area have been discussed by various authors. Many interacting factors are important and they should be considered as a whole rather than in isolation. This paper discusses appropriate technology of particular relevance to the rainforest zone of southern Nigeria, bearing in mind that such discussion should centre on resource endowment among the small farmers within the region and the appropriateness, divisibility and sustainability of the technology.

Problems and requirements

According to unpublished data from a study I conducted in Nigeria, awareness about animal traction was low (30%). Another 5% of sampled farmers were completely ignorant about animal traction. No farmer in the study area had adopted animal traction for whatever purpose. This suggests that animal traction is alien to farmers in the area, at least, at the farm level. Economic factors accounted for differences in the responses of farmers. Potential adoption of animal traction was correlated with farmers' income. These findings are consistent with earlier works on animal traction in the humid zone by Reynolds (1986) and Jaeger and Matlon (1990).

The primary requirement of draft animal power is animals. A steady supply of mature animals would be needed for a successful animal traction scheme. Cattle are currently scarce in southern Nigeria due to the presence of tsetse flies. The parasitic disease trypanosomiasis carried by these flies infests 90% of cattle in the humid zone. A recent survey in southern Nigeria indicated the presence of 0.3 million cattle, compared with 12.0 million cattle in the whole country (Akinwumi and Ikpi, 1985). However, some breeds, such as N'dama and West African Shorthorns, can tolerate trypanosomes. In the forest regions very few cattle are found but they are present in larger numbers in the savannah where dense rainforest has been cleared.

The suitability of the cattle breeds available in Southern Nigeria depends upon what work is to be performed, and the power required. Work output is related to body size and a small breed would be more limited in its usefulness where heavy soils, which require more effort for land preparation, are found. Farmers can overcome the lower power capacity of small animals by increasing the numbers in a team.

Fodder is available throughout the year. However, the nutritional value of grasses falls rapidly as plants mature and become upalatable. Browse maintains its feeding value over a long period and is therefore a valuable supplment to grass. Leguminous browse in particular has the potential to provide high quality feed at low cost throughout the year.

Maintenance of draft animals in good health is partly dependent on husbandry and hygiene. Farmers unaccustomed to dealing with cattle could find it difficult to recognise health problems at a sufficiently early stage to allow simple remedies to be effective. Accessible and well-trained extension staff would be necessary to provide advice. Prophylaxis against diseases such as rinderpest, trypanosomiasis, and contagious bovine pleuropneumonia requires medication from sources external to the farm so a veterinary system would be required. There is evidence that stress, which can arise from work, poor nutrition, other concurrent diseases, pregnancy and lactation, increases susceptibility to trypanosomiasis. These factors have not been quantified but they constitute additional constraints on the use of draft power in a tsetse-infested zone.

An important issue is the integration of cattle into a farming system where at present livestock are absent. Small ruminants exist but they are usually free-ranging and are not specifically cared for. If farmers are to own draft cattle there must be drastic changes to the present farming systems.

Animals are of limited use without complementary equipment for land preparation. Animal traction is not likely to be profitable at the early stage of its adoption because of the high set-up costs required, particularly for the purchase of animals and implements. Access to adequate credit on favourable terms is a prerequisite for the adoption of animal traction (Munzinger, 1982). The potential adoption of animal traction in the study area was found to be positively and significantly related to the amount of credit available to the farmer ($p<0.05$). This is the major reason for soliciting government involvement to ensure that farmers do not get discouraged before the technique starts being profitable. Extreme caution should be exercised at the introduction stage because there is always the tendency for a resource-demanding intervention to be targeted at wealthy or elite farmers, with whom poorer farmers may not be able to identify (Kalb, 1982). What really matters in the adoption of any innovation is the way the project is perceived by potential adopters. Thus target farmers should be able to observe animal traction in operation over a period of time and be able to try it out for themselves.

Farmers in the study area have not been practising animal husbandary within their current farming systems hence they consider the fact that they need to keep cattle a major decision to take. The majority of farmers do not know and cannot even comprehend how animal traction operates, let alone consider committing their finances to the fixed investment its acquisition demands. Culturally and socially they suspect they might find work oxen a bit difficult to handle. Demonstration units would be needed, and pioneer farmers would have to be identified, trained and provided with necessary animals and implements so that others can see the innovation and relate it to their own circumstances. This should be closely followed by training programmes for animals and farmers.

According to empirical results in the study area (Daramola, 1987 unpublished), potential adoption of animal traction in the study area was found to be positively and significantly related to farm income (at the 0.10 level), the amount of credit available to the farmer (at the 0.05 level) and the distance of farm to input source (at the 0.01 level). A negative and significant relationship was found to exist between potential animal traction adoption and distance of product market (at the 0.05 level). Some of the problems also discovered to be confronting farmers in the study area included the escalating cost of hired labour, lack of hired labour during peak periods and diminishing family labour.

Since most farmers in the target area are resource-poor, the initial animal-traction acquisition cost is the most important economic factor that will limit possible adoption of animal traction. Prejudices and, in exceptional cases,

cultural factors will also be constraints to adoption. In the light of farmers' lack of knowledge of animal traction, educational intervention directed to the possibility of animal traction in the study area will be necessary.

Being rational decision-makers, farmers are willing to embrace any innovation that will assist them in overcoming labour shortages on their farms, more so if it contributes to profitability. In this direction, many empirical studies have established that animal traction can be a profitable innovation for small farmers if the production inputs are not supply-constrained and the necessary infrastructure is provided. Previous studies have also shown that farmers broke their adoption chain due to input supply problems (Daramola, 1987; unpublished). This suggests that adequate preparation is important to ensure successful introduction and adoption. Intervention is expected to be facilitated by the World Bank-assisted State-wide Agricultural Development Projects in Nigeria.

Conclusion

Although animal traction is not an easy technology to introduce because of its many interacting facets and the complexity of the infrastructure required, it is important to state that this study indicates that none of the difficulties of introducing animal traction to the humid zone is insurmountable, as was also found by Reynolds (1986).

From the empirical information gathered, favourable consideration will be given to animal traction in the region by farmers as long as its adoption will not seriously disrupt existing farming systems, jeopardise farmers' subsistence nor impose additional strain on their limited resources. The majority of farmers have very little knowledge about animal traction and this is the major challenge facing animal traction introduction in south-western Nigeria.

Acknowledgement

My gratitude goes to the International Foundation for Science, Sweden, for sponsoring this research.

References

Akinwumi J A and Ikpi A E, 1985. *Trypanotolerant cattle production in Southern Nigeria.* ILCA Research Report, Internation Livestock Centre for Africa, Ibadan, Nigeria.

Daramola A G, 1987. *A quantitative analysis of factors influencing the adoption of improved food production technology in Oyo State, Nigeria.* Unpublished PhD Thesis submitted to the Department of Agricultural Economics, University of Ibadan, Nigeria. 127p.

Delgado C L and McIntire J, 1982. Constraints on ox cultivation in the Sahel. *American Journal of Agricultural Economics* **63**:188–196.

Jaeger W K and Matlon P J, 1990. Utilization, profitability, and the adoption of animal draft power in West Africa. *American Journal of Agricultrual Economics* **72**(1): 35–48.

Kalb D, 1982. Sociological aspects of the use of draft animals on African smallholdings. pp341–372 in: Munzinger P (ed) *Animal traction in Africa.* GTZ, Eschborn, Germany. 490p.

Munzinger P, 1982. Economic aspects of using draft animals. pp 267–338 in: Munzinger, P (ed) *Animal traction in Africa.* GTZ, Eschborn, Germany. 490p.

Niaz S M, 1978. Appropriate technology for achieving full potential of small farms. In: *Technology for increasing food production.* Proceedings of the second FAO/SIDA seminar on field crops in Africa and near East, Lahore, Pakistan 18 Sept - 5 Oct 1977. FAO, Rome, Italy.

Norman D W, Simmons E B and Hays H M, 1982. *Farming systems in the Nigerian savanna: research and strategies for development.* Westview Press, Colorado, USA. 275p.

Reynolds L, 1986. The relevance of animal traction to the humid zone. In: Starkey P and Ndiame F (eds) *Animal power in farming systems.* Proceedings of the second workshop of the West Africa Animal Traction Network, held September 19–25, Freetown, Sierra Leone. GATE/GTZ, Eschborn, Germany. 363p. ISBN 3-528-02047-4

Evolution of farming systems and the adoption and profitability of animal traction
A case study from the savanna highlands of North West Cameroon

by

Kizito Langha

Management Attaché, North West Development Authority, PO Box 442, Bamenda, Cameroon

Abstract

The study examined the effects of agricultural intensification on the adoption and profitability of animal traction under four farming systems at various levels of evolution. It was found that while intensification could increase the pace of mechanisation, adoption depended more on the agronomic requirements of particular crops, and the need to expand cropped areas to take advantage of market availability. Profitability depended critically on the level of evolution in the farming system and the availability and use of complementary, fertility-enhancing inputs.

Introduction

Efforts to tractorise farming in sub-Saharan Africa to improve the performance of agriculture have generally failed (Pingali, Bigot and Binswanger, 1987). Following these failures, there is renewed interest in animal traction as an appropriate technology (Starkey, 1988), and a convergence of views that it is more suitable to the region (Langha, 1995). However, it is still far from evident under what conditions animal traction can be used to optimise farm output. Moreover, despite substantial resources devoted to promoting its use over seventy years, the whole region has less than six million of the 400 million draft animals in the world, and only some 15% of farmers use animal traction (Starkey, 1988).

Early studies examined factors limiting the supply of appropriate technologies and concluded that adoption was blocked by equipment inappropriately designed for African conditions, inadequate extension and an absence of support services (Kline et al, 1970; Le Moigne, 1978). Later field studies shifted the debate to farm-level profitability and the demand for animal traction. Delgado and McIntire (1982), for example, concluded that ox cultivation in isolation from other components of improved technologies, is not sufficiently profitable to compensate for the high opportunity costs of farm resources tied up in the technology. Crawford and Lassiter (1985) criticised this work for overestimating labour costs for animal maintenance and underestimating supply-side constraints. More recently, Jaeger and Matlon (1990) concluded that high levels of utilisation are crucial for profitable adoption; levels of use depended on agroclimatic conditions and learning (experience). Panin and Ellis-Jones (1994) emphasised the linkage between profitability and adoption and suggested that profitability depends significantly on macro-economic aggregates (exchange rates, taxation, import duties etc), and that a longer-term vision needs to be taken in any promotion efforts.

A major weakness of many studies is the failure to include the dynamics of rural change in the analysis. As a result, the intermediate stages of the adoption process are often set aside, resulting in an incomplete and often hazy understanding of the adoption process. Therefore, a number of theories and some empirical studies have focused on the evolution of farming systems as a measure of rural change. First developed by Boserup (1965; 1987) and later on expanded and refined by Ruthenberg (1971), Pingali, Bigot and Binswanger (1987), Strubenhoff and Jahnke (1989) and others, the theory links farm mechanisation to increased agricultural intensification (more frequent cropping of land) which itself results from rising labour to land ratios (pressure on limited land). When labour to land ratios increase, aggregate population food needs rise while the quantity of land available for cropping falls. The increased labour available is applied to intensify farming. When population pressure leads to an increase in the frequency with which land is cultivated, there follows a diminution of the tree cover and the advent of grassy vegetation with roots that are too strong to be cleared by fire or removed by the hand hoe. These conditions necessitate the introduction of mechanical power, usually animal-drawn plows, which are usually the first stage in farm mechanisation. On the other hand, increased population leads to urbanisation and specialisation, thereby facilitating infrastructure improvements and providing farm households with crop markets and a source of improved farm inputs. These conditions combine to favour an improvement in welfare.

The chain of linkages running from population pressure, agricultural intensification, technical change in agriculture, economic growth and societal well-being is very long and complicated, and is difficult to test empirically in a comprehensive way. Turner, Hyden and Kates (1993) limited their investigations to the linkages between intensification, growth and material well-being. Langha (1995) traced the household-level linkages between demographic and socio-economic characteristics of households, population pressure, agricultural intensification, farm incomes and material well-being. This paper is limited to the relationships between intensification of agriculture and the adoption and profitability of animal traction in the savanna highlands of North West Cameroon.

Measuring agricultural intensification in North West Cameroon

Boserup (1965) identified five stages in the evolution of land use from hunting and gathering through fallowing to stationary cultivation systems. Ruthenberg (1971) developed a numerical measure of farming intensity (R) that takes into consideration the number of crop cycles per year, the number of years of fallow and the number of years of cultivation. Later on, Pingali, Bigot and Binswanger (1987) related the five stages of evolution to the R-value, identifying the types of technology at each stage.

North West Cameroon is located between latitudes 5.2° and 7.0° N in the northern part of a volcanic mountain range that extends north-eastward from the coast for 800 km on the Cameroon side of the Nigerian border. The land area is approximately 18,000 km^2, characterised by an extrememly varied relief composed of mountains, escarpments, valleys, plains and plateaux. A basic pattern of altitude zones runs along a southwest to northwest axis, with a central mountainous spine running from Western Province to Nkambe in the north of North West Province. The central axis is flanked by high altitude plains, with lower-lying plains and river valleys forming a third zone.

Farming systems are, in general terms, at an advanced stage of evolution towards intensive stationary cultivation. The diminution of tree cover, the advent of grassy vegetation and declining soil fertility mean that peasant farmers require high levels of skill and effort for survival. The advent of grasssy vegetation has attracted cattle-rearing Fulanis to settle in the region, thereby creating some crop-livestock interactions which could favour agricultural intensification. The number of fallows is falling rapidly towards the annual cropping stage, and crop rotation on family holdings is a feature on many farms. With farming intensities (R-value) averaging 60 to 100, and population density exceeding 70 inhabitants per sq km, this region falls squarely within the short fallow stage of evolution when animal power begins to be used side-by-side with the hand hoe.

However, this general picture is somewhat misleading, as the varying relief, climate, soil types and infrastructure characteristics result in a number of subsystems at various stages of evolution. These subsystems are characterised by different levels of input use and farming intensities, farm output and and returns to the factors of production, and agricultural commercialisation. The two case-study villages were selected to reflect this variation.

The first locality, Wum, represents a zone of low intensity of animal power use and farmers using the technology were geographically very widely dispersed. On the contrary, animal power use in Bamessing village was very intense, although it was introduced in 1988, barely four years before the study. Both areas, located at about 1000 m above sea level, have similar temperatures, rainfall, vegetation and soil types.

In each of the two villages, two subsystems were identified. In Wum, there were marked differences between Fulani farmers in the hilly countryside and the sedentary population in the low-lying ares, with the Fulanis cropping more intensively and using higher levels of animal power. In Bamessing, the same farmers practised different systems on rice and non-rice fields, with the rice fields receiving multiple cropping and high levels of animal power and labour while non-rice fields were less advanced. Hence, four subsystems can be classified, in order of evolution of the farming systems, beginning with non-Fulani farmers in Wum through non-rice farms in Bamessing through Fulani farms in Wum to rice fields in Bamessing.

All ox farmers in each community were interviewed between February and October 1992. There was a total of 59 farmers in both communities, 23 from Wum and 36 from Bamessing. Twelve of these farmers failed to respond regularly, thereby reducing the effective number of respondents to 47.

Some of the important data on farming intensities in each village are shown in Table 1. In Wum, the mean total input was 94 traction days per farm (or 62 traction days per ha). Animal power in Wum was used entirely for land preparation (mainly tilling and, to a very limited extent, ridging to form contour bunds), and for no other operation. Compared with the entire sample (105 traction days per farm) and with Bamessing, the intensity of animal power use in Wum was very low. The seven Fulani farmers owning oxen

and equipment, however, had a mean total animal power input that was more than double the mean animal power input for the entire sample, and far above mean use in Bamessing. Mean use on non-Fulani farms was less than half that on Fulani farms. Even when cropped areas were considered and utilisation was calcualted on a per ha basis, Fulani farmers continued to display a more intensive use of this technology.

As Fulani farms were on average much larger than other farms in the region, whether in Wum or in Bamessing, it is tempting to conclude that animal power was used mainly to expand cropped areas. Per ha utilisation levels on Fulani farms were much lower than those in Bamessing, indicating that cropped area expansion probably accounted for much of the increased level of use on Fulani farms. However, per ha utilisation levels on Fulani farms were still more than double those on non-Fulani farms in Wum. Therefore, the level of animal power use on these farms is likely to have been determined by a mixture of agricultural intensification and crop area expansion.

The structure of animal power use on non-rice fields in Bamessing was very similar to that in Wum: all the animal power on these fields went for land preparation. While the mean level of use on all of these fields combined was just a little over half of that on rice fields alone, the intensity of use on the latter was about four times the intensity on the non-rice fields. Animal power use

Table 1: Farming intensities and other characteristics of farms in Wum and Bamessing

	Wum			Bamessing			Entire sample
	Fulani	non-Fulani	Mean	Rice	Other	Total	
Size of holding (ha)	8.3±1.4	3.8±2.1	6.8±4.9	0.6±0.5	4.9±3.5	5.4±3.7	6.0±4.3
Cropped area (ha)	1.9±1.3	1.3±0.7	1.4±0.9	0.6±0.5	1.2±0.7	1.8±0.9	1.7±0.9
Farming intensity	96±27	84±38	88±29	167±18	78±35	103±24	94±22
Traction days per ha	116	57	65	128±11	33±30	89±65	62
Traction days per farm	219	75	94	75±61	41±42	124±95	106
Labour days per ha	220±219	125±93	208±113	288±122	110±78	167±101	175±117
Labour days per farm	417±311	166±122	301±276	167±80	137±93	304±105	301±234

Note: Where samples are large enough figures are given ± standard deviation

on rice fields in Bamessing was far more intensive that anywhere else. Most of it (48%) was used to transport people and farm produce from the distant fields in the valleys to storage structures in the village and the roadside market at Ntenka. Tilling and puddling the swamps took up a further 43%, and the rest went for weeding (9%).

While rice fields were cultivated more intensively than any other fields, the structure of animal power use on rice fields would seem to suggest that utilisation levels were determined more by transportation needs (to reduce the burden and time required to move persons, farm inputs and produce to and from the distant valleys) and the agronomic requirements of rice cultivation (need to till and puddle swamps). It does not seem that animal power is required either to displace or to supplement labour at peak periods: incremental labour requirements are met by hiring more labour.

These results were tested further by constructing a *primary mechanisation function*, a multiple regression model. The dependent variable in this model was total traction days. Independent variables included:

- farming intensity, capturing the level of evolution of the farming systems

- household resource base, captured by housing standards and household possession (scored following the methods used in the region in MIDENO, 1984), farm income levels and the availability of off-farm sources of income

- commercialisation of agriculture, captured by the level of crop sales the previous year and the value of purchased inputs used on the farm

- farm labour input and years of experience.

The results demonstrated that, contrary to the common logic of farming systems evolution theory, the relationship between agricultural intensification and primary mechanisation is not at all straightforward. While intensification is likely to increase the pace of mechanisation, the latter will not necessarily result from the former. On rice fields in Bamessing, for example, primary mechanisation appeared to be driven along by the agronomic requirements of rice cultivation. On non-rice fields, it was determined by the desire to expand the area under crops in order to increase

marketed output. The role of agricultural intensification in these cases was marginal. In Wum, the experience of users was the important factor, perhaps driven along by the need to expand cropped areas.

Agricultural intensification was not found to be a precondition for primary mechanisation; intensification alone would not result in mechanisation. Increasing intensification will only accelerate the pace of mechanisation, provided that the agronomic conditions require mechanisation, that there are crop markets, and that labour is available to meet the increasing requirements of crop area expansion or increased cropping frequencies. Hence, mechanisation will succeed only if there are gainful interactions between mechanised farms and the rest of the economy.

Evolution of farming systems and the profitability of animal traction

Yield response to animal power (not reported here), confirmed the findings of previous studies elsewhere: improved tillage does not necessarily increase yields. Cropping patterns and farm output were more varied in Bamessing than in Wum as Tables 2 and 3 show.

With very low levels of inputs and near total absence of diversification, output in Wum was correspondingly low, although Fulani farmers had a farm output almost twice that obtained by non-Fulani farmers. Cropping patterns and farm output in Bamessing were more varied and more interesting. Intercrop densities (not reported here) were much higher in Bamessing, while maize densities were lower than those in Wum. On non-rice fields, the pattern was very similar to that in Wum.

While a greater variety of crops was produced on the rice fields in the valleys, total crop value from these fields was less than half of the crop value from the other fields. Indeed, the crop value from rice fields represented barely 31% of total crop value, while the rice crop iteself constituted a mere 20% of the value of all crops. Yet more than 60% of all animal power and nearly 55% of all farm labour in Bamessing were used on rice fields alone. None of the rice produced was kept for home consumption, while less than half of the maize produced on the non-rice fields was sold.

Table 2: Crop production and crop value on Fulani and non-Fulani farms in Wum

Crop	Fulani (n=7)	non-Fulani (n=11)	Mean (n=18)
Maize			
Quantity (kg)	2700 ± 2759	1609 ± 1300	2141 ± 2003
Value (000 FCFA)	257 ± 262	153 ± 124	203 ± 190
Groundnuts			
Quantity (kg)	-	52 ± 137	52
Value (000 FCFA)	-	6.2 ± 16	-
Total crop value (000 FCFA)	257 ± 262	159 ± 143	219 ± 195

Notes:
1) Figures ± Standard deviation
2) US$=FCFA500 approximately

Therefore it seems that increased labour and animal power were used on the rice fields because of the *cash income* to be derived from cropping them. In other words, *crop marketing* appears to have been the driving force of (both) intensification and mechanisation on the rice fields, and that primary mechanisation is not necessarily a consequence of agricultural intensification.

To investigate the profitability of animal traction and its impact on the efficiency of resource use, a production fucntion was estimated, and marginal value products (MVPs) of all inputs calculated. The results are shown in Table 4.

Table 3: Crop production and crop values on rice and non-rice fields in Bamessing

Crop	Rice field	Other fields	Total
Maize			
Quantity (kg)	673 ± 438	5054 ± 2345	5727 ± 3458
Value (000 FCFA)	64 ± 42	480 ± 223	544 ± 329
Beans			
Quantity (kg)	59 ± 129	84 ± 95	143 ± 107
Value (000 FCFA)	7 ± 14	9 ± 11	16 ± 12
Potatoes			
Quantity (kg)	200 ± 189	-	200 ± 189
Value (000 FCFA)	6 ± 6	-	6 ± 6
Rice			
Quantity (kg)	2786 ± 2100	-	2786 ± 2100
Value (000 FCFA)	139 ± 105	-	139 ± 105
Total crop value (000 FCFA)	216 ± 167	489 ± 233	705 ± 451

Notes:
1) Figures ± standard deviation
2) US$ 1 = FCFA500, approximately

From the MVP values it seems that farm income response to primary mechanisation depended largely on the level of evolution of the farming system. In systems such as that in Wum at the start of evolution, farm incomes depended on the area under crops and the level of use of purchased inputs rather than on animal power. Crop area expansion and the use of fertilisers and good quality seed would be the more appropriate policy options to raise farm incomes. Expansion of cropped area need not necessarily be done using animal power. While a lot of animal power was used to expand cropped area on non-rice fields in Bamessing, the study results were not conclusive as to whether it would be profitable to use animal power for crop area expansion in Wum.

In systems with high cropping intensities, such as the rice fields in Bamessing, farm income response to animal power input depended critically on the use of complementary factors of production, principally land augmenting, fertility-enhancing inputs such as fertilisers. In these systems, some cropped area expansion would be profitable, but as this is virtually impossible, increased intensity of

use of purchased land-augmenting inputs (ie increased use per unit of land) was the key to raising farm production in conjunction with animal power.

In other, less highly-evolved but progrssive systems such as that on non-rice fields in Bamessing, marginal returns to animal power were approaching zero. Analysis of the production function indicated that increasing the level of use of animal power per ha would lead to heavy losses in farm incomes. The key to increasing farm incomes in such a system using animal power lay in expanding the area under crops. As in the other subsystems, the absence of land augmenting, fertility-enhancing inputs appeared to be the most limiting factor.

The interpretation of the ratio MVP:MFC is not at all straightforward. A ratio of 1.0 is accepted in previous studies as evidence of allocative efficiency (Shapiro, 1976). If this criterion is accepted then it is tempting to conclude that, with the exception of purchased input use (per ha) in Wum and non-rice fields in Bamessing, there is considerable inefficiency in the use of animal

Table 4: Marginal value products (MVP) and ratios of MVP to marginal factor costs (MVP:MFC) for animal power, human power and agricultural inputs

Factors	Wum MVP	Wum MVP:MFC	Bamessing (rice) MVP	Bamessing (rice) MVP:MFC	Bamessing (non rice) MVP	Bamessing (non rice) MVP:MFC	Sample MVP	Sample MVP:MFC
Animal power								
per farm	-	-	-107	-0.06	14	0.008	-326	-0.19
per ha	856	0.51	624	0.37	-640	-0.38	1571	0.94
Labour								
per farm	-	-	775	3.16	402	1.64	531	2.16
per ha	-525	-2.14	-467	-1.91	-	-	-	-
Inputs								
per farm	116	0.12	-122	-0.12	355	0.36	-281	-0.28
per ha	1052	1.05	1077	1.08	3500	3.50	6879	6.88

Notes:

1) - indicates variable was not significant in the primary mechanisation function

2) There was no factor market for land, and its value was therefore excluded.

3) MVPS are in FCFA; US$ 1 = FCFA 500, approximately.

4) Cropped area was a significant variable in the model, but is omitted from MVP estimates because of the absence of a factor market for land in the region.

power and the other factors of production in North West Cameroon. In the absence of comparable data on farming without animal power, it is impossible to say whether this is better or worse than the predominant hand-hoe-based system. However, it is noteworthy that the ratio MVP:MFC for the entire sample of animal power users is very close to one when considered on a per ha basis. What can be said for sure, therefore, is that at current levels of use per farm there are considerable inefficiencies, but that the situation can be improved by increasing the intensity of use, subject to respecting the location-specific differences in farming systems.

Shapiro (1976) provided a formula for calculating the relative change in marginal value product (D) that is required in order to equate MVP to MFC thereby optimising output. The required change is the absolute value of D in the equation:

$$D = \frac{MVP - MFC}{MVP}$$

The D values (in percentage terms) for the ratios in Table 4 are presented in Table 5, with the exception of land for which there is no factor market.

The D-value for animal power is highest for non-rice farms in Bamessing (input use per ha), and lowest for rice fields in Bamessing (input use per farm). The interpretation of D=168 is that to be

efficient, farmers should have been obtaining about double the marginal value product of animal power at the current intensity of use on rice fields. With a production elasticity of 0.21 (ie a 1% increase in animal power input would result in an income change of only 0.21%), and with MVP increasing less than proportionally to MFC, it would seem that the appropriate adjustment would be to reduce the intensity of animal-power use. Considerable inefficiencies also existed in the intensity of use in Wum and Bamessing.

In the case of labour, there were large inefficiencies in Wum, Bamessing (rice and non-rice) and for the entire sample. Purchased input use was far more efficient in Bamessing rice fields that on all other farms, whether considered per ha or per farm. In Wum, MVP could be increased up to seven-fold at current levels of input use per farm. On non-rice fields in Bamessing, MVP could be increased two-fold. In general terms therefore, the levels of efficiency could be said to be lowest in the more advanced farming system in Bamessing and highest in the more remote areas in Wum.

Conclusions

The adoption of animal power does not depend solely or even mainly on the level of farming intensity. Farmers will adopt the technology in order to expand cropped areas to increase aggregate output so as to take advantage of market

Table 5: Percentage change in the marginal value products required to equate them with the marginal factor costs

	Wum	Bamessing (rice)	Bamessing (non rice)	Entire sample
Animal power				
per farm	-	6	118	19
per ha	59	168	38	6
Labour				
per farm	-	216	39	54
per ha	214	119	-	-
Inputs				
per farm	762	12	65	28
per ha	5	7	250	588

Notes:
1) - indicates that the variable was not significant in the regression model

availability, or to meet the agronomic requirements of cultivating a specific crop such as swamp rice, provided there is a ready market for the crop.

Animal traction is a profitable technology if used under conditions of double- or multiple-cropping, accompanied by application of soil fertility-enhancing inputs such as fertilisers. Therefore, to improve profitability and increase the chances of adoption and sustainability, promotion efforts should be accompanied by measures to reduce fallow periods and introduce stationary cultivation. Development of rural infrastructure to link farmers to markets for crops and inputs is an essential accompanying measure.

References

Boserup E, 1965. *The conditions of agricultural growth.* Allen & Unwin, London, UK.

Boserup E, 1987. *Economic and demographic relationships in development.* Johns Hopkins University Press, Baltimore, USA and London, UK.

Crawford E and Lassiter G, 1985. Constraints on oxen cultivation in the Sahel: comment. *American Journal of Agricultural Economics* 67: 684–685.

Delgado C and McIntire J, 1982. Constraints on oxen cultivation in the Sahel. *American Journal of Agricultural Economics* 64: 188–196.

Jaeger W K and Matlon P T, 1990. Utilisation, profitability and the adoption of animal draft power in West Africa. *American Journal of Agricultural Economics*

Kline C, Green D, Donahue R and Stont B, 1970. *Agricultural mechanisation in equatorial Africa.* Institute of International Agriculture Research Report no 6. Michigan State University, East Lansing, Michigan, USA.

Langha K, 1995. *Mechanisation of peasant farms and agricultural growth in sub-Saharan Africa: a farming systems evolution approach.* PhD Thesis, Wye College, University of London, UK.

Lassiter G C, 1982. *The impact of animal traction on farming systems in Eastern Upper Volta.* Phd Thesis. Cornell University, Cornell, USA.

le Moigne M, 1978. *Culture attelee en Afrique francophone.* Centre des Etdues et Experimentation du Machinisme Agricole, Antony, France.

MIDENO, 1984. *Baseline economic survey of North West Province, Cameroon.* Bamenda, Cameroon.

Pingali P, Bigot Y and Binswanger H P, 1987. Agricultural mechanisation and the evolution of farming systems in sub-Saharan Africa. Johns Hopkins University Press for the International Bank for Reconstruction and Development, Washington DC, USA.

Ruthenberg H, 1971. *Farming systems in the tropics.* Clarendon Press, Oxford, UK.

Shapiro K E, 1976. *Efficiency differentials in peasant agriculture and their implications for development policy.* Discussion paper 52. Centre for Research on Economic Development, University of Michigan, Michigan, USA.

Starkey P, 1988. The introduction, intensification and diversification of the use of animal power in West African farming systems: implications at farm kevel. In: Starkey P and Ndiamé F (eds), 1988. *Animal power in farming systems.* Proceedings of workshop held 17-26 Sept 1986, Freetown, Sierra Leone. Vieweg for German Appropriate Technology Exchange, GTZ, Eschborn, Germany. 363p. ISBN 3-528-02047-4

Strubenhoff H W and Jahnke H E, 1989. Animal traction and agro-ecological zones in western Africa. *Quarterly Journal of International Agriculture.*

Turner BL, Hyden G and Kates R W, 1993. *Population growth and agricultural change in Africa.* University Press of Florida, Florida, USA.

The effects of war on animal traction in Sierra Leone

by

Abu B Bangura

Programme Manager, Sierra Leone Work Oxen Programme, PMB 766, Freetown, Sierra Leone

Abstract

The Sierra Leone Work Oxen Programme has passed through two distinct phases in its development efforts, namely the research phase (1979–94) and the development/extension phase (1985–95). The programme is at the verge of moving to a self-reliant phase but, due to the war, this is unlikely to be realised in the very near future.

The rebel conflict has generally affected the economic life of the country. Agricultural infrastructure and livestock have been damaged seriously and depleted. The farming population and activites have been disrupted and animal traction activites and development have been affected seriously. This state of insecurity is a serious constraint and has affected donor support and placement of field staff. Farmers are the worst off, since they have lost all their life-savings to the gunmen. A large proportion of the population are refugees who have been forced to flee their land.

A rehabilitation programme for animal traction should be effected immediately after the war, including schemes to assist in resettlement of farmes and to restock the cattle herds. This requires the assistance of the animal traction networks in sensitising potential donors of the need to rehabilitate animal traction activities in Sierra Leone.

The status of animal traction in Sierra Leone before the war

The Sierra Leone Work Oxen Programme is the sole institution charged with the responsibility of developing animal traction in the farming systems of the country. The programme has gone through two distinct phases, namely the research phase and extension/development phase. The research phase lasted from 1979 to 1984 and included on-farm and on-station trials, surveys, testing and modification of animal traction equipment, all geared towards tailoring the technology to the agro-socio-economic circumstances of the users (Starkey, 1981; Bangura, 1990; Starkey, 1994).

The research phase showed that work oxen have a great potential in Sierra Leone for the small-scale farmers who still depend on hoes and machetes but are the principal producers of rice, the country's staple food. Equipment such as the plow, harrow and cart were developed with high versatility for use in the varying ecologies.(Starkey, 1981, 1994).

With research conviction, the Sierra Leone Work Oxen Programme moved to the development phase in 1985. During this period the programme strengthened the equipment production aspect at Rolako Work Oxen Technical Centre and adopted a catalytic strategy by working through development agencies to promote the technology among the various target groups. This strategy resulted in the spread of the technology in various parts of the country. The Work Oxen Programme also intervened in areas where there were no development projects, but conditions for ox traction were favourable (Bangura, 1990).

During the research and development phase the programme rendered its services free of charge and the goods were heavily subsidised. This strategy was essential for the project to convince farmers and development agencies to adopt the technology. This approach worked very well and the original 30 sets of oxen and equipment in 1980 multiplied to 2000 sets located in different parts of the country.

The survival and operation of the project was mainly due to donor support and Government of Sierra Leone funding. Donor funding is not permanent and the sponsors felt that a workshop should be organised in 1995 so that local experts of different disciplines could put their ideas together to shape up a model project that was self-sustaining. The project management is currently working on the outcome of this workshop so that the programme will eventually be independent of donor support. However, the effect of the war will make this difficult to achieve in the short term.

The war and its effects on rural Sierra Leone

The rebel war started in 1991 as an incursion of Liberian rebels to Sierra Leone, which was regarded by the Sierra Leone government as aggression. With the presence of Sierra Leoneans resident in Liberia, the rebel leader Charles Taylor took advantage of this situation and forciby conscripted and trained youths, both male and female, to destabilise Sierra Leone. The rebel incursion then became a full-scale war which in 1995 affected almost every part of the country.

The rebel war in Sierra Leone is similar to other guerilla conflicts in Africa, since they all have common features such as:

- one section/group of society is politically disgruntled
- massive destruction of property including infrastructure, livestock and civilian life
- indicsriminate killing and massive abductions of civilians, particularly teenagers
- complete collapse of the economic activites of the country
- complete militarisation of the country.

Agriculture is the mainstay of the economy, providing livelihood for over 70% of the population, 40% to the country's GDP and about 10% of its export earnings. Prolonged disruption to agricultural activities is bound to affect every facet of the national economy seriously. Since its onset in 1991, the rebel war has resulted in a fall in agricultural production and has damaged existing agricultural infrastructure severely. In the first year alone, nearly 30% of all cash crops and all livestock including domesticated animals were lost, thereby reducing 1.5 million people to destitution. By the begining of 1995 nearly all districts except the western area had suffered from the effects of war.

The effect of war on the farming population

The total population of the country according to the last census is 4.7 million. The population of the capital city, Freetown, before the war was 500,000. By the end of 1994 over 60% of the total population had been affected by the war and most people in the affected areas had been forced to move away from their farming areas to refugee camps inside and outside the country, or to stay with relatives in the big towns and the rebel free

areas. As a result of the war, the population distribution in 1995 was as follows:

- Freetown has about 1.5 million inhabitants
- refugee centres outside the country have about 600,000 inhabitants
- refugee centres inside the country have about 1 million inhabitants
- the remainder of the popoulation live in the so-called safe areas.

Over 60% of this total population are farmers, who have been moved away from their farming areas. They all survive on food aid provided by international bodies.

Farming activites are limited to the rebel-free areas, predominantly in the northern and western areas of the country.

The effect of the war on livestock

Since the war has affected almost every part of the country, the first victimis of the war after dislodging the people were the livestock. All livestock including domesticated animals in the rebel-affected areas have been slaughtered for food by the armed men.

The effect of the war on animal traction

Animal traction technology has been affected seriously by the war, particularly as follows:

- work oxen farmers were forced to flee their farms, with most of them leaving their oxen behind
- the donor-funded projects that promoted work oxen stopped abruptly as a result of the rebel attacks and the heavy presence of soldiers
- the Work Oxen Programme was temporarily closed at one stage. One of its officers was abducted by the rebels and two others were reported killed. In January 1995, the project manager and a good number of his staff were refugees in the Republic of Guinea
- appropriate cattle for oxen are difficult to find due to the scarcity created by the conflict
- livestock are among the first casualties in any rebel attack on a village. As a result no farmers, even in the rebel-free areas are prepared to invest in animal traction for fear of losing the animals to the rebels or government soldiers

- staff morale in the Work Oxen Programme is very low due to the state of insecurity in the country.

Conclusion

The rebel war has had a devastating effect on the general economic life of the country, and in particular animal traction development. The Sierra Leone Work Oxen Programme has scaled down its activities considerably. The work oxen population has been reduced drastically. Government and donors should put in place a serious work oxen rehabilitation programme after the war. This should include provision for resettlement of farmers and replacement of the cattle herds. The animal traction networks should assist in this respect to sensitise donors on the need to rehabilitate the animal traction development activities in Sierra Leone.

References

Bangura A B, 1990. Constraints to the extension of draft animal technology in the farming systems of Sierra Leone. pp. 324–327 in: Starkey P and Faye A (eds), *Animal traction for agricultural development.* Proceedings of the Third Regional Workshop of the West Africa Animal Traction Network, held 7-12 July 1988, Saly, Senegal. Technical Centre for Agricultural and Rural Cooperation (CTA), Ede-Wageningen, Netherlands. 475p. ISBN 92-9081-046-7

Starkey P, 1981. *Farming with work oxen in Sierra Leone.* Ministry of Agriculture, Freetown, Sierra Leone. 88p.

Starkey P, 1994. The transfer of animal traction technology: some lessons from Sierra Leone. pp. 306–317 in: Starkey P, Mwenya E and Stares J (eds), *Improving animal traction technology.* Technical Centre for Agricultural and Rural Cooperation (CTA), Wageningen, The Netherlands. 490p. ISBN 92-9081-127-7

Meeting the challenges of animal traction

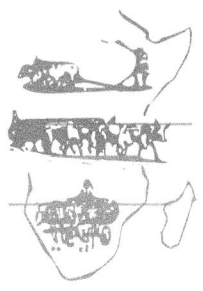

National challenges and perspectives

The challenges of animal traction in Tanzania

by

Hussein Sosovele

Institute of Resource Assessment, University of Dar es Salaam
PO Box 35097, Dar es Salaam, Tanzania

Abstract

After almost a century of activities related to the development and use of animal traction in Tanzania, many farmers are no closer to adopting the technology. What is more disturbing is the realisation that neither the farmers nor the government are near to finding lasting solutions to many pressing problems confronting the use of this technology in Tanzania.

This paper identifies the major challenges facing the technology of animal traction, including constant wavering in government policies and support towards agriculture in general and animal traction in particular, low adoption of the technology and institutional matters. It is argued that most of the challenges facing animal traction in Tanzania are a result of national policies that have given too little attention to the needs of smallholder farmers. Current government practices on economic recovery programmes reflect a shift of priorities away from the agricultural sector. The paper concludes that the challenges facing animal traction in Tanzania can be overcome if agriculture is regarded as the leading sector in the national economy and enough support is provided for its development.

Introduction

As the twenty-first century approaches, it is becoming more obvious that the agricultural sector is not given adequate attention as Tanzania's key economic sector. Statements such as "agriculture is the backbone of the country's economy" are no longer the popular slogans they used to be in the 1960s and 1970s. This is reflected in the current government's shift of policy emphasis to issues such as trade liberalisation and privatisation.

With the formal introduction of a market economy to Tanzania and the foisting of harsh economic reforms, the agricultural sector which was, until the time of the Economic Recovery Program (ERP - 1986), very much underdeveloped, has been left to slump into a technological quagmire.

The Economic Recovery Programme *inter alia*, has meant a drastic cut in government expenditure in various sectors of the economy (including research, development and training). Consequently, the needs of resource-poor farmers are gradually and 'conveniently' forgotten by decision makers. As the government reduces its commitments to agriculture-related activities, the development of animal traction is left in the hands of individual farmers and some donors who may not be able to address all the pertinent issues concerned with its development. This situation raises concern. How do these changes impinge on the development of animal traction? How will agriculture be returned to the centre of development policies in Tanzania?

A brief political economy of animal traction in Tanzania

In order to understand the challenges that face animal traction in Tanzania, it is imperative to put its development within the context of the changing political economy. Animal traction as a production tool is closely related to the policies, strategies and changes that have characterised agricultural production in Tanzania.

The period from 1884–1918

German colonialists continued, intensified and diversified the various uncoordinated attempts and experiments that were started by the missionaries in pre-colonial times to harness animal power for production and transportation. The earliest attempts to use animal power were for transport. These included experiments such as the use of zebras, elephants, horses and camels. Results were disappointing due to diseases and inefficiency of some of the animals for example, elephants were slow and could not carry more than what 28–40 porters could transport (Koponen, 1994). Camels were good in dry desert-like conditions but useless in moist climates. Nonetheless, some encouraging results were achieved with oxen and donkeys.

In 1876–77, the London Missionaries tried to use ox-wagon transport across Tanzania to Ujiji on Lake Tangyanika. The results were not reassuring because all the oxen were lost on the way due to overstrain (Koponen, 1994). After the construction of the roads, experiments on ox wagons produced encouraging results. By 1903, wagons drawn by six oxen became a common sight in Tabora, Mwanza, Dar es Salaam and Lake Nyasa. They were used more extensively in Arusha were Boer settlers began regular ox-wagon traffic to the coast and to Uganda. Nonetheless, oxen were used only on roads that were free from tsetse flies. The most reliable beast of burden continued to be the donkey and the mule as it was in the pre-colonial days.

Alongside transportation, experiments on animal power in production were also undertaken as part of the campaign to increase production. For example, in Dar es Salaam, an Asian company owner used camel-powered machines in the manufacture of oil (Mascarenhas, 1966). However, most hopes were placed on the ox plow, although opinion amongst colonial officers was not convergent. Whilst some officers believed the plow could revolutionise agriculture in Africa, others were more sceptical. Among those who took a more sanguine view were the British Church Missionary Society who believed that the means to "regenerate Africa were the Bible and the Plough" (Koponen, 1994). Some missionaries went even further and argued that the introduction of the plow would reduce the incidence of polygamy (Koponen, 1994).

Others argued that African farmers could quintuple their yields by exchanging the hoe for the plow. Officials decided "the replacement of the primitive hoe cultivation with the plough must be considered the most important means of opening up Africa since the railway" (Koponen, 1994).

Other colonial officers were sceptical about the plow. Koponen (1994) quotes Stuhlman, who was then responsible for the Amani Research Station, calling the campaign for the ox plow as Utopian, and saying that it was based on the imperfect understanding of the dynamics of hoe cultivation. In his view, hoe cultivation was not inferior, but merely different. Stuhlman warned that Germans should not imagine that they could overthrow such a system in a few years.

These views influenced the way in which animal traction was introduced. The first attempts to introduce it were scattered, but in the early 1900s efforts to spread the use of the plow gained a measure of continuity as the colonial office decided to promote it. Plows were distributed to missionaries and European farmers and to some "intelligent natives".

A more concerted state-directed attempt to introduce plows to African cultivators was made in 1910. This push came from Berlin as demand for agricultural produce from the colonies was increasing. By the end of 1910 plows and harrows were sent to almost all the district offices, and like previous efforts, the results were disappointing.

Animal traction failed to become established during German colonialism because (i) efforts to introduce it had been scant, unsystematic and poorly organised, (ii) draft animals were absent in some parts of the country, (iii) in some local areas, the plow was found to be useless or damaging to the environment and (iv) some social and economic factors imposed limitations on the adoption of the plow (Koponen, 1994).

The period 1919–1961

British colonial officers, like the Germans, also attempted to expand and intensify the use of animal power in production and transportation. Efforts that were put into the expansion of roads and railway transportation and marketing helped to increase the adoption of the plow. Transport enabled cash crops to reach the market, and the earned income was invested in plows. By 1945 plows were in use in areas that produced rice, cotton and maize for a wider market because they were cheaper to use than hiring agricultural labour (Iliffe, 1979; Kjærby, 1983; Sosovele, 1991).

After the war most of the ox-plow pioneers were shifting to tractors. Encouragement for the use of the tractors was provided by the colonial state through the promotion of large-scale farming projects. The tractor also became popular among African farmers in the cash-crop areas. The shift of interest to tractors was taking place when the use of animal power had not developed beyond pulling a few plows and carts (Sosovele, 1993). Animal traction therefore, continued to spread where its use was considered profitable and feasible. For example, in Sukumaland cattle and cotton provided the basis for oxenisation to develop widely. The adoption was so high that the colonial office had to introduce the so called 'Sukuma Plough Rules' which confined plowing to heavy soils in order to control soil erosion that was caused by careless ox plowing and overgrazing.

Towards the end of the British colonial period, restrictions imposed on ox plowing were removed and the technology began to get encouragement from the authorities. However, neither an elaborate institutional mechanism nor a policy framework were provided in order to guide the development of animal traction more systematically. Consequently adoption was scant.

The period from 1961–1995

After independence, although the government appeared to favour animal traction, in practice it emphasised the use of tractors through policies such as the settlement schemes, mechanised block cultivation, villagisation, parastatal state farms and special donor-funded projects (Coulson, 1982; Sosovele, 1991). The period between 1961 and mid 1970s was thus characterised by stagnation in the adoption of animal traction. However, the tractorisation drive was difficult to sustain either by the government, the donors or individual farmers. Gradually, the government began to pay more attention to the use of animal traction. This was done through a combination of measures including campaigns, policy formulations (eg The Agricultural Policy, Agricultural Mechanisation Policy), infrastuctural support etc.

In 1970 the Ubungo Farm Implements factory (UFI) went into commercial production of ox plows, hoes and plow parts. Other manufacturers were also encouraged to step up production and distribution of their produce. By late 1984, the Mbeya Farm Implements factory (Zana Za Kilimo or ZZK) started with an installed capacity of 10,000 mouldboard plows per year.

Yet other measures which also contributed to an increase in the adoption of animal traction include projects funded by donors. These include Iringa Oxenisation Project with support from the European Union; Mbeya Oxenisation Project supported by the Canadian government; Tanga Oxenisation Project funded by German government and several others. The Iringa Oxenisation Project was successful in establishing

ox-training centres (OTC) and introducing the farmers to implements other than the plow, However, because of financial limitations, and poor dissemination strategy, most farmers failed to adopt them and continued to use the plow in much the same way as before the project was started. Information about the Mbeya and Tanga oxenisation projects is scant, but these two projects attempted to focus on women users and weeding bottlenecks. The Mbeya Oxenisation Project has already been stopped.

On the economic front, performance from 1978 to the late 1980s was poor, culminating to a major crisis. As a response to the emerging crisis, the government adopted a Structural Adjustment Program (SAP 1983–1985). With regard to animal traction, the SAP policy meant, among other things, giving attractive producer prices to the farmers. In Shinyanga region, this led to a spontaneous rise in the adoption of ox carts for transportation of cotton from the farms to the markets (Sosovele, 1991). Overall, however, SAP policies did not help the economy to any great extent because they were hinged on foreign capital inflow which was not sufficiently delivered.

The failure of SAP, the sheer magnitude of the crisis and external pressure all combined to persuade Tanzania to adopt an Economic Recovery Program (ERP). The overall objective of the economic recovery programme was the gradual attainment of sustained growth in real incomes and output. Specifically, the measures undertaken included the following policy changes: trade reform; agricultural sector reform; monetary and credit policy reform; parastatal restructuring and private sector development and; civil service reforms (Bagachwa et al, 1995). There is no doubt that policy changes implemented on producer prices, crop marketing, input distribution and restructuring of the cooperative unions have stimulated and streamlined production. However, the economic recovery policy in general does not address specific problems that have affected the adoption of animal power in production and transport. In effect, measures such as monetary and credit policy reforms, parastatal restructuring and trade sector reform have negatively affected the adoption of animal-power technologies.

For example, a recent survey (Bagachwa et al, 1995) indicates that one of the major constraints to animal traction is the expense required to purchase a plow (about Tsh 30,000). Many farmers cannot possibly afford this level of investment. Prices have gone up because among others, subsidies have been removed. Credit has become tighter and difficult to get. Consequently, smallholder farmers are affected more than the others because they can not provide collateral for the loans.

Although one of the objectives of the Economic Recovery Programme is to ensure that industries operate at full capacity, most of the industries are now closing due to lack of capital to run them. It is not known whether the closure of most of the textile industries (eg Musoma textiles, Kilimanjaro textiles, Sungura textiles, Ubungo Garments etc) has had any effect on cotton production and consequently on the adoption of animal traction.

Until 1983, UFI factory was getting some of the capital from the government through bank borrowing. Following the implementation of the monetary and credit policy reforms, bank borrowing has been severely curtailed. This has affected the ability of the factory to obtain raw materials, and consequently production has been affected too (see Table 1). According to the General Manager, while they are struggling to run

Table 1: Target and actual production of plows by UFI 1985–1995

Year	Target	Actual production
1985	20,000	1,840
1986	–	6,622
1987	–	13,871
1988	20,000	2,482
1989	20,000	10,186
1990	30,000	12,094
1991	30,000	19,753
1992	15,000	11,438
1993	25,000	4,744
1994	20,000	4,190
1995	10,000	1,899 (to September)

Source: UFI annual reports 1985–1995

the factory under difficult conditions, they expect that by 1996, 50–60% of the steel requirements will be obtained locally. In his view, lack of capital might have affected the ZZK in Mbeya more severely than UFI because UFI had accumulated some savings before the changes were introduced.

Following the adoption of the economic recovery policies, the government is withdrawing from productive activities and instead is encouraging growth in the private sector. Productive sectors (eg agriculture) now receive a declining proportion of government funding (URT, 1994).

Animal traction is likely to suffer most because private investors can invest only where the potential for high margins is great. Hitherto, the adoption of animal-powered technology has been very low and also the agricultural sector has not been given adequate support from the government so as to motivate farmers to produce more. Thus there seems to be very little incentive to attract private investors in this area.

Although there has been very little research and development done in this area, and also while extension has contributed very little to the adoption of animal traction, government withdrawal from the direct productive sector will affect research, development and extension of animal traction.

The challenges ahead

The challenges facing animal traction in Tanzania can be classified in three categories: policy matters, institutional issues and matters related to the technology itself.

Policy issues

In the Rolling Plan and Forward Budget, one of the plan documents (URT, 1994), the government has identified the main tasks facing Tanzania's agriculture in the 1990s and beyond. These include: to achieve self sufficiency in food production; to raise incomes of all Tanzanians; to promote sustainable production and environmental protection; to increase foreign exchange earnings; to produce raw materials for the industries.

One area that will require a new sense of urgency in order to achieve the above objectives is the use of animal power. It will be important first to bring agriculture back to the centre stage of the national development policies. Currently, the government is not paying enough attention to this sector. Policy makers ought to realise that Tanzania cannot be developed by trade liberalisation and similar changes which have produced many 'bare-foot traders'. These energetic men and women can best be organised to produce wealth from the land rather than walking about the whole day selling third-rate imported goods. After all, the majority of Tanzanians depend on agriculture for their livelihood. They can only enjoy the benefits of trade liberalisation and other policy changes if agricultural production improves.

The government must therefore undertake a serious reassessment of its policy priorities and pay special attention to the development of the agricultural sector. For example, government decisions to reduce its commitments to productive sectors will have damaging effects because the decisions have been taken at the time when the agricultural sector is at a low point. Most of the services that depended on government support are likely to suffer because there are not enough private investors to fill the vacuum left by the government. In addition, there are areas (eg infrastructure) which cannot be developed by private investors. Therefore the government must take a leading role in supporting this sector.

Institutional matters

A number of institutional problems must be solved if animal traction is to be used effectively. This is an onerous challenge considering that since the colonial period, well known institutional problems have persistently affected the adoption of this technology. These problems include:

Lack of coordination

Efforts to develop and spread the use of animal traction have been scant, poorly organised and unsystematic. Any future efforts must take into account past experiences to avoid similar mistakes.

Research and development

There has been very little research and development on animal traction. It must be stressed that an efficient animal traction technology cannot develop spontaneously, it must be guided by research. The challenges of the years ahead include how to make research and development on animal traction more attuned to the needs and conditions of the smallholder farmers. For example, research is required to find

solutions to many problems affecting the adoption of animal-drawn weeding technology (for some of the problems see Loewen-Rudgers et al, 1990; Birch-Thomsen, 1993; Sosovele, 1993). Research and development must also be expanded to include research on the impact of the technology on the environment, on socio-cultural issues and attitudes towards the technology, and on the division of labour in the households and the role of women in use of the technology. More importantly, research and development must be directed to how to develop more sustainable production systems using animal traction. Given Tanzania's peculiar circumstances, the government must continue to support research and development on animal traction until cooperatives or farmers' associations are able to run such institutions on their own.

Extension

Extension and diffusion of this technology are very low. Whereas plow usage can vary between 20 and 70% in localised areas, adoption is less than 20% throughout the country. Labour bottlenecks have shifted to weeding and transportation. The main challenge here is how to increase the use of the plow and to expand the adoption of animal-drawn weeding technology and transport. Extension services must be provided with the basic infrastructural support to enable them to reach the target groups and stimulate adoption of other animal-drawn implements.

Supply and distribution of implements

Production of animal-drawn equipment is currently low and mostly confined to the plow because raw materials are not adequately available and the demand for other implements is low. The national demand for plows is only 20,000 annually. If the two factories were producing at full capacity, about 30,000 plows could have been produced each year. Following financial reforms, efforts must be stepped up to ensure that alternative sources of funding are available to the factories so that production continues.

Distribution of the equipment and spare parts must also be streamlined. Currently UFI has outlets in various regions, but it will be important to ensure that a smooth distribution system is in place. More often than not, plows or carts lie idle in rural areas simply because a simple nut or bolt is missing. With the trade liberalisation programme, private traders may be encouraged to participate in the distribution and retail of equipment. The main challenge is for the government to stimulate demand and use of this technology. If demand remains so low, there will not be a sufficient incentive for private traders to invest in animal traction.

Matters related to the technology

Compared with tractors, for example, animal-powered technologies have only a limited ability to work hard soils. Farmers often have to wait for the first rains to soften the soils before cultivation starts. However, this is also the best time for sowing. In a farming system which depends on rainfall, timely planting is not only crucial in determining yields, but it is also vital if the full benefits of other improvements (eg seeds and fertilisers) are to be achieved (Sosovele, 1991). Animal power can plow 0.2–0.3 ha a day (Arnon, 1981), but this area is reduced if the soils are heavy. Some of the tools which could be used in heavy soils are too heavy, require large teams of animals to pull them and are expensive. The challenge ahead is to produce an affordable and suitable tool which can be used in heavy soils.

Similarly, it appears that animal power can be used in areas which have moderately high rainfall. However, these areas also contain tsetse flies. The use of animal traction can be limited by tsetse flies unless extensive bush clearing and other control measures are taken. Burning, use of chemicals and biological control are expensive, and may not be environmentally acceptable. The main challenge ahead is to how to direct research and development to finding solutions to this problem.

Conclusion

The development of animal traction has been influenced by the changes that have characterised the political economy of Tanzania. The colonial and post-colonial governments made efforts to increase the adoption of animal traction but, in all cases the measures were scant, unsystematic and poorly organised. In more recent times, economic reform programmes have further affected the adoption of animal traction, and the sector does not command the same importance as it did in the 1960s and 1970s.

One of the main challenges facing animal traction now and in the future is how to bring agriculture back to the centre stage of development policies in Tanzania. Once agriculture is recognised as the leading sector in the country's economy and development policies, efforts will have to be made to ensure that animal traction is given adequate support to overcome some of the institutional and socio-cultural hitches that have affected its adoption in the past.

References

Arnon I, 1981. *Modernization of agriculture in developing countries: resource, potential and problems.* John Wiley and Sons, Chichester, USA. 405p.

Bagachwa M S D, Shechambo F, Sosovele H, Kulindwa K, Naho A and Cromwell E, 1995. *Structural adjustment and sustainable development in Tanzania.* Dar es Salaam University Press, Dar es Salaam, Tanzania. 398p.

Birch-Thomsen T, 1993. *Effects of land use intensification - introduction of animal traction into farming systems of different intensities: the case of south-western Tanzania and northern Zambia.* PhD Thesis. Part 1. Institute of Geography, University of Copenhagen, Denmark. 130p.

Coulson A , 1982. *Tanzania: a political economy.* Clarendon Press, Oxford, UK. 352p.

Iliffe J, 1979. *A modern history of Tanganyika.* Cambridge University Press, Cambridge, UK. 616p.

Kjaerby F, 1983. *Problems and contradictions in the development of ox-cultivation in Tanzania.* Research Report 66. Scandinavian Institute of African studies, Uppsala, Sweden and Centre for Development Research, Copenhagen, Denmark. 163p.

Koponen J, 1994. *Development for exploitation: German Colonial policies in mainland Tanzania.* Helsinki/Hamburg. 748p.

Loewen-Rudgers l, Rempel E, Harder J and Klassen Harder K, 1990. Constraints to the adoption of animal traction weeding technology in the Mbeya region of Tanzania. pp 460–471, in: Starkey P H and Faye A (eds) *Animal traction for agricultural development.* Proceedings of workshop held July 7–12, 1988, Saly, Senegal. Published on behalf of the West African Animal traction Network by technical Centre for Agricultural and Rural Cooperation, Ede-Wageningen, The Netherlands. 479p.

Mascarenhas A, 1966. *Urban development in Dar es Salaam.* MA Thesis, University of California, Los Angeles, USA. 221p.

Sosovele H, 1991. *The development of animal traction in Tanzania: 1900–1980s.* PhD Dissertation. University of Bremen, Bremen, Germany. 352p.

Sosovele H, 1993. Constraints to the adoption of animal-drawn weeding technology in Tanzania. pp 170–177 in: Starkey P and Simalenga T (eds), 1999. *Animal power for weed control.* A resource book of the Animal Traction Network for Eastern and Southern Africa (ATNESA). Technical Centre for Agricultural and Rural Cooperation (CTA), Wageningen, The Netherlands. ISBN 92-9081-136-6

URT, 1994. *The Rolling Plan and Forward Budget for Tanzania for the Period 1994/95–1996/97 Vol. 1* Government Printer, Dar es Salaam, Tanzania. 138p.

Animal traction in Mozambique: results from a survey of small-scale farmers

by

Alfredo de Toro and Alfredo B Nhantumbo

Agricultural Mechanisation Section, Faculty of Agronomy and Forestry Engineering
Box 258, Maputo, Mozambique

Abstract

In 1994 the Department of Statistics of the Ministry of Agriculture of Mozambique conducted a national survey in which 2749 households were interviewed with the goal of producing basic information on subsistence and small-scale farmers. A multistage sampling method was used as follows: 10 provinces x 3 districts x 8 villages x 12 households = 2880 households.

The main results relevant to animal traction were:

- *Cattle are the main base for animal traction in Mozambique. Less than 1% of the households use donkeys for draft power.*

- *From the estimated 2,465,000 households for the whole country, about 100,000 own some 470,000 cattle, mainly concentrated in the southern provinces.*

- *About 60% of the farmers owning cattle use them for animal traction, mainly for plowing and to some extent for transport. Animal-powered planting and weeding are rare.*

- *Gaza Province has the highest use of animal power with 16% of households using animal traction. In the northern provinces, the use of animal power is negligible.*

- *Although the farmers owning and using cattle for work purposes are a minority (about 60,000 households or 2.5%) they plowed about 550,000 ha or 12% of the country's cropped area. They cultivate a larger area (average 3 ha) than farmers using manual labour (average 1.7 ha). In addition, they hire out their services to other farmers. Farmers hiring draft animals cultivate 2 ha on average.*

- *About 1.5% of all households owned ox carts; this corresponds to 34% of farmers that own draft animals.*

- *As a consequence of the war, the cattle population was reduced drastically. Families who previously owned and used cattle lost their herds. The survey shows that 8% of farmers own plows while only 4% own draft animals.*

- *As in other countries, more men than women use animal power.*

Introduction

Animal traction was the main power source for the family farming sector in southern Mozambique during colonial times. The number of cattle was estimated to be 1,400,000 (SV, 1973), most of them owned by small-scale farmers. Oxen were used mainly for plowing and to a limited extent for transport. As at present, planting and weeding using animal traction were rare.

As a consequence of the war and the disruption of the economy, particularly in the agricultural sector, the number of cattle has been reduced drastically. At present the cattle population is recovering.

The peace accord signed in 1992 widened the possibilities of countryside development, making it possible to implement new policies and projects in the agricultural sector that were unrealisable during the war. The need for reliable data on the actual situation of the subsistence and small-scale farmers (the 'family sector', according to Mozambique terminology) was evident. However, statistics covering the whole country did not exist in 1993, so the Ministry of Agriculture restarted carrying out annual agricultural surveys. Based on a survey carried out in 1994 and analysing the data from an "animal traction point of view", this paper presents information on the use of animal power in Mozambique.

Methods

In 1994, the Department of Statistics at the Directorate of Agricultural Economics in the Ministry of Agriculture conducted a national survey in which 2,749 households were interviewed. The goals of the survey were:

- to produce basic statistical data on the family sector
- to have a reference point to evaluate future changes for field data collection and analysis
- to further refine the methodologies in the field and centrally.

The main topics of the questionnaire were:

- household composition and type
- agricultural production and its utilisation
- animal production
- resources (land, labour, farm implements, seed, fertiliser and pesticides).

The following multistage sampling method was used:

- three districts in each one of the ten provinces of the country were chosen in a deterministic way from a priority list (Figure 1)
- eight villages in each district were selected at random with a probability according to their size
- within each village, twelve households were chosen randomly.

In summary the planned sample size was:

- 10 provinces x 3 districts x 8 villages x 12 households = 2880 households.

The actual number of households interviewed was 2749, reflecting a 4.5% no-response rate. This was considered statistically acceptable.

To draw conclusions at national level, the provinces and districts were weighted as follows:

- District weight = total population in the district / sum of the number of members of the households interviewed
- Provincial weight = total population in the province /sum of the number of members of the households interviewed
- National weight = sum of the results from provincial level.

The field work of the survey was executed from August to October in 1994 by staff from the Ministry of Agriculture. The personnel involved were trained properly in the objectives of the

Figure 1. Map of Mozambique showing the ten provinces and the districts surveyed

survey and the questionnaire. The raw data were entered on computer and stored and analysed using the Statistical Analysis System (SAS) program.

Results

Main crops and cultivated area

The main crops cultivated by the family sector are presented in Table 1. The chief crops are maize, cassava, sorghum and rice. The only cash crop of some significance for the family sector was cotton (4% of the total area).

The level of farm mechanisation

The majority of farmers in the family sector still cultivate by hand. From the weighted estimate of 2,464,000 households for the whole country, 87% use only human energy for land preparation, 8% also utilise animal traction – owned, hired or borrowed – and 5% tractor mechanisation, largely through hiring (Table 2). Only 2.5% of households both own and use cattle for work (Table 3). Animal traction is predominantly practised by a minority of the households in the southern provinces of Manica, Inhambane and Gaza (Table 3). As the utilisation of tractor and animal traction is limited to plowing and to a certain extent to transport, the importance of hand labour for agriculture is greater than is reflected by the figures presented in Table 2.

Table 1: Estimated number of households cultivating maize, cassava, sorghum and rice, total area per crop for the country, mean area per household, mean yields and range of district mean yields for the family sector

	Number of holdings (000's)	Total area cultivated (000 ha)	Mean area per household (ha)	Mean yield (kg/ha)[1]	Range of district mean yields (kg/ha)[1]
Maize	1,923	1,737	0.90	350	74–1,435
Cassava[2]	1,383	786	0.57	540	147–1,003
Sorghum	701	339	0.48	400	113–747
Rice	644	302	0.47	610	88–1,054
Total	**2,464**	**4,489**			

Notes
1) 1994 survey, average yield for 30 districts, no distinction made for multiple cropping
2) Cassava yielded below average in 1994

The area cultivated per farmer owning oxen and/or plows was about 80% larger than that of households using manual labour only, but when expressed per adult equivalent the differenceasis smaller (see Table 2). Farmers owning oxen and plows cultivated an average of 3 ha while farmers hiring animal traction cultivated 2 ha (Table 2).

The total area cultivated by the family sector using animal traction for primary tillage is estimated to be about 550,000 ha or 12%, of which

Table 2: Estimates of number of farms, area cultivated and labour use for hand-, animal- and tractor-cultivation

	Manual labour only	Animal traction Own	Animal traction Hired or borrowed	Tractor hiring
Number of observations	2,247	160	124	218
%	82	6	5	8
Number of farmers (weighted)	2,134,000	109,000	105,000	117,000
%	87	4	4	5
Area cultivated, weighted				
total (ha)	3,585,000	331,000	221,000	352,000
%	80	7	5	8
Mean (ha)	1.7	3.0	2.1	3.0
Standard deviation	1.7	2.7	1.7	2.6
Labour per holding				
Adult equivalent	4.0	5.2	4.4	5.3
Standard deviation	2.1	2.7	2.3	2.5
Cultivated area per adult equivalent (ha)	0.4	0.6	0.5	0.6

Notes:
1) Animal traction and hired tractors were used for plowing only in most cases

Table 3: Estimates of cattle popoulations and use for draft in the provinces of Mozambique

				Draft cattle		Draft donkeys	
Province	Number of farmers (thousands)	Farmers owning cattle	Cattle population	Farmers owning and using	Number of animals	Farmers owning and using	Number of animals
Niassa	139	1,800.	11,700·				
Cabo Delgado	264					1,300[1]	1,300[1]
Nampula	485	1,300	1,300				
Zambezia	573						
Tete	174	8,600	58,000	1,400	5,000	358	1,400
Manica	103	11,000	70,000	9,400	30,800		
Sofala	178	10,200	42,000	2,100	4,300		
Inhambane	274	20,600	43,000	17,000	37,200	4,800	6,900
Gaza	181	43,700	229,000	29,200	85,000	3,300	4,900
Maputo	95	3,700	13,000	2,200	6,000	900	1,100
Total	2,465	100,900	470,000	61,700	168,200	10,700	15,700

Notes
1) One of these two figures is incorrect - it is extremely unlikely that all donkeys are used for draft

about 60% is cropped by farmers owning animal traction and the rest by farmers hiring or borrowing animals (Table 2).

Estimation of the number of working cattle and donkeys by province

The total number of cattle is estimated to be about 470,000 which are owned by some 100,000 farmers or 4% of the total 2,464,000 households (Table 3). As shown in Table 3, cattle are concentrated in the southern provinces, mainly in Gaza (which has about half the country's total) and Manica. About 60% of all farmers owning cattle in the country use their animals for traction, but in the provinces of Tete and Sofala only about one farmer in five uses cattle as draft animals.

Farmers using donkeys for work are a minority, about 0.4% of the total number of households. The total number of donkeys is estimated to be about 15,000. Like cattle, donkeys are concentrated in the southern provinces.

Implements

As mentioned, the plow is the most important implement utilised by farmers with animal traction. From the total sample of 2,749 households, 215 reported owning plows (about 8% of the total sample) but only 85 of them owned draft animals. About 130 households owned plows but did not use animal traction, probably as result of cattle losses during the war. On the other hand, only seven farmers who owned draft animals reported hiring plows.

Few households own ox carts, less than 2% (43 households) of the total sample, or about 34% of the farmers having animal traction. As in the case of plows, a considerable proportion (16%) of the farmers owning ox carts do not have draft animals. Hiring carts was much more common than borrowing, 40 farmers reported hiring as opposed to ten borrowing.

As planting and weeding using animal traction implements is extremely rare in Mozambique, there were no questions about this in the survey questionnaire.

Table 4: The proportion of farm households requiring paid external labour for different tasks

Number of households	Total	Primary cultivation	Planting	Weeding	Harvesting
Total in survey	2,749				
Number paying labourers unweighted	516	226	96	326	137
%	19	44	19	63	27
Estimated country total[1]	2,134,000				
Number paying labourers weighted	484,000	216,000	74,000	285,000	131,000
%	23	45	15	59	27

Notes
1) Weighted number of households based on the number of responses to this question

Animal traction and gender

As in other parts of the world, animal traction is biased to male use. Only 5.5% of the households possessing draft animals did not have male adults compared with 10% in the group without draft animals. Most of the households in Mozambique therefore have at least one male adult and this accords with the available labour per household (Table 2).

External paid labour

From the total sample (2,749 households), 19% of the households (516) required hired labour for agricultural tasks. A breakdown of the number of households requiring hired labour for different tasks is shown in Table 4. Most of the households required one or two person-weeks of hired labour, with a mean of 3.4 person-weeks and a standard deviation of 7. Weeding and primary cultivation required the most supplementary labour (Table 4). As primary cultivation is a power-demanding operation about 13% of the households (Table 2) executed it with the help of animal traction or hired tractors.

Discussion

The surveys executed by the Ministry of Agriculture in 1993 and 1994 are the largest surveys of family sector agriculture for at least 20 years. As the primary intention was to obtain statistics at national level, the level of detail is not

really sufficient to examine a specific issue, as in this study. However, they are one of the main information sources, if not the only one, on family sector agriculture with national coverage. One limitation is that the districts in each province were not selected randomly but were chosen from a list of "priority districts", which, in general, have better agricultural conditions. In this way, the surveys were biased to some extent.

Animal traction is still utilised by a minority of farmers in Mozambique and is used mainly in the three southern provinces. Gaza was the province with the highest rate of utilisation with 16% of the farmers owning draft animals, followed by Manica and Inhambane with 9% and 6% of the household possessing working animals, respectively.

The geographical distribution of cattle poses limitations to a rapid expansion of animal traction due to:

- high crop failure risks in the southern provinces due to erratic rains, combined with sandy soils which have low water retention capacity
- lack of a tradition of cattlekeeping and animal traction in the northern provinces where better agricultural conditions prevail
- the presence of tsetse flies and trypanosomiasis, particularly in zones distant from the coast, which has a significant

influence on the geographical distribution of cattle.

The supply of implements did not appear to be a constraint. Agro-Alfa, a national producer of animal traction equipment, has a good manufacturing capacity but marketing is very weak. This is reflected by absence of traders of such equipment in the provinces. It is often easier to procure plows in neighbouring countries.

Ownership of animal traction does not seem to have had a big impact on increasing the area cultivated. The survey results shows the mean area cultivated by such farmers is 3 ha (Table 2). Several factors contribute to this:

- since animal traction is used only for land preparation, labour shortage for other operations, in particular weeding, restricts the area that can be cultivated
- the need for cash motivates such owners to rent their services
- poor and/or absence of marketing services. The prolonged war was responsible for the disruption of the basic marketing

infrastructure (roads, storage facilities, transport means) and displacement of people.

Power availability is one of the main constraints for the family sector for increasing production and improving their living conditions. With the present crop yields and expected increasing population, tractor mechanisation is not a viable alternative in the foreseeable future. Although not more than 4% of farmers use animal traction they plow about 12% of the cultivated area of the country. These two figures indicate that animal traction can be a realistic alternative power source in the future for many farmers who currently cultivate by hand.

Acknowledgements

The authors would like to thank the Department of Statistics of the Ministry of Agriculture of Mozambique for providing all the raw data without which this paper would not been possible. Our gratitude to Mr Lars Carlsson for their valuable assistance in accessing 'SAS-file' and examination of the draft as well as to Dr Gilead Mlay for the English language correction and the contents. This paper was produced within the framework of the Programme AEEP/UEM, financed by the Swedish International Development Agency (Sida).

Reference

SV, 1973. Serviços de Veterinária. Ministério de Agricultura, Maputo, Moçambique.

The challenges of reintroducing animal traction in post-war Mozambique

by

Alix von Keyserlingk

Livestock Coordinator, World Vision Tete, Mozambique World Vision Tete
Private Bag 11214, Nelspruit 1200, South Africa.

Abstract

In post-war Mozambique, livestock production is still recovering from the adverse effects of the 16-year civil war. Many factors, both positive and negative, make this country a unique location for work with livestock in general and animal traction in particular. Tete Province is an area of priority due to the long tradition of husbandry in the province. However, there are four main constraints to the development of the animal traction sector: 1) an acute lack of animals due to depletion during the war; 2) a lack of implements as Mozambique's local hardware production has not reached large-scale production, and what is produced is mostly very expensive; 3) a lack of credit schemes which makes it very difficult for local farmers to buy animals or tools even when they are available; and 4) the lack of an extension network to disseminate animal traction technology.

The aims of the Ministry of Agriculture's programme, in conjunction with other organisations working in the province, are to reestablish family sector livestock production as well as to make animal traction an accessible alternative for farmers who have the initiative and the conditions for using this technology.

Introduction

As the third largest country in southern Africa, Mozambique has a broad range of agronomic zones ranging from coastal rainforest through semi-arid drought-prone lowlands to the upland sour grasslands. The traditions and cultures of the various peoples also vary, many having their own farming systems, mechanisms of production and development objectives. As the country opens up after the savage 16 years of civil war, which was preceeded by nearly 10 years of independence struggle, the population is attempting to develop the high agricultural potential of the country by re-establishing both their traditional farming systems and also the basic infrastructure which suffered so much. Positive factors include the low population pressure (20 people per km^2, with about 75% living in rural areas), the strong optimism that is found in the population due to the opportunity for a new start after the war, a relatively high soil fertility in many areas due to land lying fallow during the war. However, these are counterbalanced by negative factors, including the difficulties of recuperating a post-war economy, weak government structures due to lack of resources and lack of infrastructure such as transport and communications.

Constraints to dissemination of animal traction: an example of Tete Province

Tete Province lies in the northern central part of Mozambique. With an area of 100,700 km^2 it juts out peninsular-like between Zimbabwe, Zambia and Malawi. Geographically very varied, Tete ranges from an upland plateau 1000 m above sea level, with relatively high rainfall (800–1300 mm) and sour veld pastures mixed with *miombo* woodland to the lowland Zambezi valley (150–500 m above sea level) with a semi-arid climate (400–700 mm rainfall) and *mopane* savanna. It was once an important centre of animal production, and still has many advantages which encourage animal production and with it the use of the animal in multiple ways to improve the farming system. Due to often variable climatic conditions, the local farming systems have traditionally diversified and included livestock in their means of production in order to guarantee a food source even in years of poor rainfall. However, animal production in itself is restrained at present by factors including tsetse-fly infestation and lack of effective veterinary and technical assistance. Furthermore, animal traction is severely constrained on a larger scale because cattle, implements and credit schemes are all very difficult for rural farmers to procure.

Lack of animals

During the war, the animal population suffered greatly. For example, from 1973 to 1992 the cattle population dwindled from 196,000 to 33,000 head (see Table 1). Since then, the cattle population has been slowly recovering, partly aided by the government and other organisations, partly 'spontaneously'. However, the human population has also increased sharply to 1.1 million inhabitants due to the large number of refugees returning to Tete Province (UNHCR estimates 821,000 refugees have returned since the peace accord; UNHCR, 1995). The number of animals per inhabitant has decreased to 0.04.

Tete province is divided into 13 districts, some of which have a stronger emphasis on animal production than others. In 1973, the main cattle areas were Angónia, with 79,000 head, Changara, with 66,000 head, Mutarara, with 17,000 head, and Moatize, with 10,000 head (Estatistica, 1995). The distribution of tgoats was similar, with Changara at 54,000, Angónia at 40,000, Mutarara at 20,000 and Moatize at 4,000 head carrying the bulk of the caprine population.

The cattle registration of 1992 revealed that in Angónia, only 3% of the cattle remained, in Changara, 24%, while Mutarara was totally depleted. Moatize, being a semi-urban area near Tete City, had been a concentration point during the war so 50% of the 1973 cattle population was registered in 1992.

The poor infrastructure within the country, and the common poverty have worsened the situation to the extent that it is almost impossible for local farmers to alleviate the situation. They lack funds and means to arrange for cattle transport, it is impossible to organise credit, areas with enough cattle to be able to sell are few and far between and most cattle owners, even in areas with larger cattle populations, are not willing to sell because they too are still restocking. The veterinary authorities have limited means of transporting cattleeven if the farmers were to approach them for assistance.

The Ministry of Agriculture and its Provincial Department of Agriculture and Fisheries (DPAP) have been acutely aware of the problem, and have started various restocking schemes, which, by using a two- to four-year credit system will hopefully help to rehabilitate the cattle population in as short a time as possible.

Lack of implements

Various animal traction projects have been launched in different parts of the province. Most of these have been involved in the importation of animal traction implements into the country, rather than attempting to establish local manufacture of these tools. This, combined with the present difficulties of finding raw material in the country, has resulted in a marked lack of local manufacturers of animal traction implements. There are a few manufacturers in Maputo, the capital, which is about 1500 km to the south of Tete Province However, inquiries into the availability and prices of these implements have not led to any viable information, quantities

Table 1: Changes in the animal population of Tete Province 1973–1992

| Year | People | Animal population | | Animal density (no./km²) | | Number of animals/habitant | |
		Cattle	Goats	Cattle	Goats	Cattle	Goats
1973	526,000	196,000	143,000	2.0	1.4	0.4	0.3
1980	831,000	112,000	85,000	1.1	0.8	0.1	0.1
1992	319,000	33,000	45,000	0.3	0.4	0.1	0.1

Source: Sotomane, 1993

produced are low, and transport prices so high that in fact importation from neighbouring countries is economically more viable. In Beira, 600 km south of Tete, plows can be purchased at the relatively affordable price of 700,000 Meticals (10,000 Mts = US$ 1), but in very small quantities for the time being. In Tete itself, plows can be bought from the state-run hardware suppliers, but here the prices reflect not only high manufacturing costs, but also the high costs of transport from Maputo, one plow being sold at 1,000,000 Mts. As a comparison, the value of one head of cattle, at least 3 years old, currently varies between 800,000 and 1,500,000 Mts in the province).

Besides the standard mouldboard plow, some of these manufacturers also make ridgers and simple triangular harrows. However, no planters, diagonal harrows, cultivators or other implements are at present being manufactured on a large scale.

The prices of these locally manufactured implements are largely unaffordable for farmers in the rural areas. The majority of cattle owners find themselves incapable of obtaining tools for animal traction, although some, especially in locations close to the Zimbabwe or Malawi borders, manage to smuggle in a few implements, mostly plows and ox carts.

There is a possibility that local importers will start with the regular importation of animal traction implements from the above-mentioned countries. However, this will only become reliable once the local demand becomes more regular, which at the moment is not the case, due to the lack of liquidity of the farmers. This system would at any rate result in animal traction elements priced significantly above those of the countries exporting the goods, as transport (+25%) and import tax (+20%) will easily increase the selling price by more than 50% when the merchant's margin is also taken into account.

Lack of credit schemes

Another factor impeding rapid reestablishment of animal traction in appropriate zones of Tete Province at the moment is the lack of a credit system which would enable farmers to obtain animals and tools, start work, and repay the costs as they realise the economic benefit of the improved technology. The Banco Popular de Desinvolvimento (Peoples' Development Bank),

which is the only bank accessible in rural Mozambique, and even here in a very few locations, charge very high interest rates (40–46%) due to inflation in the country. Also, they are not willing to give credit to small-scale farmers. They insist on guarantees of property, valuable possessions and the like, which such a client cannot give, since the land they farm is not theirs (the new land-ownership policy of Mozambique clings to the socialist pillar of land belonging to the state). They are more likely to concede credit to a farmers' association, if this association can prove backing from a third party. There are no alternative credit schemes, for example schemes backed by international development organisations, available in central and northern rural Mozambique. To date there has been little long-term developmental input into these parts of Mozambique, as international organisations are currently closing down their relief efforts and turning their attention to developmental issues.

It will probably take up to five years for any large-scale credit scheme geared towards the small-scale farmer and operated wholly on internal funding to establish itself. Until then, farmers requiring credit will be forced to apply for informal credit from different organisations active in agricultural rehabilitation and development.

Lack of an extension network

During much of the recent history of Tete province, the farming population has often been left to its own devices for survival and development, because there has not been a strong extension network in existence which was able to cover the major agricultural areas of the province. In terms of animal traction, this means that there has been no technical support for the rural population which would enable interested farmers with cattle to either renew their tradition of animal traction or to adopt this technology due to suitability.

The promotion of animal traction in Tete Province is at the moment very much reliant on spontaneous adoption based on neighbouring families learning from one another. However, there are now districts with a growing cattle population where animal traction is not being revived, nor is it being introduced, even though rapid rural appraisals indicate a potential for this technology.

The need for a more formalised system to make animal-traction technology known to more than just those villages that have up to now been successful in reinstating traction has been identified by the DPAP and other organisations working with livestock, yet funding and the constraints discussed above limit the activities in this sector.

Short- and long-term activities

DPAP priorities and activities

Development of the livestock sector

As mentioned above, one of the main constraints in developing animal traction in Tete Province is the lack of animals available to the farming sector in general, due to the great losses suffered during the war. Thus the DPAP has defined restocking as being a priority in rural Mozambique. Various activities related to the reestablishment of the cattle population in the Province of Tete are being promoted.

A nationwide restocking scheme that imports animals from nearby Zimbabwe has been developed. The goal is to distribute the animals to the family sector based on a uniform credit scheme managed by the government extension team. The imported animals are initially breeding stock, as 95% are female and 5% are male. It is hoped that the cattle population will in this way expand rapidly and thus put the family sector farmer in a position to maximise the use of his cattle, be it for traction, sale, manure production, banking system, etc.

Mozambique is part of the regional, European Union-funded campaign against trypanosomiasis, which is being carried out in Zimbabwe, Zambia, Malawi and now Mozambique. This year, work in identifying trypanosomiasis infestation has begun, with a team surveying cattle populations in locations of high cattle density. As the first phase of the campaign involved the charting of trypanosomiasis infestation in the cattle population, the second phase will concentrate on mapping the prevalence of the tsetse fly in the province. Results of this campaign have not been published yet, but they will shortly be made available, so that zones appropriate for increased cattle production will be identified.

The rehabilitation of animal health posts and dip tanks is also a priority, with livestock centres and dip tanks being rehabilitated slowly.

Training centre for oxen in Angonia District

In Angonia District, an upland district of Tete provincelying on the border with Malawi, the DPAP has been rehabilitating the livestock centre, building up a nucleus herd of the indigenous Nguni cattle, a small, hardy zebu. This year, a traction training programem has been started with 70 young oxen, who will be trained on-station, and then made available for purchase by the local population on a credit scheme. The training scheme is scheduled to take three months, after which the oxen will be trained to pull ox carts as well as plow. Although this means that for the 1995/96 season the oxen will not be trained and sold, animal traction in Angonia is used most commonly for transport, so that the young oxen will have a year to work in transport, and be used for plowing in the 1996/97 season.

World Vision activities

Although World Vision has recognised the need for restocking, this is a very cost-intensive activity, with a general estimate being that the importation of one head of cattle, included sanitary inspection, transport and quarantine, costs US$ 1,000. Thus it has not been possible to make restocking a project priority. However, much potential for work in animal traction has been noted by the project, and certain activities have been started, with the hope of increasing work in this area.

Animal traction workshops

In many rural areas, it has became evident that there was great interest among livestock owners in restarting using animal traction if at all possible. Many farmers used to use their cattle for traction, but though the tradition is present in the area, have either lost the skill or lack encouragement via an extension network. A need for a basic introductory workshop was identified, and, supported by the DPAP and by a Mozambican animal traction specialist, a two-week course was held. The course included training of participants in the training of oxen for one-person handling, building specialised yokes for specific traction activities, making of rope from animal hide, and other local skills relating to animal traction.

The course was considered a success by the participants, who all agreed that a follow-up course was needed in the near future to deepen knowledge in improved cultivation methods, including weeding and ridging using animal traction, both of which are not common practices in the area. The participants of the workshop will act as contact persons in their home villages, who will train other interested livestock owners in the newly acquired techniques. As development of the animal traction sector intensifies, the project will facilitate further workshops of this kind in all areas of project activities.

Improvement of availability of animal traction implements

There are many solutions to the problem of lack of implements, but none is straightforward. In a short-term project, members of the livestock farmers' groups working in the project will be able to buy plows and other implements through World Vision, who will set up a simple and an as short-term as possible credit scheme to enable farmers to buy what they require. The implements thus sold will be obtained as far as is possible in Mozambique, since this avoids the high import taxes and transport costs. However, the project will encourage local importers to import traction implements from Zimbabwe, where two manufacturers produce quality implements, one of these selling on discount to Mozambique. Since the project is in its initial phase its success cannot yet be judged.

On a longer-term scale, the project hopes to encourage farmers' associations in rural areas to establish local agricultural input stores, which would stock and sell inputs including fertiliser, seed, plows and spare parts. This too would initially involve a credit scheme supported by the project to get the programme started.

Improvement of animal health in rural areas

As one of the main restraints in introducing animal traction is the lack of animals, effort has gone into the reestablishment of the health support system for livestock. In Changara district, one of the traditionally strongest livestock areas of Tete province, livestock farmers have not had the benefit of a comprehensive animal health care for as long as 10 years. The project has used transport facilitation and technical support to encourage farmers in villages with over 500 head of cattle to build handling corrals out of local materials. These corrals have become centres of general and regular veterinary assistance. Thus, the conditions for increased livestock production are slowly evolving in these zones, and more and more farmers are finding themselves in a position to acquire cattle.

Conclusions

As conditions in post-war Mozambique improve dramatically over the next few years, efforts by the DPAP and other organisations will not only be able to stabilise the precarious situation of the cattle population in Tete province, but also to encourage the integration of cattle into the farming systems. Animal traction is seen as a priority for this area, as it possesses the tradition. However, the crucial constraints of lack of animals, implements and credit schemes must be tackled both by the national support system as well as by organisations working in animal traction in the provinces. Only in this way can farmers maximise the potential for animal traction in Mozambique.

References

Estatistica, 1995. *Estatistica 1995*. Servicos Provinciais de Estatistica - Commissiao Provincial do Plano, Tete, Mozambique.

Sotomane I E, 1993. *Estrategia de Desenvolvimento Pecuário*. Direccao Provincial de Agricultura, Servicos Provinciais de Pecuária, Ministério de Agricultura, Maputo, Mozambique.

UNHCR, 1995. *Statistical Report 1995*. UNHCR, UN Publications, Geneva, Switzerland.

Animal draft power in South Africa: past, present and future

by

Richard Fowler

Agricultural Research Council, Grain Crops Institute
Private Bag X 9059, Pietermaritzburg 3200, South Africa

Abstract

The indigenous peoples of South Africa used cattle for riding and packing. The European settlers of the 1600s introduced plows, carts, horses and later donkeys, but early in the 20th century steam and internal combustion engines became the preferred sources of draft power. Animal draft remains, however, often the most profitable and environmentally friendly power source. It is used by possibly 40% of South Africa's 1.3 million functional rural households (including a number of large-scale commercial farmers) and its wide and more efficient adoption could markedly increase farmer incomes and stabilise rural communities. Also, to quote the first basic principle of South Africa's Reconstruction and Development Programme, animal draft power could help to "harness all our resources in a coherent and purposeful effort that can be sustained into the future".

Introduction

Mechanisation does NOT necessarily mean tractorisation, or the use of fossil fuel. Animal draft can also be (and often is) *mechanised*, and on many small (and some large) farms it is by far the most economic and environmentally friendly source of power. However, its advocates, and in fact all those involved with mechanisation, face a fundamental problem.

Especially when we in South Africa discuss small-scale mechanisation, we are addressing a Third World situation which, for the past 50 years, has been bombarded with First World norms and desires. We are a society which, for the past half century, has been orchestrated and manipulated, directly and indirectly, by Hollywood hype and multi-national self-interest. We are a people who have been indoctrinated in the beliefs that big is beautiful and glossy is glamorous; that guns go bang but do not kill; that tractors are prestigious but do not cost money. As a result, even small-scale farmers want tractors and extentionists recommend their use, even when wisdom and logic often suggest they should avail themselves of animal draft power: for example, for a South African rural household cultivating 1–5 ha, the cost per hectare to obtain draft power might be R 1000 (US$1 = Rand 2 in 1994) to buy a tractor or R 80–500 to hire a tractor, but only R 25 to buy oxen or donkeys.

Past use of animal power

The Khoi and other indigenous tribes in South Africa used oxen for packing, riding and war, but it was the European settlers of the mid-1600s who started using oxen to pull wagons and plows. They also brought in horses and donkeys, and much of the tillage and transport in the next two centuries depended on animal traction. From the beginning of the 20th century, however, fossil fuel power became increasingly important for long-distance transport, mining and large-scale farming systems, and it was primarily only among small-scale farmers and for rural transport that the use of animal traction continued to expand.

Early in the 1950s, however, South African officials and farmers started to succumb to the enormous amount of direct and indirect tractorisation propaganda which flooded the country. Animal draft power was old fashioned, inefficient and time-wasting, so the hype went, and for the next 40 years animal traction was at best ignored, and often denigrated, by officials and educators in South Africa. As a result, the introduction and development of animal draft power equipment practically ceased. Moreover, there was increasing government pressure to destock communal areas, sell off 'unproductive' animals (such as mature oxen and donkeys), buy up old tractors and use subsidised tractor pools, and so the use of animal draft power among small-scale farmers declined. This trend was exacerbated by the resettlement policies and droughts of the past four decades.

Present use of animal traction

Today, few commercial farmers use animal draft power, although there are some notable exceptions: some farmers in the Eastern Cape and elsewhere continue to plow with teams of oxen and a number of farmers countrywide use animals for short hauls, especially around the farm. However, for small-scale farmers increased use of animal traction offers many cost-saving opportunities, especially on short hauls of up to 5–10 km, and on light field work such as planting and weeding.

Various estimates have been made of the extent to which small-scale farmers use animal draft power. In the early 1980s, Lea and Stanford (1982) found that 23% of the homestead gardens in Ndeleshane in KwaZulu were prepared using oxen, and 18% using donkeys, compared to 18% by tractor. Steyn (1982) found that almost 100% of rural households in the Amatola Basin in Ciskei plowed with animals. In 1988, 25% of rural households at Peddie and 32% at Alice (both in Ciskei) used animals for plowing, compared to 67% in Biyela (northern KwaZulu) and 71% at Nhlangwini (southern KwaZulu) (Auerbach, 1990). In 1987, 83% of farmers in one area of Transkei plowed with oxen, compared to none in another area (Starkey, 1988).

The 1994 animal traction rapid rural appraisal commissioned by the South African Network of Animal Traction (SANAT) concluded that more than 500,000 oxen and 300,000 horses, donkeys and mules are currently being used by 40–80% of the functional rural households in some areas in South Africa (Starkey, 1995). Draft animals are used for a wide range of operations including harrowing, seeding, weeding, mowing, raking, crop lifting, fertiliser spreading, dam building and logging, but the majority of small-scale farmers use animals for plowing and transport. There are probably 200,000 animal-drawn plows in use today (compared to 330,000 in 1964) with annual sales of 6000–8000 units/year. In addition there are about 90,000 cultivators (annual sales about 6000) and 60,000 planters (annual sales also about 6000) being used in South Africa today.

Roughly 50% of functional rural households own animal draft implements (Auerbach and Gandar, 1994). Sales of animal draft power equipment by the major distributor in the 1993/94 season are shown in Table 1. Very few carts are available commercially.

The future of animal draft power in South Africa

Starkey (1995) concluded that some 500,000 oxen and 300,000 equines were currently being used for animal draft power in South Africa. In contrast, Auerbach and Gandar (1994) estimated that, in 1990, 1.1 million bovines (25% of the total cattle) and 108,000 equines (50% of the total horses, donkeys and mules) were being used for draft power (see Table 2). It must be borne in mind, however, that donkeys have been the targets of extermination campaigns, especially in Bophutatswana, and farmers frequently do not admit to using, and even less to owning, them. The number of tractor units in the homelands (see Table 2) are *best guesses* from information presented at the 1990 Mechanization and Irrigation

Table 1: Animal-drawn equipment sales, SAFIM, 1993/94 season

Implement	*Sales in 1993/94 season (units)*	*Approximate current retail (R)*
Plows (4 types)	12,500	480
Cultivators:5 tine	2,300	550
2 and 3 tine	2,200	365
Planters	5,750	1 500
Harrows: zig zag	4,700	350
triangular	250	250
diamond	310	240

Source:Vetsak Coop Ltd (personal communication, 1995)

Table 2: Available draft power in the former South African homelands, 1990

| Homeland | Cattle (000s) | Equines (000s) | Tractor units | |
			Public	Private
Gazankulu	44	8	80	150
KaNgwane	32	1.5	40	200
KwaNdebele	8	0.5	200	50
KwaZulu	379	23	131	1733
Lebowa	126	14	316	2147
QwaQwa	4	<0.5	96	144
Transkei	366	41	2225	1507
Bophuthatswana	117	16.5	421	1039
Venda	34	2.5	147	94
Ciskei	31	1.5	431	265
Total	**1141**	**108.5**	**4087**	**7329**

Source: Auerbach and Gandar (1994)

in Developing Areas Symposium, and Auerbach and Gandar (1994) note that the number of privately owned tractors in the homelands could be underestimated by 200%.

Assuming the figures in Table 2 to be correct, however, and ignoring the possible contribution of equines, in 1990 oxen alone could have plowed all of the area estimated to be cultivated by small-scale farmers in the former South African homelands (see Table 3).

Pimental and Pimental (1979) found that a pair of oxen could plow 1 ha in 65 hours; Auerbach, Nichol and Gandar (1991) found that in KwaZulu a team of four oxen took 18 hours. Pimental and Pimental (1979) found that a 38-kW tractor took four hours to plow 1 ha, whereas Auerbach, Nichol and Gandar (1991) found that in KwaZulu this was taking 6.5 hours. It is fair to deduce, therefore, that the un- or under-utilised bovine draft power in the small-scale farming areas of South Africa is at present equivalent to that which could be provided by 18,000–50,000 38-kW tractors, and that assumes that the 300,000 plus equines are not being used at all.

In other words, animal draft power is a major national asset, whose utilisation could markedly increase both employment (in a country with 40–60% unemployment) and profitability (in a sector in the past primarily financed by direct and indirect subsidy); moreover, it uses renewable energy (grass) and has the potential to appreciate (especially where cows or equine females are used) rather than depreciate in value. Replacing this asset with tractors could cost South Africa R 0.5–1.5 billion in foreign exchange, and these tractors would be dependent on non-renewable (and also largely imported and environmentally unfriendly) fossil fuel.

Certainly major constraints exist, notably:

- the time taken to perform operations (however unemployment is currently 40–60%)

- the often poor condition of animals when maximum draft is required (eg, for plowing), a situation Starkey (1995) believes can and is being partially remedied by *complementarity* – tractors performing the heavy task of plowing, and the more available animal draft power being used to perform lighter tasks whose timing is more critical, such as planting and weeding

- the gender issues influencing bovine use; however women are allowed to handle equines, eg, donkeys

Table 3: Areas which could be plowed using available draft power in the former South African Homelands, 1990

Homeland	Cultivated area (000 ha)	Percentage which could be plowed[a] by	
		Animals	Tractors
Gazankulu	119	82	12
KaNgwane	50	142	26
KwaNdebéle	35	49	34
KwaZulu	589	143	15
Lebowa	345	81	37
QwaQwa	10	80	150
Transkei	534	152	47
Bophuthatswana	334	78	25
Venda	88	86	18
Ciskei	67	103	70
Total	**2171**	**99.6**	**43**

Source: Auerbach and Gandar (1994)
[a] Assuming:
1. 25% of all cattle are used for draft
2. each animal works 50 4-hour days
3. 90 animal hours per hectare plowed
4. operational homeland tractors plow 80 ha/year
5. 46% of homeland tractors could only plow 10 ha/year

- the lack of suitable equipment for cartage, but designs do exist in other countries and development of, for example, the 1-t semi-tipping golovan has commenced locally
- conservation tillage, but systems are evolving or being developed in other countries, notably Brazil, Zambia and Zimbabwe, and the Grain Crops Institute has started to develop and investigate techniques which could be employed by small-scale farmers in South Africa
- the diminishing pool of animal draft power experience and expertise as former practitioners die
- the lack of commonly accepted rental systems for animal draft power, such as those which exist for the hire of tractors
- the negative image of animal draft power.

However, if these constraints can be countered the following, currently increasing, problems will be reduced markedly:
- drain of people and money from urban areas
- premature and excessive loss of soil and water from the countryside
- pollution of oceans and atmosphere
- dependence of farmers on non-renewable fossil fuel
- use of South African-generated wealth to bolster multi-national profits and provide employment in other countries.

Therefore, the future of animal draft power, and in many ways the future of South Africa as a whole, depends on the extent and speed with which:
- the various Departments of Agriculture can be persuaded to take it seriously

- the experience and expertise of other countries (and our own) are synthesised and used
- "new" technologies are developed and introduced
- suitable affordable equipment, especially carts and weeders, is made available to small-scale farmers
- small-scale farmers are convinced that self-reliance *is* better, and that small *is* beautiful.

Finally we in South Africa are being challenged at this time to give effect to our nation's Reconstruction and Development Programme by, *inter alia*, making better use of our country's natural, labour and capital resources to improve GNP, lessen unemployment, improve personal incomes, stabilise rural communities and eradicate the violence threatening the fabric of society.

The first basic principle of the Reconstruction and Development Programme, as set out by President Nelson Mandela in a recent government White Paper (Mandela, 1994), is "to harness all our resources in a coherent and purposeful effort that can be sustained into the future". In this context, "all our resources" means labour, capital and energy, and sustainable means appreciating and renewable, ie animal traction.

If we can get off the slippery slope induced by foreign hype and hyperbole and adopt the high road of wisdom and logic, animal draft power can be a major factor in creating a happier and more prosperous South Africa for all.

References

Auerbach R M B, 1990. *Sustainable agriculture.* Proceedings of a seminar. Institute of Natural Resources, University of Natal, Pietermaritzburg, South Africa.

Auerbach R and Gandar M, 1994. Energy and small-scale agriculture. *EPRET Paper 7.* Energy for Development Research Centre, University of Capetown, Capetown, South Africa.

Auerbach R M B, Nichol G D and Gandar M V, 1991. *The tractor as a multipurpose machine in KwaZulu.* National Energy Council Report No. EOA 5. Institute of Natural Resources, University of Natal, Pietermaritzburg, South Africa. 100p.

Lea J D and Stanford P S, 1982. Crop production practices on residential and arable sites in a peri-urban area of KwaZulu. In: Bromberger N and Lea J D (eds), *Rural studies in KwaZulu.* Development Studies Research Group, University of Natal. Pietermaritzburg, South Africa.

Mandela N, 1994. Government's strategy for fundamental transformation. *White Paper on Reconstruction and Development.* Government of South Africa, Cape Town, South Africa.

Pimental D and Pimental M, 1979. *Food, energy and society.* Edward Arnold.

Steyn G J, 1982. *Livestock production in the Amatola basin.* MSc thesis. University of Fort Hare, South Africa.

Starkey P, 1988. *Animal traction directory: Africa.* Vieweg for German Appropriate Technology Exchange, GTZ, D-6236 Eschborn, Germany. 151p. ISBN 3-528-02038-5

Starkey P, 1995 (ed). *Animal power in South Africa: empowering rural communities.* Development Bank of Southern Africa, Gauteng, South Africa. 160p. ISBN 1-874878-67-6

Animal draft power challenges in Zimbabwe

by

J Francis [1], B Mudamburi [2] and B Chikwanda [2]

1) Department of Animal Science, University of Zimbabwe
PO Box MP167, Mount Pleasant, Harare, Zimbabwe
2) Institute of Agricultural Engineering, PO Box BW330, Borrowdale, Hatcliffe, Zimbabwe

Abstract

More than 85% of communal area farmers in Zimbabwe use animal draft power for tillage and transport. Oxen provide more than 75% of this power, but cows and donkeys seem to be gaining prominence.

Research and development efforts on animal draft power technology are negligible when compared with work on beef and dairy cattle. A lot of work has also been done on implements, but without much consideration of the draft animals used on smallholder farms.

The country has experienced two severe droughts since 1990, but the extent of damage caused to the draft animal population is not known. Other challenges facing animal draft power development in Zimbabwe include the non-existence of appropriate on-farm feeding strategies, poor health management (particularly for donkeys), assessing the impact of training programmes for rural artisans, and development of affordable harnesses for donkeys. Appropriate farmer participatory approaches (including gender analysis) should be used in all efforts aiming to meet these challenges.

For there to be sustainable development of animal draft power in Zimbabwe, educational curricula should be revised and tailored to promote animal power instead of tractor power. Finally, animal draft power workers should share information, engage in collaborative activities, and avoid duplication of efforts. Meagre resources would then be used more efficiently to develop animal draft power.

Introduction

More than 70% of Zimbabwe's human population of about 10.3 million live in communal (smallholder) farming areas (Chiduza, 1994). Almost 90% of these people live in marginal agro-ecological regions III–V, where the soil is characteristically loose, sandy and infertile. Rainfall is less than 650 mm annually, and its distribution in the growing season (November–March) is very erratic. In some cases, the growing season may end prematurely. Dryland farming is, therefore, a very risky undertaking in these regions, and farmers are usually advised to plow early and ideally plant maize (staple crop in the country) by mid-November. This is only possible if animal draft power is available when it is required. Yet communal farmers always face a critical shortage of animal power. Communal farmers also use low input–low output technologies in production. As a result, the gap between the rate of increase in food production (2.2%) and population growth rate (3.2%) continues to widen (Chiduza, 1994).

This paper reviews the availability of animal draft power in Zimbabwe's communal farming areas, and research conducted to date. The major challenges for development of animal draft power are also highlighted.

Sources and availability of animal power

More than 85% of the 1–1.2 million communal farming households in Zimbabwe use animal draft power. Table 1 is a summary of the composition and availability of draft animals in selected areas of the country. Oxen provide about 75% of this power and are normally worked in pairs, although four or six animals are sometimes used in a team.

About 30–45% of communal area farmers have four or more oxen, which they consider as adequate animal draft power (Bratton, 1984; Christensen and Zindi, 1991). Thus, the majority of farmers do not have adequate animal draft power. For example, Francis (1993) found that only 5–8% of the farmers in Chinamhora communal area had sufficient draft animals. Farmers who do not own their own draft animals hire them from farmers who do, but their access to draft animal power is always untimely and associated with poor crop yields (Shumba, 1984; Francis, 1993).

Table 1: Availability of draft animals in selected areas of Zimbabwe

	Makoni	*Chirumanzu*	*Chivi*	*Mberengwa*	*Nswazi*
Agro-ecological zone	II	III	IV	IV	V
Number of draft animals	537	581	407	451	499
Draft animals per household	2.2	2.3	1.6	1.8	2.0
Composition of draft					
Oxen (%)	78	69	48	51	66
Cows (%)	18	22	19	11	9
Donkeys (%)	4	9	33	38	25

Source: GFA (1987)

Farmers may supplement their stock of draft oxen with cows and donkeys if these are available. There is no reliable information on either the number and distribution of draft oxen in the country, or the extent to which cows and donkeys are used for draft purposes. Because the draft capacity of cows is only 60–70% of that of oxen (Howard, 1980), ownership of two cows is normally counted as equivalent to one ox. Farmers believe that subjecting cows to work would result in reduced reproductive performance, so the use of cows for draft could be a useful indicator of the severity of the animal draft power problem. Although donkeys are more suitable for packing and carting tasks, they are particularly important for traction in marginal agro-ecological zones where successive droughts have killed off almost all the cattle.

Major animal draft power challenges
Farmer involvement

A lot of research and development attention has been focused on crops, particularly maize (Shumba, 1984; 1986), even when diagnostic studies highlight shortages of animal draft power and manure as critical production constraints. Considerable work has also been conducted on implements. Some available technologies, such as minimum tillage techniques, have not been widely accepted by farmers, principally because farmers were never consulted to provide an input into the technology development process.

Use and maintenance of equipment

Little or no literature is available on the proper use and maintenance of agricultural machinery by communal farmers. Complex implements, such as planters, need proper calibration and maintenance, but are sold without manuals. Although farmers set and maintain machinery incorrectly, poor quality is often regarded as the major cause of inefficient performance and the ultimate use of labour-intensive methods by farmers. Farmers also persistently remove the hitch assembly of the conventional, right-hand mouldboard plow soon after purchase. This is occurring against a background of vigorous campaigns that highlight why it is necessary to maintain it intact. Also, manufacturers have done little to develop a more farmer-acceptable product; they need to be involved in the training of farmers.

Rural artisans

The Institute of Agricultural Engineering in Harare, the German Agency for Technical Cooperation (GTZ) and Intermediate Technology Development Group are running a joint programme to train rural (communal area) artisans to make and repair farm tools and animal-drawn implements. Another thrust of the programme is to provide cheaper and easily accessible spare parts for these implements and farm transport devices such as carts, wheelbarrows and four-wheeled wagons. The impact of the existing (trained and untrained) artisans on animal draft power development has not been assessed: such assessment would assist in identifying the constraints faced by the artisans so that sustainable solutions can be sought, and would also reveal the weaknesses of the current training programme for the artisans, which should be addressed.

Inadequate research

Draft animals have received negligible attention from researchers and policy-makers, compared with the work done with beef and dairy cattle. What little work has been done on draft animals has tended to concentrate on feeding management of Mashona oxen (Mupeta, Ndlovu and Prasad, 1990; Francis and Ndlovu, 1993; Francis, Ndlovu and Nkuuhe, 1994; Prasad, Khombe and Nyathi, 1994). There is little documentation on donkeys (Prasad, Marovanidze and Nyathi, 1991; on-going work funded by the UK Overseas Development Administration), probably because donkeys have a low socioeconomic status and research involving donkeys is considered backward and not glamorous by most agricultural scientists.

Although draft animals work for only short periods during the year, they need to be well fed if they are to be ready to work efficiently at the start of the cropping season. Considerable basic research has been done on nutrient requirements of draft cattle, but there is a need for much more research on how improved animal nutrition can increase efficiency under on-farm conditions. Knowledge of the needs, goals and resource endowment of farmers would play an integral part in the use of animal draft power. A thorough understanding of the range of feeding practices and various feedstuffs being fed to draft animals is also important. This, together with an idea of the nutritive value of available feedstuffs, would enable the formulation of rations which best meet the requirements of draft animals in the target farming systems.

Animal–implement disproportionality

Farmers concede that their draft cattle are becoming increasingly stunted, mainly due to poor nutrition (Tembo, 1989; Francis, 1993). Cows and donkeys are also increasingly being used for pulling mouldboard plows which have not been modified to allow for their lower liveweights and draft capacities. These trends have not been correspondingly matched by sufficient efforts to develop implements suited to weaker animals.

Need for affordable donkey harnesses

Although there have been extensive training programmes on making appropriate harnesses for donkeys, farmers continue to use inappropriate harnesses. Some commonly used harnesses injure animals during work, and it is quite common to find donkeys being neck-yoked together. Farmers argue that they cannot afford the recommended harnesses, and that the materials needed to make them are both scarce and too expensive.

Effects of droughts

The droughts of the 1991–92 and 1994–95 cropping seasons have reduced the country's animal draft power availability but the magnitudes of their effects are still not known. Furthermore, although hiring and lending of draft animals and implements between households is common during the cropping season, there is a need for systematic studies of the extent of this. Such studies should also quantify how late, when compared with draft owners, draftless farmers plant their crops and what are the related yield differences.

Health care

Intestinal parasites are a major cause of poor condition in draft animals throughout Zimbabwe. It is generally recommended that cattle should be dewormed at least twice a year (in April/May and October/November), and this empirical deworming strategy is already being used by communal farmers (Francis, 1993), although few scientific tests have been carried out to establish its effectiveness. Farmers also widely use traditional deworming remedies: scientific study of this indigenous technical knowledge may suggest ways to improve animal health care.

Information exchange

Although numerous people (farmers, researchers and trainers, implement manufacturers, non-governmental organisations, etc) are involved in different aspects of animal draft power, there has not been sufficient collaboration or communication among them. One result of this has been considerable duplication of efforts. One of the functions of the Animal Power Network for Zimbabwe (APNEZ), formed in September 1994, is to facilitate the exchange on information and coordinate all animal draft power-related work in the country.

Agricultural education curricula

All curricula in Zimbabwe's educational institutions (agricultural colleges and universities) emphasise tractor power, rather than animal draft power. Yet fewer than 3% of smallholder farmers in the country use tractors in crop production. Clearly, this anomaly should be redressed.

Gender awareness

Acceptance of technological interventions by smallholder farmers depends on how the farming family and local community perceive them. All members of a farming family (men, women and children) have their own roles in agricultural production, but the work of women and children, and their knowledge and potential in promoting animal draft power technology, are often ignored. Most technological recommendations therefore tend to strengthen male-based models which in most cases are never adopted by farmers. Animal draft power workers need to consider seriously the linkages between age, gender and labour equations. Unfortunately, socio-cultural restrictions often make women and children reluctant to talk to outsiders. Inclusion of women in animal draft power teams could possibly solve this problem (Mutimba, 1994).

Conclusion

There are numerous challenges facing animal draft power technology development in Zimbabwe. Appropriate methods should be devised so that relevant technologies can be developed. Without doubt, individuals and agencies involved in animal draft power-related activities should now adopt bottom-up approaches in meeting these animal draft power challenges. This implies that farmer-participatory approaches in technology development should be used. Such a shift in approach would accord farmers (including women and children) and extension agents the opportunity to dictate what they specifically want. Ultimately, this would increase the chances of adoption of the developed technologies.

References

Bratton M, 1984. *Draft power, draft exchange and farmer organisations.* Working Paper 9/84. Department of Land Management, University of Zimbabwe, Harare, Zimbabwe.

Chiduza C, 1994. *Farm data handbook. Crop and livestock budgets for Communal Areas: agro-ecological zones I–V.* Agricultural Services Division, FAO (Food and Agriculture Organization of the United Nations), Rome, Italy.

Christensen G and Zindi C, 1991. *Patterns of livestock ownership and distribution in Zimbabwe's Communal Areas.* Working Paper AEE 4/91. Department of Agricultural Economics and Extension, University of Zimbabwe, Harare, Zimbabwe.

Francis J, 1993. *Effects of strategic supplementation on work performance and physiological parameters of Mashona oxen.* MPhil Thesis. Department of Animal Science, University of Zimbabwe, Harare, Zimbabwe.

Francis J and Ndlovu L R, 1993. Improving work performance of Mashona oxen through strategic supplementation with locally produced feeds. *Draught Animal News.* 19:3–7.

Francis J, Ndlovu L R and Nkuuhe J R, 1994. Improving draft animal nutrition management through strategic supplementation in Zimbabwe. pp 158–161 in: Starkey P, Mwenya E and Stares J (eds), *Improving animal traction technology.* Proceedings of the first workshop of the Animal Traction Network for Eastern and Southern Africa (ATNESA) held 18–23 January 1992, Lusaka, Zambia. Technical Centre for Agricultural and Rural Cooperation (CTA), Wageningen, The Netherlands. 490p.

GFA, 1987. *A study on the economic and social determinants of livestock production in the communal areas of Zimbabwe.* GFA, Hamburg, Germany.

Howard C R, 1980. The draught ox: management and uses. *Zimbabwe Rhodesia Agricultural Journal* 77:19–34.

Mupeta B, Ndlovu L R and Prasad V L, 1990. The effect of work and level of feeding on voluntary food intake, digestion, rate of passage and body weight in Mashona oxen given low quality roughage. *Zimbabwe Journal of Agricultural Research* 28:115–124.

Mutimba J, 1994. Farmer participation in research: the imponderables. In: *Progress report on DANIDA-funded project on improvement of research–extension–user linkages.* DANIDA/CARNET/UZ (Danish International Development Agency/Cattle Research Network for Africa/University of Zimbabwe). Department of Animal Science, University of Zimbabwe, Harare, Zimbabwe. pp 19–24.

Prasad V L, Khombe C T and Nyathi P, 1994. Feeding crop residues for improved draft power. pp 164–166 in: Starkey P, Mwenya E and Stares J (eds), *Improving animal traction technology.* Proceedings of the first workshop of the Animal Traction Network for Eastern and Southern Africa (ATNESA) held 18–23 January 1992, Lusaka, Zambia. Technical Centre for Agricultural and Rural Cooperation (CTA), Wageningen, The Netherlands. 490p.

Prasad V L, Marovanidze K and Nyathi P, 1991. The use of donkeys as draft animals relative to bovines in the communal farming sector of Zimbabwe. pp. 231–239 in: Fielding D and Pearson R A (eds), *Donkeys, mules and horses in tropical agricultural development.* CTVM (Centre for Tropical Veterinary Medicine), University of Edinburgh, Edinburgh, UK.

Shumba E, 1984. Animals and the cropping system in the Communal Areas of Zimbabwe. *Zimbabwe Science News* 18(7/8):99–102.

Shumba E M, 1986. Farmer maize production practices in a high potential Communal Area environment: implications for research. *Zimbabwe Agricultural Journal* 83(5):175–179.

Tembo S, 1989. Draught animal power research in Zimbabwe: current constraints and research opportunities. In: Hoffmann D, Nari J and Petheram R J (eds), *Draught animals and rural development.* ACIAR Proceedings 27:69–77.

Surveying animal traction use in Zambia

by

Henk J Dibbits

Institute of Agricultural and Environmental Engineering (IMAG-DLO)
PO Box 43, 6700 AA Wageningen, The Netherlands

Abstract

Results of a national animal traction survey in Zambia showed that small-scale and medium-scale farmers prepare about 54% of their cultivated area with oxen. In 1990 the total number of trained oxen was 266,000. There were 132,000 working plows. The national average plowed area per pair of trained oxen was 3.5 ha. Generally there were not enough weeding implements which restricted the utilisation of trained oxen. Rural transport is not well developed, particularly in the non-traditional cattle-keeping areas where there are few sledges and ox carts per 1000 farming households. About 25% of the plows and ox carts are broken down and need to be repaired. It is assumed that the other implements have the same percentage of breakdowns. Veterinary and extension services leave much to be desired. Although credit has had little influence on the development of animal traction in the past, farmers still complained about lack of loan facilities for buying oxen and implements. The survey confirmed the poor distribution of implements and spares and the need for more repair workshops. The survey results are intended to support animal traction programmes by providing information for policy development, research and extension activities, (rural) development programmes and manufacturers and distributors of animal-drawn implements. The survey is an example for other areas where more in-depth information is needed to promote animal traction development.

Introduction

The development of animal traction can be hampered by lack of information about both its current and potential contribution to agricultural production. Without reliable data it will be hard to convince a Government and/or donor agencies of the need for support. Importantly, when reasons for successes and failures are not well known it will be difficult to analyse the situation and formulate policies to stimulate progress. In Zambia, substantial experience in the promotion of animal traction had been accumulated in different parts of the country. However, the evolution of project design and implementation had occurred without proper central direction and without a formal mechanisation policy structure. The signals of the several animal traction projects and the necessity to promote a mechanisation system for small-scale farmers without requiring large sums of foreign exchange (tractorisation for small-scale farmers had failed), made the Ministry of Agriculture and Water Development decide to investigate the status of animal traction. In 1985, a team travelled to all provinces of Zambia to describe the situation, and collect statistical data on the numbers of trained oxen and implements The latter were difficult to obtain and many assumptions had to be made. Nevertheless the study resulted in a good overview of the status of animal traction at that time, and this was incorporated in an Animal Traction Investment Plan.

Many donor agencies were willing to support a part of the Investment Plan and although not all proposed projects were funded (particularly some provincial animal traction programmes), many animal traction programmes were launched. In 1990, the National Animal Draft Power Coordination Programme initiated a review of the animal traction development since the start of the Investment Plan. By that time little new statistical information was available in some provinces, but it was considered that a national animal traction survey could provide more detailed information to be included in the review. Some analyses of initial responses to the questionnaires of the survey were made and included in the report (Starkey, Dibbits and Mwenya, 1991). However, some obstacles in carrying out the national survey caused a delay in data collection and therefore an in-depth analysis had to be done later. The results are stated in a separate report (Dibbits and Mwenya, 1993).

This paper describes the main results of a national animal traction survey in Zambia.

An ATNESA resource book

Information is also given about the methodology that was used and constraints encountered. The support of this survey to further animal traction development in Zambia is explained. The survey could be repeated (adapted to the local needs) elsewhere if more in-depth information is needed to promote animal traction development.

Objectives

The objectives of the survey were:

- to collect numerical data on animal traction from all agricultural camps in Zambia.
- to provide statistics to review the importance and potential of animal traction in all districts in Zambia as a whole based on:
  - households owing oxen
  - (trained) oxen
  - area cultivated with trained oxen
  - implements in use and repair needs
  - ox carts in use and repair needs
  - animal means of transport
  - blacksmiths.

The results of this survey were intended to assist in the development of animal traction policy. Other important beneficiaries include: district and provincial rural development programmes; manufacturers, importers and distributors of animal-drawn implements and spare parts; and researchers and extension staff involved in the development and promotion of animal traction.

Methodology

To obtain the information required for this exercise, questionnaires were distributed to all agricultural camps in Zambia (the political organisation of Zambia is based on a hierarchy of provinces, districts and camps). The survey tried to get details of farming households with and without oxen, area cultivated and common animal-drawn implements in use by farmers. This was the first time that such a specific survey had been carried out by Camp Officers. The method of data collection employed included farmer interviews and observations. The first version of the questionnaire was tested in five agricultural camps in Lusaka Province. An explanatory note was added to clarify certain questions, to prevent the Camp Officers from making mistakes and to enable them to cross-check some data. The questionnaire had provision for numerical data and comments from Camp Officers concerning changes, constraints and suggestions for improving animal traction in their camps.

The farmers surveyed were mainly small- and medium-scale farmers who normally use hand tools and animal traction. These two categories and the large-scale farmers are classified according to the Ministry of Agriculture Food and Fisheries as follows:

- **small-scale farmers:** farmers who normally plant at least 0.5 ha but less than 5 ha.

- **medium-scale farmers:** farmers who normally plant at least 5 ha but less than 20 ha.

- **large-scale farmers:** farmers who normally plant more than 20 ha.

The 1989/90 Crop Forecast classified farmers as small-scale, emergent and commercial, with similar definitions.

All the questionnaires were checked by the provincial staff and the National Animal Draft Power Coordination Programme. Where data seemed to be incomplete or doubtful, questionnaires were sent back for verification. Several provinces were also revisited to consult on the data submitted.

As not all camps submitted questionnaires, district totals had to be adjusted, based on the number of missing camps and other sources of information such as cattle censuses and crop forecasting.

Limitations and constraints

Data collected on number of farming households and area under crops did not appear to be a problem for Camp Officers as they were familiar with the data required. Most of the data submitted on farming households and area under crops coincided fairly well with that of the crop forecasting estimates. However, the area cultivated with hoes plus the area plowed by oxen is not always equal to the total area under crops. This is because in some areas tractors are used for plowing and because 'permanent' crops such as cassava are grown.

The estimates of number of oxen and number of trained oxen in a camp were sometimes the same. This is feasible in areas of introduction where oxen are brought in for traction but is less likely in traditional cattle-keeping areas. The survey did not include donkeys and work cows; however, the use of donkeys and work cows was reported by a number of Camp Officers.

The numbers of mouldboard plows and ox carts not in use were included to find out the need for spare parts and repair facilities. It was expected that the situation for the other implements in the survey would be similar.

For the implement repair and maintenance, only the number of blacksmiths or blacksmith-farmers was recorded. No information was gathered about their capabilities and work. However, the number of blacksmiths in the area is a good indication of the potential for establishing rural repair facilities.

The National Animal Draft Power Coordinating Office encountered some obstacles in carrying out the national survey. Communication with Camp Officers, particularly the ones in the very remote areas appeared to be difficult. Distribution and collection of questionnaires was sometimes delayed. Furthermore, many Camp Officers had to collect the data on foot. Despite many follow-ups to provinces and districts as well as financial support for fuel, in a number of districts the handing in of questionnaires was slow. The programme had underestimated the time and manpower required to carry out such a national survey and the difficulties that might arise.

Abbreviations

In the graphs and charts in this paper the following abbreviations are used:

NP	Northern Province
LUP	Luapula Province
COP	Copperbelt Province
NWP	North Western Province
WP	Western Province
SP	Southern Province
CP	Central Province
EP	Eastern Province
LUS	Lusaka Province

Results of the survey

Area cultivated with human and animal power

Small-scale and medium-scale farmers in Zambia prepare about 46% of their cultivated area with hand hoes and 54% with oxen. A small unknown percentage, which has not been accounted for in these figures is plowed with tractors. There are big variations in percentages between the provinces

(Figure 1), and also between the districts within provinces, because areas with few trained oxen will automatically have a low percentage of ox plowing. On average, a pair of oxen in Zambia plow 3.5 ha per growing season.

In Zambia the total area plowed with trained oxen, 468,000 ha, is about five times bigger than the planted area of all the large-scale farmers (close to 90,000 ha; 1989/90 final crop forecasting).

Number of trained oxen

The total estimated number of trained oxen in 1990 was 266,000 (Table 1). Compared with the estimates in the 1985 Investment Plan, there has been an increase of about 48%, while during the same period the cattle population in the traditional sector increased with only 7%. This implies that the percentage of the cattle population used as draft oxen has been increasing rapidly. Unfortunately, after the 1990 survey many cattle/oxen have died due to outbreaks of disease in some districts.

Number and balance of implements

Finding similar numbers of several different types of animal-drawn implements means that there is a well-balanced package of equipment available, so that oxen can be used for many agricultural operations. An even balance also implies the potential for a reduction of human power use, for example with a complete set of implements human power might be used only for in-row weeding. In practice, farmers tend to buy plows first. The purchase of other equipment, particularly weeding implements, depends very much on profitability in ox farming and availability. In some cases the balance has been influenced by project interventions, as in North Western Province where for a number of years, a district development programme has promoted plows and not weeders. In Eastern Province farmers appreciate the ridger for tillage operations and weeding.

The nationwide estimates of numbers of ox-drawn implements (Figure 2) hide the differences between provinces. However, the data collected on the numbers of available ridgers, harrows, planters and cultivators per 100 ha plowed with work oxen by province (Figures 3, 4, 5 and 6) show the variations clearly.

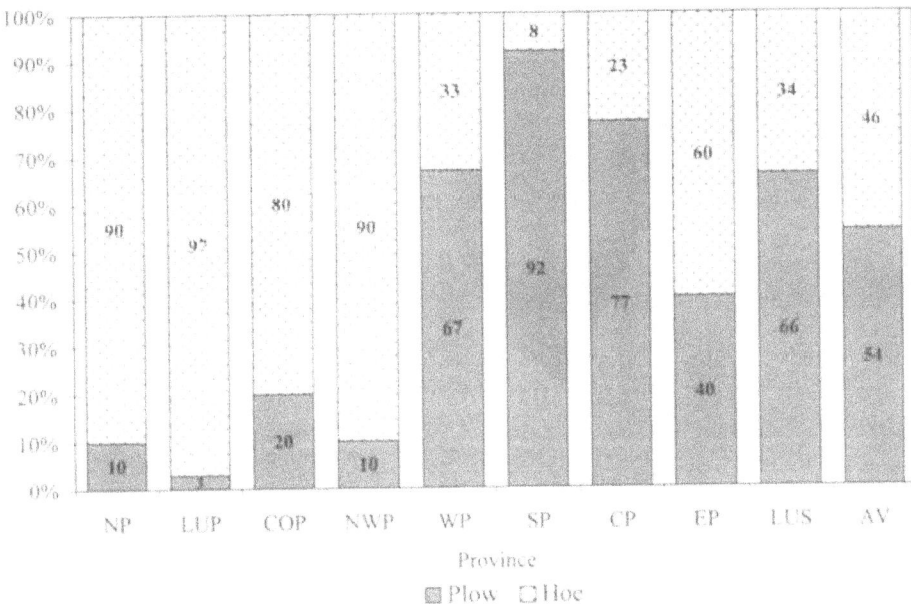

Figure 1 Percentage of area cultivated with hoes and ox plows by province and the national average

Table 1: Estimates of numbers of trained oxen and total number of cattle in Zambia in 1985 and 1990

Province	1985[1]	1990[2]	% change
Northern	4,200	4,620	10
Luapula	500	487	-3
Copperbelt	1,300	2,329	80
North Western	300	2,375	790
Western	10,000	31,700	317
Southern	96,000	126,400	32
Eastern	38,000	47,9600	26
Central + Lusaka	29,000	50,055	73
Total	**179,000**	**266,000**	**48**
Cattle (traditional sector)[3]	2,077,000	2,216,125	7

1) Estimates from 1985 Animal Draft Power Investment Plan
2) Estimates from 1990 animal traction survey
3) Livestock census figures from the Department of Veterinary and Tse tse control services

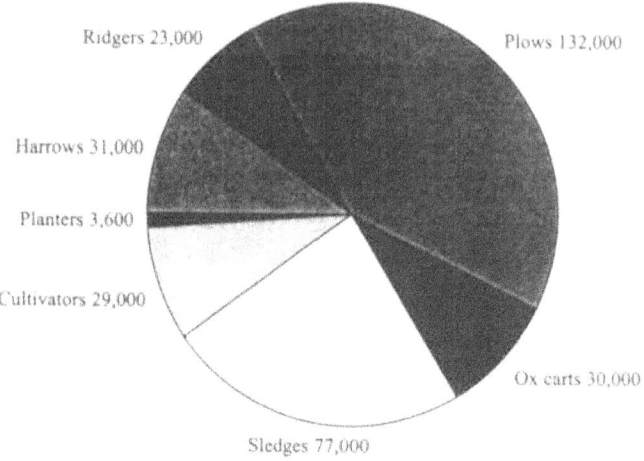

Figure 2: Estimated numbers of ox-drawn implements in Zambia

The optimum number of cultivators, harrows and ridgers seems to be about 8 to 10 per 100 ha of ox-plowed land, bearing in mind that cultivators and ridgers are interchangeable in weeding operations. From this assumption and the total area plowed with trained oxen, one can calculate the theoretically required increase in number of implements per province and per district. However, one also has to take into account local farming practices and the profitability of ox farming, as in Western and Eastern Provinces.

For example the imbalance of implements in Western Province manifests itself in the low number of cultivators and weeders per 100 ha plowed with animal power. Eastern Province is also a typical example. It has many ridgers but few cultivators and harrows. There ridgers are also used for primary tillage, particularly in groundnut production, hence the high number. In Lundazi District there are even more ridgers than plows.

Transport

Rural transport with trained oxen is not very well developed in Zambia. Although there are many sledges, 29 per 100 trained oxen, they are mainly used for on-farm transport and short distances. Ox carts are more suitable for rural transport but there are only 11 usable ox carts per 100 trained oxen (Figure 7).

The number of sledges and usable ox-carts per 1,000 farming households reveals that the rural transport situation is very bad in Northern,

Figure 3: Number of ridgers per 100 ha plowed with trained oxen

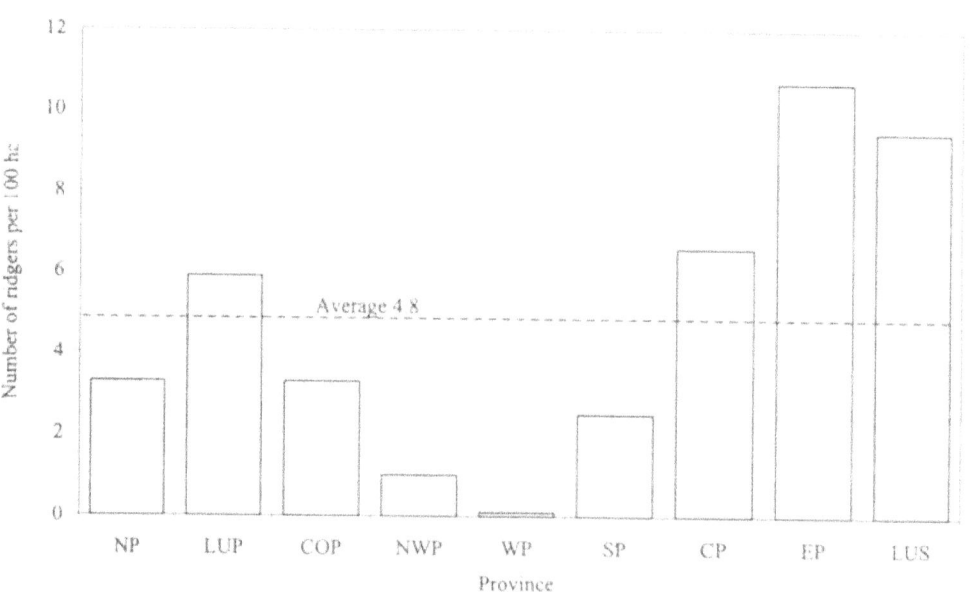

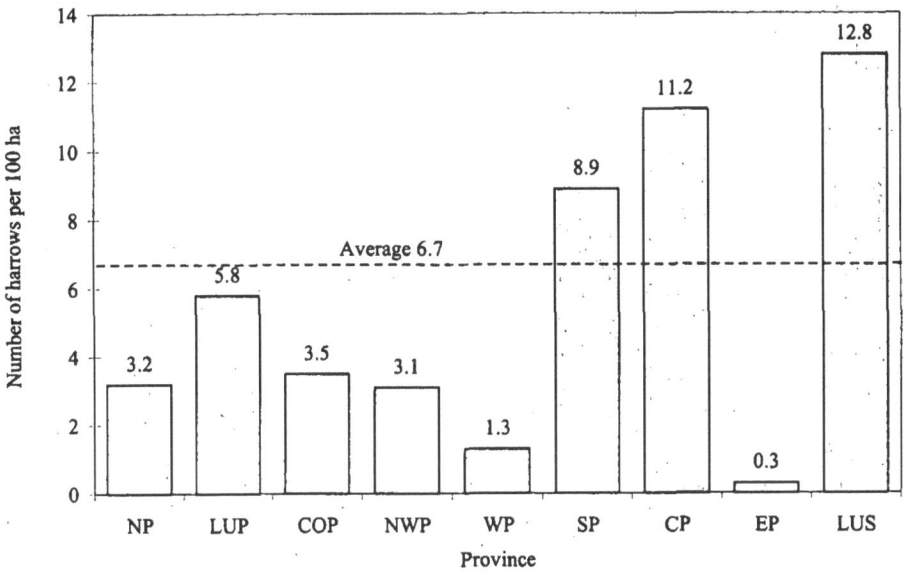

Figure 4: Number of harrows per 100 ha plowed with trained oxen

Luapula, Copperbelt and North Western Provinces. The most developed areas in terms of animal-drawn rural transport are Southern Province and Central Province. The latter has 122 ox carts per 1,000 farming households (Figure 8), the highest density in Zambia.

Repair needs for plows and ox carts

About 25% of the plows and ox carts are broken down and need to be repaired (Figures 9 and 10),

ie 38,000 plows and 11,000 ox carts. The percentage of broken plows is almost equal in all provinces, except for Western Province, which also has the highest percentage of broken ox carts. North Western Province, where many ox carts are relatively new, has few broken carts. Many ox carts with standard roller bearings were sold during the last five years and the North Western Integrated Rural Development Project replaced

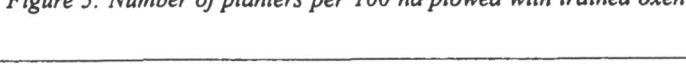

Figure 5: Number of planters per 100 ha plowed with trained oxen

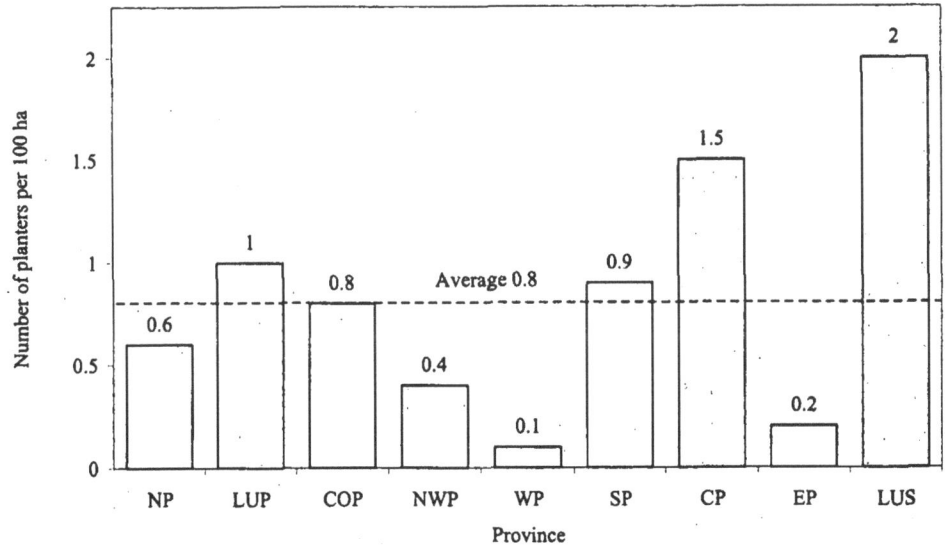

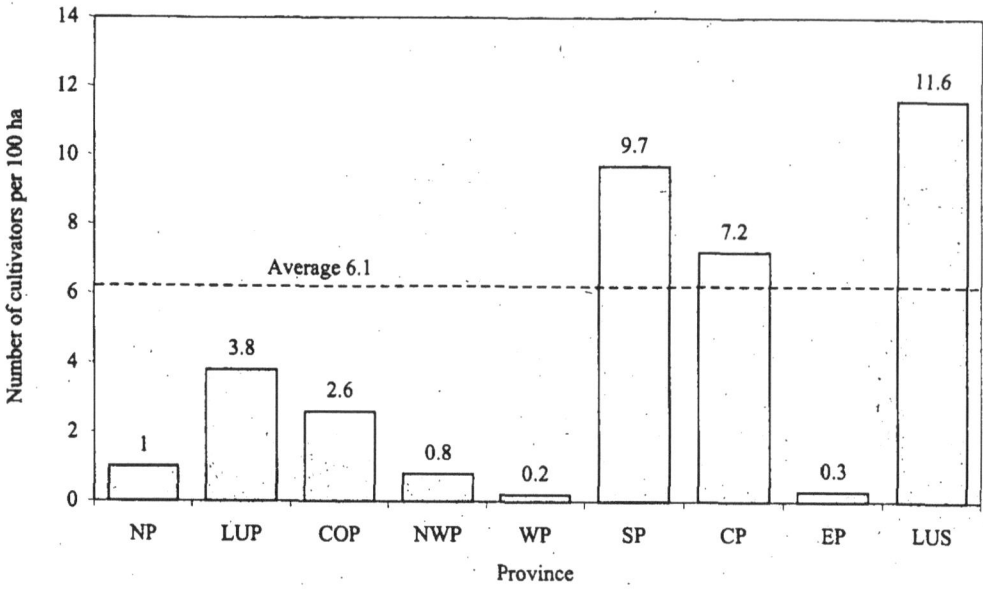

Figure 6: Number of cultivators per 100 ha plowed with trained oxen

failed axles (with bronze, nylon and wooden bearings) by axles with roller bearings.

One can assume that the other implements have the same percentage of breakdowns. In total this means that thousands of different spare parts are needed for the several makes of implements and ox carts.

Draft power utilisation

The importance of animal traction in farming also depends on the rate of utilisation. The survey gives information about the area plowed per pair of trained oxen and the potential for carrying out weeding operations and transport (balance of implements, number per 100 ha plowed with trained oxen or per 100 trained oxen). A high rate of utilisation means that the oxen will be used many days per year, not only for plowing, but also for weeding and transport. Plowing with a team of six oxen where four or two oxen can do, indicates an ineffective use of oxen. Other important aspects in achieving good utilisation of oxen are profitability in farming, rural transport and availability of implements and spares. Figures 3 and 6, indicate that oxen are under-utilised for weeding in Northern, Copperbelt, North Western and Western Provinces.

The rate of utilisation in transport is hard to estimate. At least we know that many oxen are hitched to ox carts and sledges. The average number of ox carts plus sledges in use per 100 trained oxen is 40. It varies from 29 in Western Province to 50 in Lusaka Province (Figure 7).

The average area plowed with a pair of trained oxen is 3.5 ha in Zambia. It is notable that in Northern, Luapula and North Western Provinces, where most of the introduction of animal traction has taken place during the last 10 years, the trained oxen plow more than the national average of 3.5 ha. The reason may be the relatively long rainy season and the fact that farmers plow with one pair of oxen. The Copperbelt Province has the smallest area of plowed land per pair of trained oxen: 2.8 ha. This is because of the very small areas plowed per pair of oxen in the urban districts. In Western Province (where oxen are usually hitched in teams of four or six), the area plowed is 3.1 ha.

Remarks by extension staff
Veterinary and extension services

One can imagine that the farmers experience the services of the Veterinary Assistants and the Camp Officers differently. Loss of an animal is perceived as being far more serious than insufficient

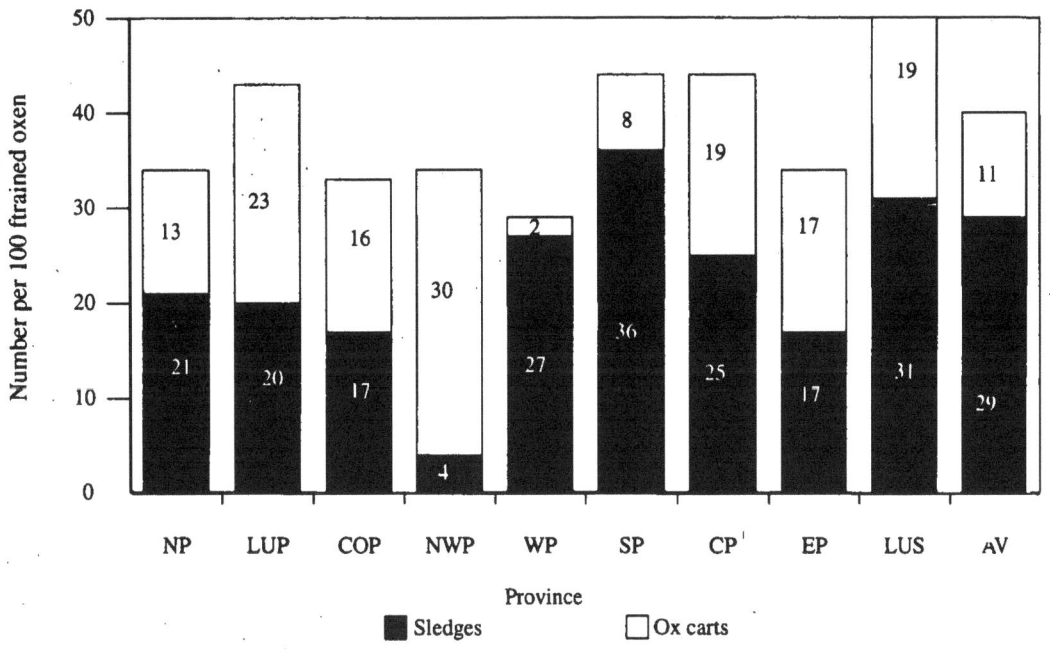

Figure 7: Number of sledges and ox carts per 100 trained oxen by province and national average

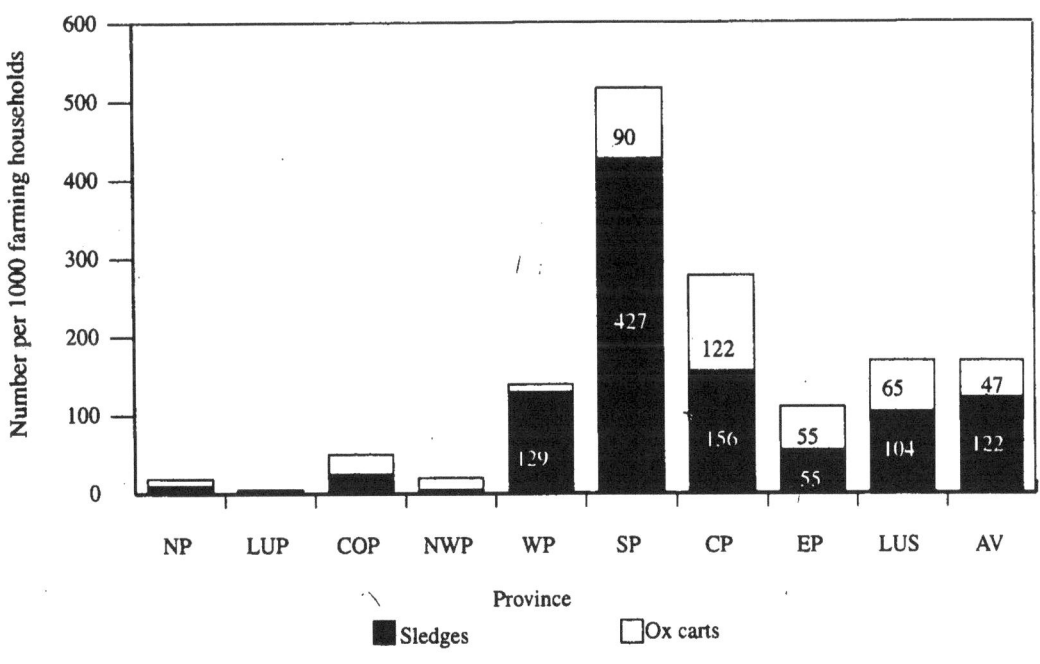

Figure 8: Number of sledges and ox carts per 1,000 farming households

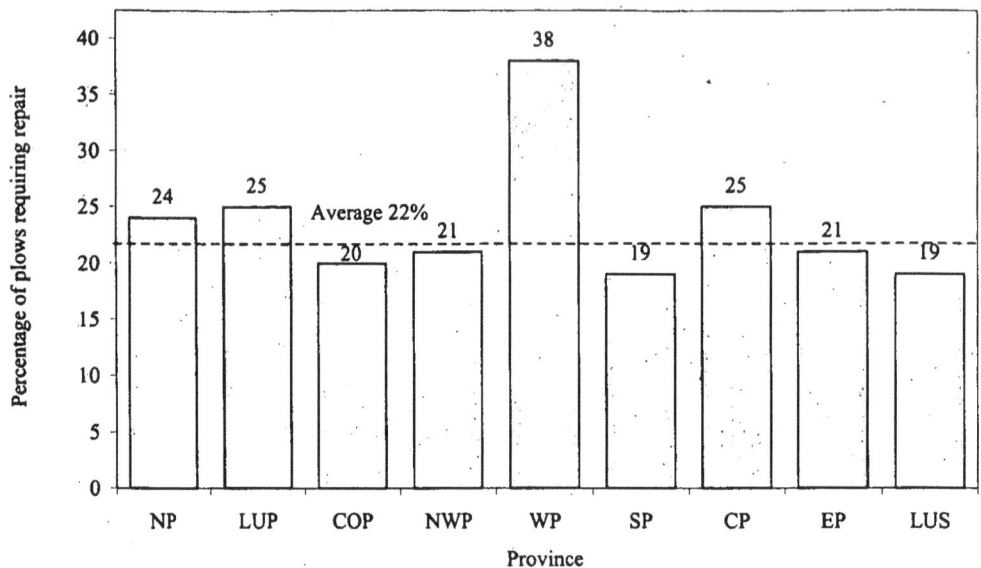

Figure 9: Percentage of plows requiring repair

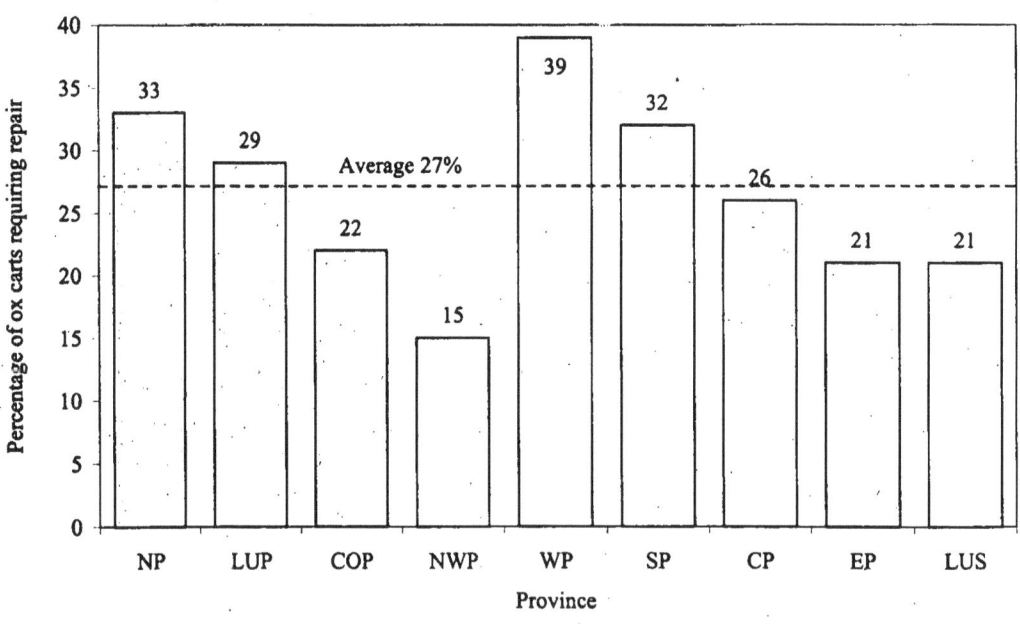

Figure 10: Percentage of ox carts requiring repair

extension services from a Camp Officer. The former has immediate financial consequences for the farmer, whereas the damage of insufficient extension is difficult to estimate. Therefore farmers' complaints about veterinary services and the rise of drug prices (to more realistic prices) are understandable. However, the circumstances under which Veterinary Assistants and Camp Officers have to work leave much to be desired.

The extension staff are hampered in their animal traction extension work by lack of basic needs like transport, protective clothing, and extension/teaching materials. Some have to walk long distances and are unable to visit all farmers. Proper housing and allowances stand much higher on their list of priorities. Camp Officers are critical about their own functioning. It is difficult to transfer knowledge or to convince farmers if you have insufficient knowledge and experience yourself. Camp Officers expressed a desire for extension pamphlets and in-service courses. However, the effect of training will be insignificant if the other problems are not solved. They also feel that farmers need more education and training. Some staff members have communication problems with farmers and others feel left to their own devices. They would like to see more coordination and communication.

Credit

In the past, credit was available and farmers were keen to obtain seasonal and medium-term loans for agricultural inputs and investments in oxen and implements. However, the percentage of farmers that obtained credit has always been low and has had little influence on the development of animal traction. In practice, the credit system was a kind of subsidy, particularly during crop failures, when repayment was poor. Parastatal lending institutions which were running at a loss were subsidised.

With inflation higher than interest rates, credit-giving organisations have seen the value of their capital reduced and have foreseen a cut-back in government support. As a result the credit available has been reduced greatly, both in terms of Kwacha and value.

The farmers' past experience in obtaining credit and the present tight financial situation of the lending institutions explain the many complaints from farmers concerning lack of loans for buying oxen and implements.

Supply and distribution of implements and spare parts

Complaints about shortages of implements and spares are reported in all districts, even in the districts along the line of rail (railway from Livingstone to Lusaka and the Copperbelt) where they are usually available. Some Camp Officers reported that farmers have to travel long distances in search of implements and spares. Sometimes farmers have to buy a complete new implement due to lack of spares.

Using the survey data, suppliers are now able to make rough estimates of the demand for various types of implements in the districts.

The demand for spares is difficult to estimate. A more detailed survey in a number of camps spread over Zambia is necessary, to map out the different makes and types of implements and ox carts, together with the repairs and spare parts required. Moreover, the distribution network should be improved, to bring the implements and spares closer to the farmers. Ideally, a number of rural blacksmiths or repair workshops should be involved in the distribution.

Blacksmiths

The survey indicates that there are quite a number of blacksmiths in Zambia. The questionnaire does not specify the minimum requirements a person must satisfy to be called a blacksmith and therefore many farmers who occasionally make a hoe may be included. Generally, artisans have no external support and mainly use simple tools and scrap materials. This means that the implement repair and maintenance service to the farmers is limited (Starkey, Dibbits and Mwenya, 1991).

Many Camp Officers recommend further training of blacksmiths and the establishment of more repair workshops. This is in line with the enormous repair needs of implements and ox carts.

Use of cows

An increase in the use of cows for draft was confirmed by a number of Camp Officers, particularly in Western (Sesheke District), Southern and Central Provinces. Loss of trained oxen has forced farmers to use cows in their teams.

Gender

With reference to a question asking if there had been progress in animal traction in the camp, some Camp Officers reported that an increasing number of women are able to work with draft oxen, particularly in Western province. A considerable number of women use animal traction in Southern Province as well.

Support to animal traction development

The results of the survey are intended to assist in the development of animal traction policy. Other important beneficiaries include district and provincial rural development programmes, manufacturers, importers and distributors of animal-drawn implements and spare parts and researchers and extension staff involved in the development and promotion of animal traction.

Animal traction policy development

Limited financial means force governments to set priorities in their support to the agricultural sector. To be able to set priorities one needs to have at least a number of areas to choose from. To get animal traction on the priority list of the Ministry of Agriculture, one needs to have sufficient information about the role animal traction can play in the agricultural development, its constraints and prospects. A widely distributed animal traction status report and national or regional survey can be of great importance to convince politicians and decision makers of the need for support.

Once animal traction is accepted as a priority area, different policies can be pursued to promote its development. The policies could involve activities of the Ministry of Agriculture (eg further data collection, research, training and extension). Others are measures to make it more attractive for the private sector to be engaged effectively in activities which will stimulate the use of animal traction technologies. One of these measures could be duty-free importation of steel for manufacturing animal-drawn implements or marketing support.

A status report or a survey will also be helpful to differentiate between the different requirements of each region. In areas where there is sufficient experience in animal traction and there are many

oxen and implements, the private sector may be able to manage most of the required businesses. In areas with little experience in animal traction the government could facilitate animal traction development in assisting in extra training and extension and also in the supply and distribution of implements.

District and provincial development programmes

Similar to the animal traction policy development, no policy can be developed in districts and provinces and no actions can be taken unless enough information is known about the benefits of animal traction and the constraints to further development.

Manufacturers, importers and distributors of animal-drawn implements and spare parts

It is almost impossible for an individual entrepreneur to gather information about the marketing possibilities for animal-drawn implements and spares. For most entrepreneurs, manufacturing or selling of these items is a minor part of their total business. To play safe, they usually order the same numbers as they did in previous years. Consequently, there are generally few implements and spares available for the small-scale farming community. In addition, implements and spares are usually not well distributed as suppliers are not aware about the demand in certain areas. Statistical data about the number of farmers using draft animals and the available implements can give more information about the potential market.

Research, training and extension

Through detailed information about the constraints in animal traction development the programmes can be adjusted and extra attention can be given to problem solving. Regions with similar problems can be grouped paving the way for more cost-effective tailored approaches in research, training and extension.

Conclusions

The national animal traction survey in Zambia has produced much in-depth information for policy makers, researchers and extension staff, (rural) animal traction development programmes, private enterprises, etc. It has showed clearly the importance of animal traction in agricultural production and development in Zambia. However, it also confirmed many constraints to the spread of animal traction including:

- poor distribution of implements and spares
- need for more repair facilities
- under-utilisation of trained oxen
- need for better veterinary and extension services.

Last but not least, it may be advisable to carry out such a survey by province. One should not underestimate the finances, supervision, communication and particularly the time required to collect and analyse the data.

References

Dibbits H J and Mwenya E, 1993. *Animal traction survey in Zambia*. Ministry of Agriculture, Food and Fisheries, Lusaka, Zambia in association with the Institute of Agricultural and Environmental Engineering (IMAG-DLO), Wageningen, The Netherlands. 119p.

Starkey P H, Dibbits H J and Mwenya E 1991. *Animal traction in Zambia: status, progress and trends.* Ministry of Agriculture, Lusaka, Zambia in association with the Institute of Agricultural and Environmental Engineering (IMAG-DLO), Wageningen, The Netherlands. 105p.

Possible initiatives for increased utilisation of animal traction in Malawi

by

M L Mwinjilo

Senior Lecturer, Faculty of Agriculture and Natural Resources, Africa University
PO Box 1320, Mutare, Zimbabwe

Abstract

Malawi is small country with a limited resource base, high birth rate, high infant mortality, low life expectancy, and a high population density in relation to arable land. The economy is predominantly agricultural, accounting for 35% of GDP in 1991 with the smallholder sub-sector contributing 26%. The smallholder sub-sector is important to the country's economy since it supplies about 85% of the country's food requirements and accounts for about 80% of the agricultural output. However, the sub-sector operates almost at subsistence level due to low farm sizes, low levels of technology use and lack of adequate draft power. Provision of draft power would improve yields through improved timeliness, and hence increase farm incomes. Due to the small farm sizes and knowledge and skills of smallholder farmers, the most appropriate power source is animal draft. However, there is low utilisation of animal traction due to limited access to credit, inadequate supply of implements and spare parts, diseases, and land pressure. Increased utilisation of animal traction could result from improved access to credit, identifying appropriate tools and implements, and the use of animals not traditionally used for traction such as cows and donkeys.

Introduction

Malawi covers an area of 11.8 million hectares, 2.4 million of which are water. In mid-1988 the population was estimated at 8 million, with almost 85% rural small-scale farmers.. A preliminary analysis of the 1987 census suggests an annual growth rate of 3.7% and a projected population of between 10.2 and 11.9 million people in the year 2000, depending on fertility, mortality rates and the impact of AIDS (UNICEF, 1992). Twenty percent of the population is under 5 years old and 45% is below the age of 15. The population density ranges from 10 to 292 people per square kilometre with a national average of 85 people per square kilometre (Statistics, 1987).

Smallholder production

Agriculture is the backbone of the Malawi economy. In 1991, it accounted for 35% of GDP with the smallholder sub-sector contributing 26% and 87% of export revenue (Planning, 1991). Of the total labour force, 81% were in agriculture. This dominance of agriculture in the economy has not changed much since independence in 1964, when agriculture accounted for 58% of GDP, 90% of domestic exports, and 90% of the resident labour force was engaged in this sector.

Thus, high priority has been given to agriculture in the country's development programme. Agriculture was seen, and is still seen, as offering the greatest potential for increasing income and thus improving the welfare of the population. Achieving self-sufficiency in food production and expanding exports have been at the centre of agricultural development policy.

Table 1 shows the distribution of household farm size. In 1990, 26% of households cultivated less than 0.5 ha. A further 30% cultivated between 0.5 ha and 1.0 ha. Thus, about 56% of rural households cultivated less than one hectare and the overall mean farm size is 1.1 ha. Holding sizes vary by region, depending on a combination of population density, population growth, quality of land and level of marketing activity. The prevalence of smallholdings within the smallholder sub-sector emanates from population growth resulting in high densities and low technological and income levels.

Given a fixed amount of village land, population growth will result in falling average holdings as land is subdivided into smaller family units. At the same time the remaining uncultivated land is brought into production. Land may be limited not only quantitatively but also qualitatively. This problem can be resolved by using

Table 1: Number of households with different farm sizes in Malawi in 1992

Farm size (ha)	Number of households ('000s)
<0.5	338
0.5– <1.0	389
1.0– <1.5	265
1.5– <2.0	143
2.0– <2.5	66
2.5– <3.0	35
>3.0	62
Total	1298

Source: UNICEF, 1992

productivity-increasing inputs such as fertiliser and high-yielding seed. However, not only must the producer know how to apply these, they must be available at the right time, in the right quantities and at an affordable level. Similarly, use of efficient agricultural tools is necessary. Government policy must be such as to confine provision of such technologies to a specific group of producers. Even so, since such technology has to be paid for, the producer must have sufficient income to afford them. If he/she does not have the means, an agricultural credit system could solve the problem. However, credit facilities might not be available to all household producers. More particularly, those households whose holdings are too small to produce more than what is required for food may consider it unwise to receive credit since they would not be able to repay it. Further, there may be certain institutional procedures acting as a constraint. Where credit is provided only to a farmer group or club, those outside may be ineligible for credit. Credit may also be tied to a repayment record.

Smallholder production is derived almost entirely from family labour. Although labour is not considered a constraint for most farmers, labour shortages do occur during land preparation and harvesting periods, especially on larger holdings. Levels of animal traction technology use are very low with virtually all cultivation done by hand using a tongued hoe set in a wooden handle. Only 5% of households own draft animals, 4% own animal-drawn plows, and 2% own animal-drawn ridgers. The use of intensive inputs is also low; in 1980 only 6% of households applied fertiliser, and 6% of the cultivated area was under improved groundnuts (Statistics, 1984). Very few farms have adequate storage facilities, and significant post-harvest losses occur. The majority of farmers have very low cash incomes estimated at 300 Malawian Kwacha in 1984/85, of which about 80% was from crops and livestock and 20% from off-farm activities (World Bank, 1987).

Mwinjilo (1987) reported that farm power was limiting smallholder production through the failure to meet crop labour requirements during critical periods (planting, fertilising and weeding). He also found that use of draft animals and animal-drawn implements reduced labour requirements, resulting in increased labour productivity. Coupled with improved inputs, this led to improved land productivity through increased yields.

Utilisation of animal traction

It is widely believed that there is a low level of utilisation of draft animal power in Malawi. The major reasons put forward include land availability, availability of or access to credit, supply of equipment and spare parts, and diseases (Mwinjilo and Kasomekera, 1989; Mwinjilo and Ng'ong'ola, 1990). Mwinjilo and Kasomekera (1989) reported that 50% of all draft animal owners used their animals for tillage and transport but the type of use varied somewhat: there are areas where farmers use their animals predominantly for transport, notably the Central Region, whereas in the Northern Region, farmers use their draft animals predominantly for tillage. However, Mwinjilo and Matenje (1993), who carried out a survey of labour-saving technologies for rural women, reported that all respondents used a hoe for weeding. Mwinjilo and Kasomekera (1989) also reported that farmers without draft animals hired animal traction for land preparation (26%) and transport (53%).

Contracting of animal traction is constrained by poor timing of operations which leads to unnecessary pressure on the resource. Farmers usually wait until the onset of rains to start land preparation. With proper training and extension this approach to land preparation can change, possibly leading to greater contracting.

The predominant animals used for traction are cattle (oxen). The number of donkeys was 2200 in 1991, 2450 in 1992 and 1950 in 1993. The cattle numbers have not changed much over the years. However, the human population has been growing resulting in much grazing land being converted to cropping. The end result has been poor nutrition of animals, overgrazing and subsequent land degradation. The supply of draft animals is limited due to competition with demand for beef.

There have been efforts to introduce donkeys into smallholder agriculture since colonial days. Soon after independence, the government intensified the effort but has met with minimal success. In areas where donkeys are prevalent, it is only because of cattle theft that farmers resorted to donkeys and they are used only for transport.

Due to limited holding sizes, most smallholder farmers have low incomes which do not permit them to purchase draft animal power packages with their own funds. The government has credit funds for the purchase of draft animal power packages, among other things, but these are extended to individuals who can show ability to service the loan, which usually means those with holdings greater than 3 hectares.

Improving utilisation of animal traction

The major constraint to access or ownership of draft animals is finance. As indicated above, the government provides seasonal credit to farmer groups and credit for the purchase of draft animals and implements to individuals who can show ability to service the loan, meaning those with more than three hectares under cultivation. Since the majority of smallholder farmers cultivate less than three hectares, most of them would not qualify for a loan. The only way to assist such farmers is to create credit groups to be issued with loans to purchase animal traction packages, on the understanding that they would provide contract services to other farmers. The government should consider weaning farmers off seasonal credit to make funds available for animal traction packages.

The availability of draft animals appears to be a constraint as butchers offer better prices than farmers. Consideration should be made of the use cows for traction. Studies in Malawi (Chamdimba, 1991) and elsewhere have shown that it is possible to use cows for traction as long as management takes into account the needs for reproduction and lactation. For farmers to accept the use of cows for traction, intensive efforts to educate farmers would be needed. Donkeys would offer an alternative source of draft animals for transport and possibly for tillage if appropriate tools and implements can be identified and made available to farmers. This would require education of farmers on the benefits of using donkeys for traction.

The implements currently in use are for conventional tillage, typically requiring a high draft force. There is a need to identify tools and implements for minimum tillage with light draft which could be drawn by comparatively large single ox or a pair of donkeys.

References

Chamdimba E A T, 1991. *The effect of draught work on some productive characteristics of the Malawi Zebu* (Bos indicus) *heifers and their calves.* Unpublished MSc thesis. Faculty of Agriculture, University of Malawi, Lilongwe, Malawi.

Mwinjilo M L, 1987. *Farm power and mechanization for smallholder production systems in Malawi.* Unpublished Ph D thesis. Silsoe College, Cranfield Institute of Technology, Silsoe, UK.

Mwinjilo M L and Kasomekera Z M, 1989. *Survey of animal power utilisation in Malawi.* Report submitted to the Food and Agriculture Organisation (FAO) of the United Nations, Project UNDP/FAO MLW/86/002 Animal Power Utilisation. FAO, Rome, Italy.

Mwinjilo M L and Ng'ong'ola, 1990. *An investigation of the constraints restricting the uptake of medium-term credit by smallholder farmers in Malawi.* A report submitted to the Smallholder Agricultural Credit Administration, Ministry of Agriculture, Lilongwe, Malawi.

Mwinjilo M L and Matenje F, 1993. *Labour-saving technologies for rural women in SADC countries: Malawi.* Report submitted to the Ministry of Women and Children Affairs and Community Services, Lilongwe, Malawi.

Planning, 1991. Economic Report 1991. Department of Economic Planning and Development, Office of the President and Cabinet. Government Printer, Zomba, Malawi.

Statistics, 1984. *National Sample Survey of Agriculture 1980/81 (customary land in rural areas only).* National Statistical Office. Government Printer, Zomba, Malawi.

Statistics, 1987. *Malawi Statistical Yearbook 1985.* National Statistical Office. Government Printer, Zomba, Malawi.

UNICEF, 1992. *Poverty in Malawi: Situation analysis (Draft).* Malawi Government/UNICEF, Lilongwe, Malawi.

World Bank, 1987. *Malawi: Smallholder Agricultural Credit Project (IDA/IFAD Credit).* Staff Appraisal Report No. 6886 - MAL, Southern Africa Department, Agriculture Operations Division, World Bank, Washington DC, USA.

Workhorses in Norway

by

Martin Aeschlimann

SJH N-5745 Aurland, Norway

Abstract

Norway is an industrialised country with a highly mechanised agriculture so few draft animals are used by farmers. However, the income of Norwegian farmers is falling and there is a growing focus on the ecological impact of farming. This has caused some farmers to start using work horses again. The government supports projects that lead to a wider public acceptance of workhorses as an alternative to tractors. One agricultural school includes the use of workhorses in its compulsory curriculum. This is only a beginning: major demands have to be met before a significant number of farmers begin to use workhorses again.

Introduction

In northern Europe, Norway has an area of 320,000 km^2 and 4.3 million inhabitants. The country is more than 1500 km long and has many glacial mountain ranges. The majority of the population live along the coastline. The climate is influenced by the gulf-stream:. the coastline has a high rainfall and fairly mild winters. The east gets less rain and has cold winters with snow from November to April. Some areas of mountainous valleys are fairly dry and here farming is only possible with artificial irrigation. Only 3.6% of the total land surface is farmed, with 22% productive forest. Only 2% of the working population are farmers. Norway has one of the highest levels of agricultural subsidy in the world. It is a political aim for the government to maintain a widely spread population. Supporting farming is a key step towards achieving this aim.

In the east of the country the landscape comprises rolling hills and wide valleys and the farms are much bigger than the Norwegian average of 10 ha. The main crops in this region are grains and potatoes. In the west, mountains, lakes and fjords dominate the landscape. The farms are small and often situated on steep hills. Milk production is dominant in this area. Where the land is too steep for cows to graze, goats are kept for milk. In the summer the animals are grazed in the mountains on summer farms. Until the end of the 1950s the majority of the farms used horse power. By the mid 1970s, 15 years later, tractors had almost completely replaced the workhorse. Today most of the remaining workhorses are used on smaller farms and in the mountains.

Why use workhorses in Norway today?

Since the introduction of tractors to Norway, workhorses have been regarded as old-fashioned and reminiscent of a past era. Farmers working with horses were looked upon as crazy or as hopeless romantics. Working with horses was not considered seriously. Around 1980, a new era started with a government-supported logging research project. Mechanisation of forestry with ever bigger machines was resulting in severe damage to the soil and remaining vegetation. At the same time these machines were not well-adapted to steep hillsides. Some farmers began consider using workhorses again and started a research programme. Old horse-logging equipment was modernised, and intensive field tests proved that it was efficient, causing a small-scale renaissance of the workhorse in forestry. The use of workhorses has not increased greatly since the end of the project; nevertheless, it was successful in changing the image of horse power. The public became informed about the possibilities of the workhorse in forestry. New equipment was developed and production started. Books and videos about horselogging were produced and courses held on the subject. For some years the number of professional horselogging contractors increased. Today, with new harvesting machines, prices are too low for most horseloggers to be able to compete. However, with a growing demand for focusing on ecological aspects of logging, horselogging can be expected to become more popular in the near future. Nowadays many farmers buy new equipment and use horses for logging in their own forests.

The increased use of horses in forestry has stimulated their use in farming, though most of the farms working with horses also have a tractor. The main reason why people want to work with horses is their interest in animals. With the modern farmer working most of the day on his own, having a social partner in the workhorse is another aspect. Often the whole farming family is interested and involved in using and caring for the horse. Ecological reasons are also important. With a tougher economic situation, many farmers are looking for ways of cutting expenses. Some find a solution in replacing one tractor with a horse. Although the vast majority of farmers are still using tractors, workhorses are gaining ground and small farmers in particular are beginning to recognise the advantages of horsepower.

Reintroducing the workhorse at Sogn Farming and Gardening School

Sogn Farming and Gardening School is the only college level school for organic farming in Norway. Rather than concentrating on a particular direction within organic farming, for example biodynamic or organic-biological farming, it emphasises the qualities from the established organic movement. The education is organised so that during the two years, the students will experience an entire growing season.

The 26 ha organic school farm is the main classroom at the college. It is managed after the IFOAM regulations on organic farming. The students are involved in the daily chores. The barn houses 15 dairy cows, 20 dairy goats, 40 sheep and 4 workhorses. Besides animal husbandry, horticulture is an important part of the farm (1.2 ha vegetables, 1.2–1.6 ha potatoes, 0.8 ha fruit and berries, glasshouse production).

The education at the Sogn college is a solid, professional education, which qualifies students for the title 'Agronomist in organic farming'. When finished, the students should be able to manage an organic farm on their own. The education is both practically and theoretically orientated. In recent years more than 50% of the students have been women. The course also teaches social science, where organic farming is seen in a social, historical and political light. At present the school has 53 students and 9 teachers.

The Workhorse Project at Sogn

For some years there had been discussions about whether to reintroduce workhorses to the school farm. The final decision was taken in 1990 when a government-funded project made it possible to reintroduce them, after 30 years of absence. Since there was no suitable equipment left at the farm, the project had to start from scratch. After 6 months of training horses and acquiring and repairing machinery, teaching the use of horses began. Horses are now used in many operations at the farm. Learning about horses is compulsory at the same level as mechanics and field work with tractors. All students receive both theoretical and practical training in the use of workhorses.

In the theoretical part of the course the students gain an introduction to the use and care of horses. They learn about feeding and husbandry, about the possibilities and limits of the workhorse, and about planning the use of workhorses on their own farm. In the practical part the students undergo basic training on harnessing, hitching, driving and safety before they start field work. During field work they learn to harrow, drill, cultivate and to use the toolframe for weeding. Other aspects they learn are plowing, tedding and raking hay, transport and horse-packing. In the winter there is also a course on log skidding with horses. The majority of students (50–80%) do not have any experience with horses before they come to the college. They are eager to learn and many students who have completed the course have started to farm with horses. The agricultural school also arranges shorter courses (2–3 days) on the subject.

Developing machinery

The production of horse equipment had come to a standstill in the 1960's. One of the main tasks of modern animal traction is the supply of adequate, modern machinery. There are a lot of horse-drawn machines available in the USA, but most of these are not appropriate for a small-scale Norwegian farm. Equipment is required for both single and pairs of horses. A government-supported project on developing horse-drawn equipment started in 1993. A first step was to produce a team-drawn self-loading wagon for loading loose hay and grass. A lightweight tractor-drawn pick-up wagon was equipped with a front axle and a motor with pto to drive the pick-up unit. To obtain maximum flexibility, both the front axle and power unit can easily be disconnected and used for other equipment such as a flat-bed wagon, a manure spreader and a hay tedder. The self-loading wagon has proven to meet expectations. The capacity for loading of loose hay is 30–60 m^3 per hour.

In a second step, a new Swiss multipurpose toolframe/forecart will be used as a basis. Equipped with a new hydraulic power unit, it will be used to run a hayrake, a haytedder and a sprayer. The motor drives a hydraulic system, so providing a flexible alternative for driving a wide range of pto equipment, including winches, wood saws, and milking machines.

Field research

One argument often used to support working with horses is that they have a lower impact on the soil compared to working with tractors. Yet there is little evidence to support this statement. In 1995 Sogn college started a 3-year field research project on the subject. In a hay/grass field one part will only be driven on by tractor, another part only by horses and horse-powered equipment. On a control patch no machines will be used at all. In each of the three plots there will be five fixed survey lines along which vegetation and soil characteristics are recorded. The first results are not expected before the end of 1996.

Discussion and conclusions

There is a growing interest in the use of workhorses in Norway. A lot of young people wish to learn more about how to farm and work with horses. However, people who wish to use workhorses face many problems including: getting the know-how, finding/constructing the needed machinery and getting spare parts, finding qualified help experienced with workhorses and replacement of a trained workhorse in case of sudden casuality.

Knowledge about working with animals has traditionally been handed down from generation to generation, with little written down. 'Everybody' knew how to handle a horse until the 1950s. This traditional handing down of knowledge has been broken. New ways have not yet been established. There is an increasing need for education and information. Farming schools can be involved successfully as described above for the college in

Sogn. Good books need to be written on the subject. Practical 2 to 3-day courses are held at different places but there is a need for better and longer courses tailored to farmers' needs.

It can be difficult for farmers to acquire the equipment necessary for working with horses. There is still a lot of old machinery available, but as Jean Nolle had already realised 40 years ago, new machinery has to be developed in order to keep the workhorse competitive on small farms (Nolle, 1986). Factories producing modern farming machinery are not interested in the development and production of horse equipment because the demand for such machinery is too low. Some machine shops in other European countries are manufacturing horse equipment in small quantities. However, they do not have the resources to offer all the machines needed. This lack of equipment may have a major effect on the future of the workhorse in Norway.

The negative aspects of overmechanised agriculture are becoming increasingly visible. Research about farming with workhorses will lead to a wider acceptance of animal traction. The experiences from the logging research project show this clearly.

The ecological impact of farming is of increasing concern to farmers and the public. At the same time farmers are facing lower income and have to cut down on expenses and investments in order to survive. It is also a well-known fact that the modern farmer faces major physical strains, such as back problems due to long hours with tractor work. Loneliness, as a result of industrialisation of agriculture, is another aspect of modern farming.

The ecological, economic and social aspects of farming mentioned above are arguments for greater use of workhorses. To achieve a wider use of workhorses politicians must to encourage horse power by supporting education, consulting, and the research and development of machinery. More importantly, people actively working with horses must share their knowledge of the benefits workhorses can give to modern agriculture.

Reference

Nolle J, 1986. *Machines modernes à traction animale: itinéraire d'un inventeur au service des petits paysans.* Edition L'Harmattan, Paris, France.

Improving animal traction technology in Cuba

by

Arcadio Ríos

Scientific Director, Instituto de Investigaciones de Mecanización Agropecuaria (IIMA)
Carretera Fontanar, Km 2.5, Abel Santamaria Boyeros, AP 19240, La Habana, Cuba

Abstract

Over the past 25–30 years animal traction in Cuba has decayed due to the massive introduction of tractors, mainly of Soviet production. The number of tractors increased from less than 7000 in 1960 to more than 7000 in 1990, while oxen decreased from 500,000 to 163,000 in the same period. In recent years the decline of the socialist states in Europe has produced an acute shortage of fuel, spare parts and hard currency in Cuba. The government was concerned about maintaining food production and so various measures were taken, among them a drastic reduction in the number of tractors in operation, and consequently an increase in the use of animal traction.

Another important change has been conversion of the huge government-run collective farms into self-operated cooperatives and the encouragment of the existing private sector. To these small-scale farmers animal traction is much more attractive than tractor power, and a wide-ranging project has been undertaken to finance the introduction of 200,000 oxen, the training of draft animals and their handlers, and the fabrication of new and more productive implements.

Introduction

The Cuban economy is heavily dependent on agriculture which accounts for 14% of the Gross Domestic Product but 70% of export earnings. The main products are sugarcane (15% of the agricultural land), citrus (5%), roots (5%), vegetables (4%), and rice (6%). Livestock are farmed on 55% of the agricultural land. Agriculture employs more than 22% of the country's population of about 11.2 million people. Only 16% of the population lives in the agricultural areas (CEE, 1994).

Cattle have been used as draft animals in Cuba since its discovery and colonisation five centuries ago, and agriculture was traditionally based on animal traction, mainly from oxen of rough races. Equines are not used for plowing, but horses are used for light carting and riding. In montainous areas donkeys are commonly used as pack animals to carry coffee and other products.

In the years following the revolution, a massive introduction of tractors was undertaken as part of the governmnent's strategy for transformation and modernisation of agriculture. Over the three decades from 1960 to 1990 the number of tractors increased ten fold from 7,000 to 70,000 (see table 1). The increase was not only in quantity but in quality, since the average power of the tractors grew from 40 to 75 HP. In the same period the number of oxen decreased from 500,000 to 163,000.

The growing adoption of tractor technology was encouraged by the concentration of agriculture following a socialist policy: the aim was that all farming would be carried out on mechanised collective or state farms. Only a limited

Table 1: Number of tractors and work animals in Cuba (1960–1995), figures in thousands

	1960	1970	1980	1990	1995
Tractors	7	52	68	70	40
Oxen	500	490	338	163	376
Horses	800	741	811	235	300
Donkeys	5	4	4	4	4
Mules	30	29	25	30	30

cooperative and private sector remained. Soviet aid and soft credit backed the import and introduction of tractors, implements, spare parts and fuel at low prices. This contributed to a strong socialist sector, but with a weak long-term basis.

In recent years the decline of the socialist states in Europe has produced a dramatic shortage of fuel, spare parts and hard currency in Cuba which has made the high level of mechanisation achieved in agriculture unsustainable.

Maintaining food production at a reasonable level in such conditions became a huge problem. New agricultural policies and strategies were developed and these include:

- a drastic reduction in the number of tractors in operation
- rapid increase in the use of animal traction and its infrastructure
- transformation in the forms of land title.

Animal traction in the new conditions

To face the new conditions realistically, the most important measure taken was the turning of the state farms into cooperatives, putting the organisation of production under the control of the workers and operating under the principles of self-support. A part of the production is sold to official agencies under contract and the remaining part is offered on a free market.

The number of tractors in operation was reduced from 70,000 to 35,000 and the idle ones were dismantled or prepared for long conservation.

A vast programme was undertaken to encourage animal traction at the expense of tractor mechanisation, but the task was (and is) not easy, for many reasons. A farmer trained to operate a tractor, can only learn to handle a pair of oxen with difficulty. Not only farmers but technicians and many officers see tractorisation as a symbol of progress and animal traction as a 'backward step'.

Draft animal power, engine (tractor) power and human power should be seen as complementary power sources for agricultural production, not as mutually exclusive ones. The optimum mix will depend upon the requirements of each individual farm (Inns, 1994). The new smaller-scale cooperatives and their internal division into farms are much more suitable for the use of animal traction than the huge state farms.

In 1992, the Ministry of Agriculture established a set of recommendations to encourage the use of animal traction, backed with a serious control of the amount of fuel provided for use in tractors and in the distribution of the spare parts obtained with the government's scarce foreign exchange.

Selection and supply of oxen

One of the greatest constraints to the adoption or increase of the use of animal power was the limited infrastructure in this field. For the Ministry of Agriculture a logistical problem had to be faced in obtaining oxen to be supplied to the cooperatives. This was especially difficult as cattle were not in abundance. For the cooperatives the problem was in training oxen, selecting and training ox handlers and creating basic but acceptable conditions for feeding, health care and protection of cattle.

Practically all healthly oxen in Cuba were selected and supplied to the cooperatives and the remaining state farms. The first programem in 1991–1992 involved the supply of 100,000 oxen and a second programme in 1993–1995 supplied some 100,000 more. The programme continues and now there are about 376,000 working oxen in Cuba, nearly 2.5 times the figure in 1990.

Selection and promotion of new implements

For a long time the most commonly used animal-drawn implements for crop production in Cuba were mouldboard plows and wooden ards similar to those widely used in other regions of the world. Harrows, cultivators, wooden sledges and a few basic models of two-wheeled carts were the only others available. No animal-drawn planters, fertilisers or wheeled toolcarriers were available.

A wide-ranging programme was started in 1992 for the development, selection and promotion of new, more suitable implements and the fabrication and supply of traditional and new models, also encouraging the adaptation of tractor-drawn implements such as sowers, fertilisers and sprayers to be used with animal traction.

The main change in the pattern was the development of a new model of plow patented by the Institute of Agricultural Mechanization Research (IIMA) under the name of *multiarado* (multiplow) in versions for tractors, a one-row animal-drawn implement and independent elements for toolcarriers. The *multiarado* is designed for soil plowing, cultivation and weeding without inversion of the soil. With additional shares it can be used for ridging, pre-seeding, furrowing and for covering seeeds or stacks with soil after planting.

Wheeled toolcarriers amd other similar three-row implements have been promoted with a satisfactory adoption by the farmers, making a balance among the known advantages and constraints of these implements (Starkey,1988). The recommended model is a version of the multicultor adapted to the working conditions of Cuban agriculture, with many improvements.

Regional workshops and on-farm demonstrations

In the initial stage of the implementation of the project of re-oxenisation it became evident that no satisfactory results could be achieved without an intense programme of on-farm, regional, provincial and national workshops and demonstrations. At these events the objectives to be achieved were:

- demonstrations of newly-designed or unknown implements
- selection of the most suitable implements for promotion
- training of ox-handlers, blacksmiths and artisans makers of yokes and harnesses
- discussions and interchange of experience on animal traction.

From 1993 to 1995 over 450 workshops were held which were attended by more than 30,000 farmers, technicians and others. A National Workshop was held in November 1994 and an International Congress on animal traction was held one month later.

Local manufacturing of implements

A group of implements has been selected for national promotion including a mouldboard plow, the *multiarado*, a tine harrow, an ox cart, the multicultor, a grain seeder and a yoke with width

regulation. In each of the 15 provinces a workshop for the local manufacture of the recommended implements was set up. This seems to be more convenient than centralised manufacture by the government, due to the possibility for the provinces to exploit unused reserves of steel and the ability to adapt the designs to local requirements and preferences.

Another important aspect has been the organisation and training of local blacksmiths equipped with forge, anvil and tools for repairing and even manufacturing non-complex implements and spare parts. In 1995 three were a total of 250 blacksmiths.

Conclusions and future directions

Animal traction is a suitable technology for Cuba, especially in the new conditions of an increasing number of small-scale cooperatives and as a means of saving scarce fuel. The Ministry of Agriculture and the Institute of Agricultural Mechanisation Research faced the task of improving animal traction on the basis of ensuring the manufacture of suitable implements at local or provincial level, supplying sufficient oxen, creating an infrastructure for repairing and maintaining the implements and the training of ox handlers, agricultural technicians and farmers. Much further work is required. Animal traction is not easily adopted by those familiarised with tractorisation and the constraints must be overcome to reach a sustainable balance between tractors and oxen.

References

CEE, 1994. *Anuario estadistico de Cuba*. Comite Estatal de Estadisticas (CEE), La Habana, Cuba. 600p.

Inns F M, 1994. Animal draft tillage systems: the need for an integrated approach. pp 214–217 in: Starkey P, Mwenya E and Stares J (eds), *Improving Animal Traction Technology*: Proceedings of the Animal Traction Network for Eastern and Southern Africa (ATNESA) held 18–23 January 1992, Lusaka, Zambia. Technical Centre for Agricultural and Rural Cooperation (CTA), Wageningen, The Netherlands. 490p.

Starkey P, 1988. *Perfected yet rejected: animal-drawn wheeled toolcarriers*. Vieweg for German Appropriate Technology Exchange, GTZ, Eschborn, Germany. 161p. ISBN 3-528-02053-9

Photo (opposite): Three men plowing with oxen as one man plants in the furrow in Western Zambia

Meeting the challenges of animal traction

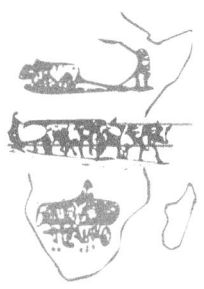

Photo Paul Starkey

Addressing the challenges:
project experiences

Animal traction development in Iringa Region, Tanzania: project approaches and future requirements

by

R Mwanakulya

Regional Agricultural Mechanisation Officer, PO Box 389, Iringa, Tanzania

Abstract

This paper reviews the performance of animal traction activities carried out by different projects, past and present in Iringa Region, Tanzania. Animal traction is not well developed in Iringa Region, although there has been limited spread of the technology through natural diffusion. The performance of the projects has been poor, mainly due to the adoption of top-down non-participatory approaches which led to inappropriate selection of implements and poorly-designed training schemes. Animal traction has the potential to increase food security in the region and should continue to be promoted actively. Natural diffusion is an effective process but is very slow and cannot keep up with the rapid increase in population. A participatory approach that combines bottom-up and top-down approaches to create a dialogue is necessary. The use of single animals for traction should be promoted. Since income from cash crops is generally low, privatised self-supporting training programmes would not be viable so government/donor intervention is still necessary.

Introduction

The use of animals, particularly oxen and donkeys, in Iringa Region can be traced back to the early 1930s. Donkeys have been in use in the semi-arid zone of Iringa District since well before this time. This is because donkeys were mostly common in the central part of the country and because the region has an agroecological zone (zone 16) where donkeys do best so that it was easy for them to be used much earlier. The donkeys were originally used as pack animals; only much later were they used for tractive purposes. Ox traction started in the 1930s when it was introduced to parts of Mtwango (Njombe District) and Malangali (Mufindi District) by Christian mission posts. Transfer of the technology was by natural diffusion, ie people learning from others by seeing the animals working on *shambas*. Spread of the technology was rather slow and took a long time to reach distant areas. However, it is effective because it is demand-driven. Use of animal traction continued to increase slowly up to and after independence. In the effort to increase agricultural production one way was to bring more land under crop production and human labour was a limiting factor. At the time the population was about half of what it is today and there were still large areas of uncultivated fertile land. The limiting factor was power availability. Previous attempts at tractorisation had failed to achieve the desired goal.

When the government shifted its emphasis from tractorisation to animal traction in the mid 1970s, the districts within the region used development funds to establish about six Ox Training Centres (OTC). These training centres operated very briefly and then closed due to shortage of funds. The method of transfer of technology was top-down with the government extension workers as active players and the clientele as passive players. In 1978, the European Economic Community (EEC) funded Iringa Region Agricultural Development Project (IRADEP) started working again on oxenisation development in the region. Activities included the rehabilitation of old ox-training centres which involved the construction of staff houses, cattle bomas, buildings for implement storage and classrooms. An extra ten training centres were constructed. Other activities included procurement of implements, and carrying out farmer training. IRADEP made a significant impact on animal traction. In terms of funding, it was the largest single animal traction drive in the region since independence.

In 1994, the International Fund for Agricultural Development (IFAD) started the Southern Highlands Extension and Rural Financial Services Project (SHERFSP) to train 600 farmers in 6 years. Emphasis was placed on training women.

Sasakawa Global 2000 has also begun a farmer training programme. HIMA-Iringa has a project trying to reduce the workload of women by using donkeys as pack animals and CONCERN also has a small project. It is hard to say whether the activities of the different projects are coordinated.

Review of animal traction projects

IRADEP Project

Success of the project

The IRADEP project created a great deal of awareness of animal traction in most rural areas, especially in the use of the plow. A total of 10 ox-training centres were constructed, although one was not completed. Each training centre had a trained carpenter and blacksmith. According to the project report 7,000 farmers received training from 1977–1981 and 1982/83–1985/86. The frequent trips to the villages enhanced the awareness raising since at that time most of the vehicles in the area belonged to the project.

Shortcomings of the project

In many ways the project seemed busy for nothing. It should be noted that even before the IRADEP project had begun, there was a good number of farmers using animal power and the project appears to have helped to increase only a smaller proportion of the total. Moreover, most farmers adopted ox-drawn plowing only, a technique which most already knew.

The project brought a variety of implements which included ridgers, weeders/cultivators, multipurpose toolbars, plows, planters etc. which were placed at the ox-training centres so that farmers could learn how to use them and eventually purchase them. Except for the plow, none of the implements has been used effectively by farmers. In a few places where animal traction is fairly well-developed, the plow is used for weeding instead of a cultivator, simply because cultivators are not available. In places where ridging is usually preferred over flat cultivation, ridging using animal-drawn implements was not adopted and farmers switched to flat cultivation despite all the advantages of ridges to soil and moisture conservation.

Planters were placed at ox-training centres for training purposes even though it was irrelevant to do so because planters of proven performance were unavailable in the region. Moreover, the planters at the ox-training centres were very expensive, and the planter components broke easily since most were made of cast iron. They were complex and could not be easily repaired by the farmers. The decision to bring in planters was unnecessary. It may have been as a result of the 'complete package' approach.

The Ariana multipurpose toolbars were unnecessary since it was an irrelevant and expensive implement which farmers were not ready to adopt.

Follow-up on the project activities was poor, so that the reported numbers of farmers that were trained is hard to accept. Farmers trained at the training centres were expected to carry out similar training in their villages which were then regarded as an Ox Training Unit (OTU). There was a system whereby Tsh 50/- was paid to the trainer for each pair of oxen trained. This became an incentive to cheat in collusion with unscrupulous village leaders. For this reason the figure of 7000 farmers trained is unrealistic.

Courses at the ox-training centres lasted 30–90 days. This is a very long time for any farmer and in particular the poorer ones. Effective training was between September and late February which was the time when farmers were most busy. The result was that primary school leavers were sent for training and farmers themselves were not trained.

The project did not give high priority to women. It is apparent that women, the main producers of food, would be excluded from such a long training session.

Ox carts were manufactured at a government district agricultural workshop (Mafinga in Mufindi district). The body of the cart was made of hardwood timber, and the wheels were made of steel. The workshop had problems ranging from managerial and administrative problems to the cart design. Farmers did not like the metal wheels and the wooden bearings were inappropriate. The metal wheels performed poorly in sandy areas. It may be that the decision to use metal wheels and wooden bearings was in accordance with a misguided misconception in which appropriate technology was equated with substandard technology. There are few metal-wheeled ox carts being used.

There was a general weakness in the management of the animal traction project. The whole project was under the direction of a regional oxenisation officer who apparently was not trained adequately. The minimum training at that level should have been a Diploma.

IFAD-SHERFSP

Success of the project

This project started in late 1994 so that it is difficult to fairly judge to its performance as yet. The project has shifted its emphasis to women by stipulating that women must constitute at least 30% of participants in each training session. This is a positive move in the right direction.

Shortcomings of the project

The project trains only 100 farmers per annum so over 6 years it will have trained about 600. The figure of 100 farmers per year is very low. It is possible to train up to three times this figure without budgetary constraint.

Farmers are brought to a central point, exposed to very brief theoretical and practical training for about 3 to 4 days per year and that is it. Additional training sessions are not offered, nor are follow-ups carried out to assess the effectiveness of the training sessions. The course content is mostly textbook notes which have little relevance to practicality. The training is done once at the beginning of the season and is repeated the next year for new farmers. The short duration of the course means that farmers have to learn much new information in a short time which means that farmers can absorb very little of the course content. The interest of the project appears to be in quantitative successes rather than in both the quantitative and qualitative ones.

Future strategies for animal traction in Iringa Region

As discussed above, there have been a number of shortcomings in the way animal traction projects in Iringa Region have been and are being implemented. This explains why the activities are not sustainable when the projects have ended. For future development and sustainability of animal traction in the region, there are two alternative approaches available. The first is to leave things as they are and hope that the technology will spread through natural diffusion, as it has done in the past. The role of the government should be to introduce new technologies which have been proven in areas where animal traction is well-developed. This may be supported by the fact that Rukwa region is one of the leading areas in the use of animals for tractive purposes although it is a product of natural diffusion with little external intervention.

The weakness with this approach is that with the low productivity to land and labour and increasing population, the region may continue to face unreliable food security. Natural diffusion cannot keep pace with the ever-increasing demand for food. The number of animals being used for tractive purposes is very small and with natural diffusion the technology will take a long time to reach the remaining farmers.

The second approach is a deliberate one in which the government, through its extension workers, engages in a dialogue with farmers and together they identify problems related to farm power issues and develop solutions to these problems. The farmer has a great deal of experience and the extension worker, in liaison with the researcher, has new technology and practices to offer the farmer. The new technologies are only good in as much as they solve the identified farmer problems. The second alternative would be preferred by the author since it offers an opportunity for accelerated development.

Suggested animal traction strategy

Participatory approach

Experiences elsewhere have shown that participation of farmers in the whole process of planning and implementation has led to sustainability of project activities. The top-down approach alone has not worked and instead a bottom-up approach with a little top-down intervention, in other words a dialogue, should be used. During a dialogue with the farmer the available resources, the real problems and their prioritisation will be learnt.

Selective technology transfer approach

All animal traction technology packages are suitable, but some have a higher priority. For instance, while mechanising seed sowing is good, mechanising land preparation is more important from the point of view of the farmer. Land preparation is important because it requires more time and labour than sowing. In semi-arid areas farmers would like to reduce the number of hours spent for primary cultivation since it can lead to timeliness penalties related to low moisture availability in the soil. Using animal traction to perform this operation can make the difference between a bumper crop and a poor harvest. Another important operation is weeding, since it is usually carried out more than once. Transport is another important operation for moving inputs to the farm as well as moving crop harvests to the home or market. These are the most important operations which should be targeted as a priority.

Farmer training

The timing of training should make sure that real farmers do attend. Emphasis should be on practical training rather than theory. Plowing and carting training may be carried out during the dry season. Weeding may be carried out during the dry as well as the rainy season. The implements used during training, especially weeders, should be readily available in the area at the time of training.

Training of village extension workers

Most village extension workers know very little about animal traction technology. Most do not know even the basics of animal traction technology. There is a need to improve the curriculum of the agricultural training institutes that offer extension certificates.

Single animals and promotion of donkeys

In all the projects animals have been used in one or more pairs. It is thought that the time has come when single animals have to be used. The following reasons call for greater emphasis on use of single animals:

- for operations such as weeding single animals perform better than pairs
- the number of animals, especially cattle, in the region has been affected due to a combination of deaths from disease and inadequate feeding and a high demand for beef which has led to large numbers going for slaughter. Cattle numbers have fallen in absolute terms as well as in relative terms when compared to population growth.

- the price of animals, especially oxen, is very high so it would do a great service to the farmer if single animals were used
- women as individuals or in groups can easily adopt and handle single animals
- use of donkeys must also be also be encouraged since they are cheaper and are easy to maintain than cattle. They can be used as pack animals in highly dissected mountainous areas.

The role of ox-training centres

Currently the central government is in the process of rationalising the mechanisation services and is considering whether to continue owning ox-training centres. The alternatives may be, in the short- to medium-term, collaboration between the government and farmer organisations/cooperatives on modalities of ownership and running the centres. Farmer organisations and cooperatives are still not strong enough to carry out this difficult task.

Most farmers are subsistent to a large extent, although some have moved substantially towards production of cash crops. Therefore, to run the centres based on demand may not work. At the moment, the centres are designed as centres of innovation and try to offer free services. To run them sustainably the organisations have to charge a small fee to cover expenses. For the same reason they cannot be run as private entities, at least at present. The central government still has a role to play in ox-training centres and if accelerated development in animal traction is felt important it should not relinquish ownership of them for the time being. If the central government has to stop owning them, then each district local government could own one ox-training centre.

Conclusion

There is still a high potential for developing animal traction in Iringa Region. More has to be done and it must take into consideration the farmers' situation and their priorities. Extension workers must be well trained in the area of animal traction and follow-ups to farmer training must be carried out.

A note on the development of animal traction in Tanzania

by

R N Mtunze and M G Lyimo

Agricultural Engineers Mechanisation Unit, Ministry of Agriculture, Dar es Salaam, Tanzania

Introduction

There are an estimated 12 million cattle in Tanzania, of which about 10% are used as draft animals. It is thought that cultivation using draft animals was introduced into the country about 100 years ago by settlers from South Africa along with the use of single-furrow steel mouldboard plows.

In 1970, the Ministry of Agriculture initiated the establishment of about 12 animal traction training centres in ten regions of the Tanzanian mainland. By the end of 1990, 80 training centres had been established. Although these centres were used mostly for training oxen, those with full board and lodging facilities enabled both farmers and their animals to be trained at the centres. Farmers were able to exchange their untrained animals for trained animals at the centres. However, this approach had limited success: familiarisation with the animals at work requires that farmers are trained with their animals. Because of financial problems affecting both farmers and government, training is now preferably carried out on the trainees' farms.

The two prominent Tanzanian factories for the production of agricultural implements and hand tools are Ubungo Farm Implements Limited and Zana za Kilimo Limited Mbeya. Ubungo Farm Implements, which started in 1970, reached maximum hand-hoe production in 1984 when it produced about 2,175,000 hand hoes, well above the installation capability of 195,000 hand hoes. The same year the production of animal-drawn plows peaked at 24,600, also above the installation capacity of 20,000.

The Zana za Kilimo Ltd Mbeya factory produces only hand tools. These two large factories do not satisfy the national demand for hand hoes or animal-drawn implements.

Animal power projects

Between 1990 and 1995, two famous projects were underway: the Mbeya Oxenisation Project and the Tanga Draught Animal Project. Both projects, in different environmental conditions, have been very successful. External finance for both projects has come to an end. Much experience has been gained from the projects.

Since the availability of implements and spare parts is a major problem for farmers, the Mbeya Oxenisation Project tried to facilitate this by developing a marketing system. The system had the following objectives:

- to distribute relevant and appropriate implements to farmers
- in co-operation with the project engineering section, to promote the use of village workshops where farmers could obtain spare parts, have repairs done or even purchase complete implements
- to establish a reliable and permanent distribution system for draft implements.

However, it was very difficult for the Mbeya Oxenisation Project to achieve these objectives. Two distribution systems were used:

- Direct selling to farmers in the villages by the project. This was a short-term strategy aimed at making farmers aware of the project and the type of implements available.
- Selling to farmers through village-based sales agents. Most of the agents were government employees, cooperative societies, trading companies and private businessmen.

However, these approaches proved inefficient: the implements were not in the right place at the right time. The marketing system needs to be strengthened to benefit farmers, perhaps by a mobile sales system.

The Tanga Draught Animal Project aimed to improve the utilisation of draft animals by farmers in selected areas of the region for agricultural and transport activities. This project was faced with working in a difficult rural environment with weak infrastructure and lack of determination by all parties to overcome the organisational and human constraints associated with the project. Project activities proved unsustainable when the donors finished their contract. The Ministry of Agriculture is trying to find ways in which the project team can resume technical advisory services. This was supposed to be considered before the external finance stopped.

Another project which came to the end of external funding was the Usangu Village Project. In its terminal report it recorded that 70 farmers participated in draft animal power training courses in 1992 and 1993. The response was quoted to be good and demand for plows and carts increased.

Current projects

Sasakawa Kilimo

The Kilimo SG 2000 oxenisation programme is being implemented in seven regions of Tanzania. The main objective of the programme is to train farmers in the use of draft animal technology. The aim is to train farmers to be efficient in using the implements and to observe improved timeliness by using the animal-drawn implements. This will in turn reduce drudgery and raise their productivity. Presently, groups of 10 farmers are trained in each centre by village extension officers.

IFAD

The International Fund for Agricultural Development (IFAD) oxenisation programme covers the Southern Highland regions of Tanzania. The approach used is different from that of Kilimo SG 2000. IFAD has selected one animal traction training centre in each region. The centre will be strengthened with implements, equipment and animals. Farmers are trained at the centres.

The problem of this approach is whether farmers will come to the centres at their own expense, which is difficult for them. The other alternative of the project, to provide full board and lodging, can only be done when there is external finance, otherwise it will be difficult in the future.

FAO

The Special Programme for Food Production of FAO has pilot phases in Morogoro and Dodoma Regions. It is aiming at increasing production of rice and maize. The project aims to facilitate the availability of agricultural and mechanisation inputs such as animal-drawn implements.

Several interest groups, for example an oxenisation interest group, will be formed to avoid dependence of farmers on central government and the project. Project staff will then act as advisors to the groups. Some of these will help farmers purchase various animal-drawn implements through a bank or credit societies.

Netherlands assistance

In the lake zones The Netherlands Government is carrying out a feasibility study on how agricultural development can be improved. In Kagera Region, the project will concentrate on the improvement of the use of animal power.

Government policies

In the past, the Ministry of Agriculture tried to do too much through the public sector, assigning tasks to institutions which were ill-fitted for the task. Some activities like distribution of agricultural inputs, research and extension still remain in public hands although conditions and resources no longer justify this. The government ought to limit itself to critical services for which it has a comparative advantage in delivering and should leave the rest to the private sector.

Due to inadequate financial support from the government there is a need to reduce the number of animal traction training centres. The Ministry of Agriculture will advertise to run the centres jointly with NGO's or other associations which have an interest in developing draft animal technology. Some will be sold directly to interested groups or associations, but the Ministry of Agriculture will retain one training centre in each region. They will be strengthened in terms of the package of implements and animals available. Several approaches to contact farmers will be applied to make farmers familiar with the full package of draft animal technology. Farmers near the training centres will be visited easily, while the mobile unit approach should be used for distant farmers.

Through extension, the Ministry of Agriculture will continue to advise farmers to form various associations which in turn will raise the productivity of land and labour. Formation of draft animal farmers' associations will help the technical personnel explain their ideas. As a group, farmers will discuss ideas given to them before implementation. The associations are expected to run commercially.

A note on reviving cotton production in eastern Uganda: the animal traction perspective

Alex Nyugo and John Olupot

Smallholder Cotton Rehabilitation Project, Ministry of Agriculture, PO Box 102, Entebbe, Uganda

Introduction

The use of animal traction in Teso zone dates back to the introduction of the cotton industry in the area in 1909. Before the civil strife in the 1980s the cattle population was about 355,000. This was reduced to about 18,000 by the turmoil. Since animal traction was the major means of land opening, this had a negative effect on agricultural production in the area. The cotton industry was affected worst. After the use of animal traction for over half a century, the technology had become part of the farming system. The use of this technology was not confined to men, but extended to tyoung people and women.

Animal traction was used mainly for land opening and transport using sledges. A few farmers were attempting to weed their cotton, maize, groundnuts and beans planted in rows. There were also some attempts to use mouldboard plows for groundnut lifting. An assortment of animal-drawn equipment was received from the USA, South Africa, India and the UK. This equipment was tested in the Serere Agricultural and Animal Research Institute (SAARI), but their take-up by farmers was low for socio-economic reasons, for example the preparation of a fine seedbed for row-cropping was considered time-consuming, since farmers could not afford to hire extra labour.

Another factor that hindered the development of animal traction technology was the introduction of tractor hire services in the 1960s. By 1990, a total of 3,021 tractors were imported into the country, 136 of them into Teso. However, this scheme was short-lived since farmers could not afford the hire costs and there was a lack of technical know-how in the area of maintenance. The farm holdings which averaged 2.5 ha per family were too small to justify the ownership and use of tractors. Due to high hire charges for tractors, the farmers reutrned to the use of animal traction and hence the scheme collapsed. Teso continued to lead in the use of animal traction in Uganda until the 1980s when civil strife set in, resulting in massive depletion of the working herd and equipment.

The revival strategy

To revive the cotton industry in the Teso farming system and boost national production which fell from the annual average of 360,000 bales to 11,000 bales in the 1980s, the government of Uganda recognised the importance of promoting animal traction alongside the revival of cotton production.

This was made more important by the fact that Teso was the centre of the SATU cotton breeding and multiplication centre. Hence in 1993 a 'smallholders cotton rehabilitation project' was conceived and launched with funding from the International Fund for Agricultural Development (IFAD). This project has a large component for promoting animal traction, especially credit to facilitate purchase of animals and equipment by farmers on 'soft' terms. SAARI received support to enable it to test and develop equipment and produce extension packages for farmer use.

The strategy adopted to produce quick results in reviving animal traction technology includes:

- organising farmers into animal traction credit groups to enable them to aquire loans for animal traction
- establishing Livestock Animal Traction Extension teams in each district to facilitate farmer training
- establishment of animal traction training centres for testing new technologies and training of staff, artisans, farmers and their animals
- encouragng the development of prototypes by SAARI for fabrication by the Soroti

Agricultural Implements and Machinery Manufacturing Company for farmer use.

The project is gaining ground as can be seen from the following indicators:

- a majority of farmers who benefited from credit gave priority to procurement of oxen and ox equipment and the purchases were made as soon as the loan was received
- many groups were formed for credit provision and extension services. This group approach has also assisted in fast loan recovery: 53% of matured loans have been recovered so far
- women's participation in all project activites is now 37–40% against the project target of 30%. This has been enhanced by rapid formation of women's groups
- many farmers benefited from intensive training on-farm and at training centres
- the cropped area under cotton has increased with the cropping area index moving from 33% to 103% .
- farmers have been able to comply with the optimum planting date of cotton, increasing their yields from 400 kg/ha to 800–900 kg/ha
- there has been a reduction in peak labour requirements
- farmers have moved from subsistence farming to limited semi-commercial farming
- the food security situation and the social status in the area have improved greatly
- Serere Agricultural and Animal Research Institute has developed the SAARI master plow and the SAARI versatile weeder. These are being fabricated for on-farm testing before large-scale production starts
- Soroti Agricultural Implements and Machinery Manufacturing Company has achieved record sales, especially of ox plows.

This is mainly due to the credit available for these inputs.

Challenges ahead

The following challenges need to be met:

- there is need to emphasise local manufacture of durable traction equipment rather than the current import of weak implements
- the promotion of rural blacksmiths and artisans is important in sustaining the supply and repair of fast-wearing parts
- there is a need to develop a planter suitable for local conditions. At present most of the planters used in the region are not suitable
- the cattle used for traction (East African Zebu) are small and cannot provide the required tractive power. There is a need to develop a breeding programme to improve on their size and strength
- credit is a critical factor in sustaining steady agricultural production. There is therefore a need to identify workable financial intermediation for the rural areas
- though the National Agricultural Research Organisation has been established, research that is orientated to solving farmers' problems should be encouraged
- the formation of an animal traction network for Uganda is overdue.

Conclusion

The role animal traction has played in reviving cotton production in the Eastern block of Uganda is an indication of its importance in the national economy. If the challenges discussed above are addressed positively, then animal traction will play its rightful role in the economic and social development of the region.

Hiring and lending of oxen for plowing in Kaoma, Zambia

by

Kennedy Kakwaba and Martin van Leeuwen

Western Province Animal Draft Power Programme, PO Box 940007, Kaoma, Zambia

Abstract

This paper discusses lending and hiring arrangements of oxen for primary tillage in Kaoma District, Zambia based on the results of interviews with 152 ox-owning and non-ox-owning households. Only 13% of rural households own oxen but approximately half the cultivated area in Kaoma District is plowed under hiring and borrowing arrangements with ox owners. Most (51%) of oxen exchanges were on a sharing basis, with 26% of transactions conducted for cash and 15% hired in return for fertiliser. A few transactions were conducted for labour, maize or were lent free of charge. More than half of hiring transactions were to relatives, and relatives, especially women, had a high priority in terms of time of plowing. This study found that farmers were optimistic about the continued importance of the hiring market. However, a more recent study indicates that removal of subsidies for fertiliser and hybrid seed may have a negative impact on the hire market.

Introduction: lending and hiring animal draft power

Hiring and borrowing of oxen has been mentioned in animal draft power literature since the 1960s (Nicolas, 1968; Dronne, 1969). In some cases lending has been explained by owners' interest to be assured of labour for weeding and harvesting (McCown, Haagland and de Haan, 1979). In other countries with a long history of using animal power, for example Ethiopia and Bangladesh, farmers appear to have a range of words for different hire arrangements (Helmrich, 1986b; Gryseels, 1988). A few aspects of hiring and borrowing oxen in other districts of the Western Province of Zambia have been described by other authors (Nambayo and Vierstra, 1988; Vijfhuizen, 1987; Beerling, 1988; Beerling and Mwenda, 1988; Beerling, 1985; Sutherland, 1984).

In the Western Province of Zambia, Kaoma District has the highest proportion of oxen (20%) in the total cattle population (Muntali et al, 1995). If one assumes that each pair of oxen plows 8 ha in a season, all 5,600 oxen could plow over 20,000 ha, which is more than the total area cultivated. Futhermore herd analysis shows that in Kaoma only 12% (250) of the oxen required each year need to be imported. However, a district-wide survey and contacts with farmers showed that only 13% of all rural households actually own oxen (Kakwaba, 1995).

The Western Province Animal Draft Power Programme (WP-ADPP) estimated that 30–40% of the rural households would have access to animal draft power and so in 1994 initiated research into the arrangements of hiring and borrowing draft animals. The aim of the research was to establish whether access to draft power could be increased by project intervention. After some exploratory interviews, 152 households of four types in the maize belt of Kaoma were interviewed:

- farmers owning more then two oxen (n=44)
- farmers owning two oxen only (n=39)
- farmers who hired and borrowed oxen (n=39)
- farmers tilling the land by hand (n=30).

Results

On average, farmers with oxen kept bigger households (mean 7.7 people) than the farmers who used manual labour (mean 5.3 people) or borrow oxen. Of households owning oxen, 4% were female-headed, while 32% of households using manual labour or hiring oxen were female-headed.

Some farmers from all four categories used hired labour, especially farmers with only two oxen, of which 28% employed hired labour more or less continuously. About 70% of all respondents used fertiliser during the period 1992–1994 for maize production, without significant differences between the four categories. External financing of fertiliser by agricultural lending institutions increased during that period for farmers using manual labour and for ox borrowers (while self financing decreased). For ox owners the source of fertiliser

Table 1: Area of maize grown by respondents

	Area of maize planted (ha)		
	1992	*1993*	*1994*
Ox owners with >2 oxen (n=44)	6.1	3.7	4.6
Ox owners with 2 oxen (n=39)	2.2	2.4	2.5
Ox hirers (n=39)	1.4	2.0	2.3
Hoe users (n=30)	0.7	0.8	0.8

Notes:
Differences between categories are statistically significant (t-test at p<0.05)

Table 2. Quantities of maize sold by respondents

	Maize sales (tonnes)		
	1992	*1993*	*Mean*
Ox owners with >2 oxen (n=44)	4.1	5.7	4.8
Ox owners with 2 oxen (n=39)	1.3	1.9	1.6
Ox hirers (n=39)	0.7	2.0	1.4
Hoe users (n=30)	0.3	0.5	0.4

did not change significantly and 20–35% financed fertiliser from their own resources.

Maize cultivation and sales were considerably different between the categories (Tables 1 and 2). In each category, some farmers do not grow maize every year, but there is a higher percentage of non-maize growing among the borrowers and hirers of oxen. During 1992 and 1993 only about 55% of these households plowed for maize cultivation. Analysis of the 3-year average area plowed, using the t-test, showed statistically significant differences between all categories.

The differences in maize sales between the categories are less marked. The difference between ox owners and ox borrowers is not significant, like the difference between ox hirers and hoe users. The analysis generally confirms the findings of WP-ADPP in 1992 (van Agt, 1992).

Plowing for others

The 82 owners of oxen plowed an average of 4.1 ha on their own fields, and 3.6 ha on other farmer's fields in 1993/94. The total area plowed per span of oxen was lower than estimated previously (Dibbits and Mwenya, 1993) and there was no difference in total plowed area between farmers with one and farmers with more than one span of oxen. Farmers with high maize production (>4.5 t/year) plowed more in total (10 ha), but plowed a relatively smaller area (43%) than farmers that produced less (60%).

Table 3 shows that plowing for others is mostly done on fields of relatives. The score for priority in time shows that, contrary to the general perspective of gender and animal draft power (Sylwander, 1992), the fields of the wives and sisters have a higher priority than fields of male relatives or clients who want to hire.

Table 3: Plowed area and priority in time of plowing for different types of plow clients (n= 82 farmers renting their oxen)

Clients	No people	Area plowed (ha)	% of total area	Area/client (ha)	Time priority
Hirers	105	92	31	0.9	5
Wife	37	68	23	1.9	1
Son	35	40	13	1.1	3
Daughter	29	34	11	1.2	4
Brother	23	25	8	1.1	6
Parents	19	23	8	1.2	7
Sister	17	19	6	1.2	2
Total	265	301	100	1.1	

Only 74 farmers could precisely describe the reward for their plowing services. Table 4 gives an overview of the arrangements. It appears that only 35 of the 74 ox owners hired out their oxen for cash. Plowing through hiring of oxen is provided by relatively more farmers with a high maize production. Less productive farmers tend to use sharing arrangements more often, whereby the plow or the plowing labour is provided by the clients. It is interesting that although farmers applied several barter values for plowing, most of the respondents kept 2 bags of maize or 2 bags of fertiliser as 'a standard price per ha', whereby fertiliser is underrated compared with real market prices.

Sustainability of ox hiring

The hiring and lending arrangements are mostly seasonal. Only 41% of the ox owners plow more than one season for the same clients. When asked why their clients change each year, ox owners responded with various explanations which are shown in Table 5. It suggests that clients can decide which owner they want to plow for them. Ox owners, asked to characterise clients who return yearly for plowing, mentioned farmers *who have something to offer*, like maize, fertiliser and cash. However, when asked which persons did not return, the same characteristics were given: *'those who have means to look for plowing services elsewhere'*.

Table 4: Arrangements for hiring and lending plow capacity (n=74)

Arrangement	Frequency (% of transactions)	Value
Sharing	51	
Hiring for cash	26	K16,000/ha
Hiring for fertiliser	15	2 x 50kg bags /ha
Hiring for labour	3	weeding
Lending	4	

Table 5: The major reasons for changes in clients from year to year

Reason for change	Frequency (%)
Clients do not come back	26
Clients come too late	19
Clients bought oxen	9
Clients book others	2
Not enough (trained) oxen	24
Much work on own fields	12
Avoiding hatred	2

Demand for hiring

Ox owners were optimistic about the plow hiring market; 94% of the owners expected expansion. The main arguments mentioned were:

- many farmers increase their cultivated area
- the district still receives a lot of immigrants
- there are not enough oxen for the increasing number of farmers.

This opinion was not confirmed in a 1995 survey of 25 farmers who had in the past obtained loans from WP-ADPP. This group was almost unanimously of the opinion that demand for hiring oxen was shrinking. When asked under which conditions they would plow more, most of these farmers replied: *'If I receive more fertiliser and seed'.* Conditions like sufficient rainfall, spares and stronger oxen were not considered important. The change in public opinion of ox owners could be caused by the fact that in 1994/95 farmers experienced the effect of reduced governement support to agricultural lending institutions, whereby almost 20% lost access to the use of fertiliser and hybrid seed (Kakwaba, 1995).

Stability of ox ownership

Another influence on the sustainability of ox hiring and borrowing services is the stability of ox ownership itself. When the 40 farmers who hire or borrow oxen were asked about their past ownership of draft animals 43% stated that they owned oxen in the past, while half of these former owners had plowed for less than 3 years with draft animals. Table 6 shows the period of ownership of famers with a single pair of oxen. More than 55% of the ox owners with 2 oxen acquired their oxen within the last five years. This relative instability of ox ownership was also found in a survey in 1994 (Bwalya and Leeuwen, 1994).

Conclusions

- About 49% of the cultivated area in Kaoma District is plowed under hiring and borrowing arrangements with ox owners.
- Hiring for cash is more important than hiring for barter (fertiliser, labour or maize), but borrowing through sharing is most common.
- Hiring of oxen in Kaoma District is applied by 43% of ox owners. Farmers with higher maize production and/or only one span of oxen hire more frequently.

Table 6: Period of ownership of oxen by farmers with 2 oxen (n=39)

Period of ownership	Frequency (%)
<1 year	10
1–3 years	26
3–6 years	23
6–10 years	26
>10 years	15

- Among family members, women receive plowing services at a relatively high priority in time.
- Ox ownership as well as ox-hiring relations in Kaoma District are rather unstable throughout the years.
- Ox owners expect less demand for hiring services when fertiliser and seed are less easily available, for example as a result of the removal of government subsidies.
- The research did not show obstacles for WP-ADPP's programme of support to resource-poor female-headed households, whereby the programme creates an environment for training of surplus oxen in herds of cattle owners by resource-poor female-headed households. It is likely that the demand for hiring oxen by this group has reduced recently, due to diminished availability of credit for seasonal inputs.

References

van Agt, A J , 1992. *Go ahead with oxen, priorities and possibilities of farm households in Kaoma District.* Western Province Animal Draft Power Programme, Zambezi Livestock and Lands, Mongu, Zambia and RDP-Livestock Services, Zeist, The Netherlands.

Beerling M E J, 1988. *The advantage of having cattle. Distribution of cattle and access to benefits derived from cattle in the WP of Zambia.* 178p.

Beerling M E J, 1985. *A socio-economic survey of Namakuyu Kalabo District.* RDSB-University of Lusaka, Lusaka and Project Planning Unit, Mongu, Zambia. 51p.

Beerling M E J, and Mwenda M Mumbuna, 1988. *Distribution of benefits derived from cattle: a community study at Looma, Mongu District.* Department of Veterinary and Tetse Control Services, Mongu, Zambia.

Brouwer B O, Schoonman L and Wagenaar J P, 1992. *Cattle production parameters from Western Province,*

Zambia. Department of Veterinary and Tsetse Control Services, Mongu, Zambia and RDP-Livestock Services, Zeist, The Netherlands.

Bwalya G M and van Leeuwen M, 1994. *From supply to facilitation of farmers' trade in oxen for Kaoma District*. Paper presented at the 4th annual workshop of Palabana Animal Draft Power Development Programme, Palabana, Zambia.

Dibbits H and Mwenya E, 1993. *Animal traction survey in Zambia*. National ADP Coordination Project, Agricultural Engineering Section, Department of Agriculture PO Box 50291, Lusaka, Zambia and Insitute of Agricultural and Environmental Engineering, Wageningen, The Netherlands. 119p.

Dronne M, 1969. Problemes humains du developpements de l'elevage en zone sud du Tschad. In: *Colloque sur l'elevage Fort-Lainy*. Institut d'Elevage et de Médecine Vétérinaire des Pays Tropicaux (IEMVT), Paris, France.

Gryseels G, 1988. *Role of livestock on mixed smallholder farms in Ethiopian highlands*. Agricultural University of Wageningen, Wageningen, The Netherlands.

Helmrich H, 1986. *Animal husbandry in Bangladesh. Conditions, functions and development potential*. Herodot, Gottingen, Germany.

Kakwaba K, 1995. *Preliminary household enumeration report*. Department of Agriculture, Kaoma, Zambia. 9p.

McCown R L, Haagland G and de Haan L, 1979. .Interaction between cultivation and livestock production in southern Africa. pp 297–332 in: Hall A E et al (eds). *Agriculture in southern African environments*. Heidelberg, Germany.

Muntali et al, 1995. *Livestock Census 1994*. Working paper 95/5, Department of Veterinary and Tetse Control Services, Mongu, Zambia.

Nambayo G S and Vierstra G A, 1988. *Ox mortality and its effects on draught power use and other agricultural activities in Senanga West*. Adaptive Research Planning Team, Mongu, Zambia. 25p.

Nicolas G, 1968. *Problemes poses par l'introduction de techniques au sein d' une societe africaine*. Faculté des Lettres, University of Bordeaux, Bordeaux, France.

Sutherland A, 1984. *Draft power and other socio economic aspects of farming systems in Senanga West District, Western Province: Preliminary report*. Adaptive Research Planning Team, Mongu, Zambia. 33p.

Sylwander L, 1992. Women in animal traction technology. pp 260–265 in: Starkey P, Mwenya E and Stares J (eds) *Improving animal traction technology*. Proceedings of the first workshop of the Animal Traction Network for Eastern and Southern Africa (ATNESA) held 18–23 January 1992, Lusaka, Zambia. Centre for Agricultural and Rural Cooperation (CTA), Wageningen, The Netherlands. 490p.

Vijfhuizen C, 1987. *Borrowers of oxen: a description of the local cultivation system of rural farmers in Loanja Ward, Sesheke District*. Department of Veterinary and Tsetse Control Services, Mongu, Zambia. 23p.

An ATNESA Resource Book

Challenges to animal draft technologies in North Western Province, Zambia

Ivor Mukuka

Agricultural Engineer, Department of Agriculture, North Western Province
PO Box 110041, Solwezi, Zambia

North Western Province of Zambia is not a traditional cattle-keeping area. Over the last 10 years two donor-supported projects, the Integrated Rural Development Project (IRDP) and the Agricultural Development Project (ADP), have supported animal traction development by providing credit to farmers through established lending institutions within the province.

The use of animal-drawn technology became common within the IRDP Area. However, this project came to an end in December 1990. The ADP, sponsored by the International Fund for Agricultural Development (IFAD) continues. This paper dicusses the approach of the ADP and suggests how it could become more effective.

Although there are many extension workers in the area, only a small proportion are trained in animal traction techology and even these can find it difficult to extend their knowledge to farmers. Extension workers should be sent on specialised courses to improve their knowledge.

There is a shortage of draft animals in the project area. Importing animals is a temporary solution. The project has set up a programme to promote small-scale privately-owned cattle-breeding centres to increase the number of animals available. The provision of credit to support these centres should be speeded-up: the IRDP succeeded because credit was an integral part of its scheme.

Tick-borne diseases are very common in the area and trypanosome infection is also widespread. A comprehensive control campaign targeting both parasites and vectors is required. This should use low-cost integrated pest management.

Recent increases in the price of seed and fertiliser have made maize less atrractive as a cash crop. Farmers should be encouraged to diversify to more profitable crops such as mixed beans and groundnuts.

The North Western Province Agricultural Development Project has been in operation since 1984. It is sponsored by International Fund for Agricultural Development (IFAD). The project area comprises the districts of Kasempa (21,100 km^2), Mwinilunga (21,070 km^2) and Solwezi (29,800 km^2). These districts constitute 67% of the total area of the province. The project area is one of the high rainfall areas in Zambia with an annual precipitation of over 1200 mm October–April. The heavy rainfall causes considerable leaching of nutrients from the top soil. Consequently, soil acidity is an important factor limiting productivity both in the project area and the province as a whole.

In the Agricultural Development Project (ADP) area there are three major farming systems: a sorghum-based traditional subsistence shifting-cultivation system, a cassava-based traditional subsistence system and a semi-commercial farming system. All these systems depend on human labour as a source of farm power and shortage of labour was identified as a factor limiting crop production. However, increased and efficient use of animal traction may increase crop production.

During the first phase of the project that ended in 1991, there was a pilot programme to promote animal traction. However, prior to the commencement of the second phase in 1992, there were strong recommendations that animal traction should be supported fully by making credit facilities available through an existing financial institution. It was also recommended that a livestock development component of the project be established with the objective of improving livestock production in the project area.

Project approach

The role of the project is to stimulate the establishment of a viable animal traction programme in the project area. Thus conditions favouring animal traction technology will be instituted while constraints militating against the success of animal traction would be eliminated. It is envisaged that shortage of steers would be eliminated through encouraging establishment of privately-owned cattle-breeding herds.

Generally, the number of steers available for draft may not increase without improvement in the productivity of cattle. Thus effort will be made for overall improvement of cattle in the project area. To ensure survival of work oxen, an efficient animal health-care scheme will be established. For this purpose, the ADP has set up a rural veterinary drug centre and veterinary drug kits for camp extension workers through which farmers would have greater accessibility to drugs for their work oxen.

Farmer and animal training will be done by the extension wing of the Department of Agriculture. The present system of centralised training is being re-evaluated and the possibility of decentralising to the district farmer training centres or trainees' farms is being considered.

Dissemination of animal traction technologies

Extension service

Extension is defined as a professional communication intervention done by extension workers to obtain a voluntary change in the behaviour of farmers. Its role is vital to enhance adoption of animal traction technologies.

A majority of the small-scale farmers in the Province said they adopted new technology because they heard from the extension workers (Mwila, 1990). The extension worker disseminates information to increase knowledge level about an innovation and exerts influence to alter or strengthen beliefs to the point where farmers adopt the recommendations (Rodgers, 1983). Therefore, the project area needs competent extension workers who will exert their influence to cause changes in favour of animal traction technology.

There are close to 100 Camp Extension Workers in the project area but only 17 have acquired in-service training in animal traction. Moreover, most of the extension workers find it difficult to communicate technical messages on crop production that they have been doing for a long time. Therefore, communicating technical messages on animal traction technologies which are relatively new would even be more difficult. The training background of extension workers has a bearing on their performance when they come into the field. Hence inadequate knowledge and skills in animal traction have affected the dissemination of animal traction technology in the project area.

The District agricultural engineers have been involved in dissemination through field demonstrations and the results have been very encouraging. Consequently the demand for extension has been increasing for the past four years. However, the section still needs the support of competent extension workers to do much more than has been done already. The agricultural engineering section has also been disseminating information about animal traction through publication of pamphlets and handouts to create awareness. This literature was mainly distributed during the agricultural shows at the provincial and district level. However, this method has had its limitations because of the literacy level of the majority of the target group. Effort was made to translate the literature into local languages but the impact was still very low because of the limited audience.

Communication skills

In diffusion of agricultural innovations where a change agent attempts to introduce improved technology through a particular social system such as a group or village, the communication strategies used are very important. When the Department of Agriculture introduced animal traction to the farmers, the objective was to increase labour productivity by using efficient production methods. The level of adoption that has been reached now is a result of communication methods that were used in disseminating the technology: the level of adoption is low because the message only reached a limited audience.

Operator and animal training

In the past, the majority of farmers in the project area plowed less than two hectares in a season.

Worse still, the quality of plowing was not good and this affected the crop yield because of poor root development. This can be attributed partly to poorly-trained farmers and animals. Subsequent operations like weeding were also affected because some farmers did not plant the crop in straight lines. There were incidences in the field where to carry out a weeding operation a farmer needed four helpers. This excess labour could have been better deployed elsewhere. The challenge that lies ahead is how to refine the training for both animals and farmers to improve productivity.

Adoption of animal traction technology

Adoption usually denotes the act of accepting an innovation and utilising it effectively. This is the ultimate goal of animal draft power extension activity in the project area.

The importance of the social and economic background of the farmers adopting technology cannot be underestimated. Several researchers in this field have affirmed that socio-economic problems associated with diffusion and adoption of new technology are not always inherent in the technology itself, but are a direct reflection of the social inequalities and economic disparities that already exist in the society.

Generally, in the project area wealthier farmers have been adopting the use of animal draft power more than the resource-poor farmers who have continued to use hoes in spite of their low labour output. Over the last four years there has been a rise in adoption rates of 20% and 11% in each grou,p respectively.

The use of animal-drawn technology became a traditional feature in the Integrated Rural Development Project (IRDP) area. This was attributed to the fact that the districts operated in a more or less homogeneous socio-economic environment. The majority of farmers in the project area at least knew the potential benefits of using animals. This is an indication that farmers in the Integrated Rural Development Project area appreciated the benefits of animal traction for their field operations. However, this has not been the case with the IFAD ADP that operates in a heterogeneous environment.

Shortage of steers

The province is generally deficient in draft animals. Worse still, for cultural reasons, the majority of farmers are unwilling to sell animals. Therefore, it has been difficult to acquire draft animals for new farmers and to replace old oxen given by IRDP project in the initial stages of the programme in 1985.

The herd that is found in Zambezi district alone may be able to meet the total animal traction requirements of the other two districts (Kabompo and Mufumbwe). Mwinilunga District also has the potential of meeting the required number of draft animals for Kasempa and Solwezi districts. However, lack of better animal husbandry practices among the farmers is another limiting factor. For instance, in Mwinilunga animals are not herded and are left alone to graze in the plains. Consequently, the majority have become semi-wild (Dipeoulu, 1994). However, farmers have begun to realise the importance of animal traction in crop production and the situation is slowly changing for the better.

Adoption may be speeded up by using the animals available within the province and supplementing with imported animals from other provinces. The ADP has set up a base to alleviate this problem by establishing a cattle breeding centre where small-scale farmers would procure breeding stock to start privately-owned breeding centres. It is envisaged that these small farms would provide a sustainable source of steers.

Animal health

Maintaining animal health requires an efficient veterinary service backed up with adequate drugs. The disease challenge posed by tick- and tsetse-borne pathogens is critical, especially for draft animals that are exposed to rigorous field conditions.

Areas infested with tsetse flies include parts of Kasempa, Mwinilunga and Mufumbwe Districts. During the livestock diagnostic study that was conducted in Kasempa and Mwinilunga it was established that tick-borne diseases are far more prevalent than trypanosomiasis. This was in complete contrast to earlier reports. With this background information, the Department of Agriculture will be better placed to advise the lending institutions and farmers on what kind of drugs are suitable in these areas.

In Kasempa District the surveillance took place in selected livestock-producing areas which share borders with tsetse-infested areas (Dipeoulu, 1994). A sample of 144 cattle was examined and 139 had blood parasites. Of the infected cattle, 50% were infected with tick-borne parasites only, 48% had both tick-borne parasites and trypanosomes with the remaining 2% infected only with trypanosomes. In Mwinilunga District 103 cattle were examined and all were infested with blood-parasites: 65% had tick-borne parasites only, 37% had both tick-borne parasties and trypanosomes, with 1% infected with trypanosomes only.

Going by these results, control of trypanosomes and tick-borne diseases is a priority. Animals which have trypanosomes in their blood and are in the pre-immunity state to tick-borne parasites are unhealthy, have low work output and ultimately low productivity. For oxen, because of the nature of their work they are more likely to suffer from exhaustion and injuries, which are inflicted during work. Furthermore, stress lowers their immunity and sub-clinical diseases are manifested. Consequently, farmers may be unable to plow their fields and transport any goods. Worse still, they will lose extra income from hiring services for plowing and transport operations.

For small-scale breeding, the long-term objective of establishing a base for making the project area self sufficient in draft animal may not be achieved, due to high calf mortality because of infections.

Farmers' use of animal traction technology

One major problem of adoption in the province and in the project area is partial- or non-adoption of the recommended technology.

Some farmers have adopted the technology for plowing, while other major operations like weeding are still done using hand hoes. Other farmers have not adopted animal-drawn technology at all because of the high initial capital investment. Still other farmers have the desire to adopt the technology but scarcity of animals and implements have prevented it. In extreme cases, some farmers lack knowledge or conviction of the benefits of using the new technology for crop production and transportation.

Farmers' use of animal traction is also affected by incompatibility of cultivation practices. Animal traction is not commonly used even in areas like Mwinilunga with a tradition of cattle keeping. This is a result of traditional cropping and land preparation methods such as mounding which is used in the cassava-based farming system. In mounding, a piece of land is cleared of its grass and small branches which are heaped together and covered with soil until a considerable heap is achieved. Sometimes the soil is upturned together with the grass and small shrubs to make a heap. The mounds are usually prepared randomly and are not usually in a straight line. Mounding is meant to improve soil fertility by incorporating organic matter. Mound-making may be difficult to achieve using animal-drawn implements.

Lack of profitable cash crops

Maize is the major cash crop cultivated in the project area. The crop has been promoted heavily through the extension system. Production of hybrid maize is reliant on use of fertiliser and certified seed. Recent increases in the price of seed and fertiliser, and liberalised marketing arrangements have made it less attractive economically hence the urgent need to diversify to more attractive cash crops that will help farmers to meet their credit obligations.

Lack of implements and spares

Since the North Western Cooperative Union (NWCU) started having liquidity problems, the lack of profitable cash crops has meant that availability of implements for various field operations and spares has been a problem. Zambia Cooperative Federation (ZCF) finance services assumed the responsibility with financial assistance from the German Agency for Technical Cooperation (GTZ), but has not been effective. The federation has concentrated more on fulfilling needs of their clients than providing services to develop animal traction technologies in the project area. As a result, scarcity of animal-drawn implements and spares has continued. It is therefore difficult to recommend implements to farmers as they are unobtainable locally in the province. This situation may have a negative effect on the adoption of the technology in the foreseeable future. The only hope is in the establishment of private hardware shops. However these are not developing rapidly.

It can therefore be said that institutional incapacity generally limits and hampers development and adoption of animal traction in the project area.

Inadequate credit

Agricultural credit, irrespective of the source, is a necessary input within the agricultural sector. Credit is often a key element in modernising agriculture. It does not only alleviate the financial constraints faced by farmers, but also accelerates the adoption of new technology, in this case animal traction, by enabling farmers to purchase inputs.

However, credit is not the only requirement for increasing agricultural output. For it to have the success desired, complementary services such as an effective extension system must be addressed adequately.

For animal traction to be adopted or enhanced requires that the risks of farming are reduced and essentials like input supply, credit and marketing should be in place.

For a long time, North Western Province has had a problem of late input delivery and very unreliable marketing. In this province there are five lending institutions namely Barclays, National Commercial Bank, Lima Bank, Credit Union and Savings Association (CUSA) and ZCF. However, only CUSA, Lima Bank and ZCF provide credit facilities for oxen. CUSA administers the credit facility on behalf of IFAD in the project area. This implies that only farmers in the project area have access to this loan facility.

Although few farmers can afford to raise money to spend on this new technology, the organisation has 143 pending applications and every month additional enquiries are being received. This rise in demand is attributed to the fact that tractor mechanisation is completely out of their reach and animal traction has become economically attractive even with the high risk involved.

The IRDP succeeded because credit was an integral part of the scheme and played an important role in establishing a firm foundation for the programme where no animals existed, as in some areas of Mufumbwe and Kabompo districts.

The possibility of adding an insurance fee to the existing ox-loan package should be further explored, so that farmers do not lose out when the animals die from disease.

Recommendations and conclusions

Farming systems

Introduction of animal traction into these farming systems requires thorough research, planning and a full understanding of the local farming system and its implications. The project should try to teach farmers how to incorporate draft animals into their farming system to raise and expand productivity.

Mechanical skills

Mechanical weeding saves time and money, but few farmers or ox owners have mastered the technique. This is because it requires more skill and training than plowing and, in most cases, farmers usually do not have appropriate implements. Another important fact is that if farmers expand their cultivated areas, more labour is required. Since acquiring labour for weeding is one of the most serious bottlenecks, farmers may not weed their expanded farms. It has thus become important to teach farmers mechanical weeding skills so that they will not have to depend on hired casual labour and will able to manage larger farms. The project should also procure training equipment to enable agricultural engineering staff to teach mechanical skills to farmers.

Training for extension workers

Vigorous in-service training for village-level extension workers would enable them to address farmers' problems adequately in the field. The knowledge acquired from colleges gives them basic knowledge but not enough to be applied in real field situations. The agricultural sector, like any other industry is changing. Hence the need to have extension workers who are adequately prepared to meet new challenges in a changing working environment.

In order to disseminate this technology effectively in the project area, the project should send some extension workers to attend specially tailored courses at the Palabana Animal Draft Power Training Centre. Although the Palabana Animal Draft Power Project has been offering training to extension workers throughout Zambia it cannot train all of them as funds are limited. Hence the need for projects like the ADP to train the extension workers in the project area.

Animal health
The control of trypanosomes and tick-borne diseases

The control of trypanosomes and tick-borne diseases in both Kasempa and Mwinilunga should be based on a comprehensive control campaign comprising pre-control and control action. In pre-control campaigns, awareness meetings with livestock farmers where the surveillance survey was held should be conducted. This is meant to inform farmers on the current situation in their area. Control campaign activities would focus on the control of the blood parasites and the vectors.

Control of blood parasites

This would be effected through a one-time curative treatment. Emphasis would be on the oxen because of their movement to and from tsetse-infested areas for plowing and transportation. This approach will be multidisciplinary involving the rural veterinary drug centre, and the extension service through block and camp workers. Farmers will have to pay for the drugs.

Control of vectors

Tsetse control will be done by setting up traps and targets, while tick control will be achieved through use of low-cost integrated tick management.

Holding facilities

Concerted efforts should be made to domesticate the existing semi-wild animals in Mwinilunga by construction of paddocks because relying on imports from other provinces will not solve the shortage of steers in the project area.

Small-scale breeding

Speeding up provision of production loans for smallholder cattle breeding will, in the long term, reduce the shortage of draft animals. Otherwise, for the next ten years or so, the province will continue to rely on imports.

The current cost of draft animals from the breeding centre in Solwezi is high, and as a result it has become difficult for resource-poor farmers to have access to this technology. Therefore, encouraging widespread breeding will lower the price of animals in the long term.

North Western Province (including the project area) is in a high rainfall area and this creates numerous perennial streams and *dambos* (wetlands) which are suitable for grazing almost the whole year, making it possible to produce livestock with little or no feed supplementation.

The fact that many farmers in the province, including the project area, are not cattle keepers is an advantage for livestock production, because in many traditional livestock farming communities of Africa, there has been strong resistance to change from traditional to improved livestock production methods. This resistance is likely to be weak in the project area, creating an opportunity to introduce improved production techniques to farmers who have no traditional alternatives. This also applies to utilisation of implements in areas where animal traction has a long history. In these areas farmers tend to remove the hitch assembly from plows to make it 'work better'.

Cash crops

The current cash crops being grown in the project area are not profitable in the present economic enviroment. Hence the urgent need to encourage farmers to diversify to more profitable cash crops like mixed beans and groundnuts. This may help farmers raise enough income to be able to meet credit obligations.

References
Dipeoulu O O, 1994. *Surveillance of animal trypanosomiasis and tsetse infestation in Kasempa and Mwinilunga districts.* Integrated Rural Development Project, Northwestern Province, Zambia. 24p.

Rodgers E M, 1983. *Diffusion of innovations.* Third edition. The Free Press, NewYork, USA.

Mwila C, 1990. *An evaluation of agricultural extension services in the Integrated Rural Development Project districts of North Western Province.* Rural Development Studies Bureau, University of Zambia, Lusaka, Zambia. 26p.

Challenges and constraints of animal traction in Luapula Province, Zambia

by

Alexander Mutali

Provincial Agricultural Engineer, Department of Agriculture
PO Box 710157, Mansa, Luapula Province, Zambia

Abstract

There has been little tradition of cattle-keeping in Luapula Province, Zambia, and the use of draft animals was introduced only recently. As a result there is a shortage of both breeding cattle and draft oxen. Most cattle breeders do not buy more animals once they have started their herd, resulting in inbreeding. This problem is being tackled by a Bull Exchange Programme.

Much land in the province is left uncultivated because of a shortage of labour. Draft animals could play a major role in allowing cultivated area to be expanded. The shortage of draft animals could be addressed by promoting the use of cows for traction and by introducing donkeys. Donkeys could be especially useful for transport and could be particularly beneficial for women.

Introduction

Luapula Province of Zambia has an area of 50,600 km^2 on the Central African Plateau. The major physical features of the province are the plateau, the valley of the Luapula River, lakes Mweru and Bangweulu and the Bangweulu swamps. The altitude ranges between 900 m and 1300 m.

In 1995 there were about 585,000 people in the province (estimate based on 1990 census), the most densely populated areas being the Northern Luapula Valley and beside Lake Mweru. Other concentrations of people are around Samfya and Mansa. The plateau is sparsely populated.

Cattle-keeping in Luapula

There is little tradition of cattle-keeping and use of animals for draft power in Luapula Province; the first cattle were introduced about 100 years ago. The use of cattle for draft was first promoted in the mid 1950s, but this was not very successful. Another attempt at introducing the technology was made in the late 1970s and early 1980s by the Integrated Rural Development Programme with support from the Swedish International Development Agency (SIDA). Support to the animal draft power programme was taken over by FINNIDA in 1988 when the Luapula Rural Development Programme was started. FINNIDA support has continued in the new phase started in 1995 under the Luapula Livelihood and Food Security Programme.

There are now a total of about 13,000 cattle in the province, of which about 800 are trained oxen. Between one and 18% of households own cattle, depending of the district. Most cattle are kept on the plateau, where there is plenty of land and no competition from fisheries for finance and labour. Concentrations are found around Mansa and the missions of Chibote and Lubwe/Kasaba, indicating the impact that the missions have had on development of cattle-keeping in the province.

As there has been little tradition of cattle-keeping and use of draft animals there is a shortage of both breeding and draft animals. The province's annual requirement for draft animals is 300, but there is currently a shortfall of about 60%. Non-cattle owners depend on hired labour and, to a smaller degree, hiring of oxen, for agricultural production. Only about 1200 ha are cultivated using oxen, compared to about 100,000 ha cultivated by hand.

The distribution of cattle between households is unequal: one farmer might have a huge herd whilst neighbouring farmers do not own any cattle. It is not possible to describe the average Luapula cattle breeder but a number of general observations can be made. Cattle breeders are older men (mostly over 50 years), who bought their cattle with cash. Knowledge about cattle was obtained by observing others (including parents) rather than from formal courses.

Inbreeding

Most cattle breeders do not buy more animals once they have started their herd, not even bulls. The use of self-bred bulls and the use of bulls for long periods is widespread. Farmers are unaware of the concept of inbreeding in cattle; it is thought to be a problem for humans only. Even when informing farmers about risks in general terms, problems like stunting, premature births and poor disease resistance surface. This indicates that the problem does exist. Poor fertility figures may in some cases also be attributed to inbreeding.

The province is addressing the problem by providing information on the difference between genetic improvement and avoiding inbreeding. This is taught to both extension workers and farmers. It is hoped that farmers will start appreciating the advantage of exchanging two bulls. Nevertheless, few farmers seem to put much effort into acquiring a (better) bull. Lack of money is often cited as a problem but, in economic terms, sale of one or two cows to buy a bull is, in these cases, very profitable. The Luapula Livelihood and Food Security Programme has introduced a Bull Exchange Programme to tackle the problem. The programme will purchase 10 bulls a year and distribute them to established livestock groups. These will be rotated after three years.

Other management issues

Another notable problem is poor feeding. Supplementary feeding, if it is practised, is limited to maize stover because nothing else is available. Herding practices may be important: farmers tend to blame herdsmen for not allowing sufficient grazing time. Herdsmen are generally poorly educated in cattle-keeping. Training herdsmen is a problem as they often do the job for only a few months.

At the level of herd management, a disease will be observed while poor condition and low fertility may go unnoticed or be accepted as a fact of life. The problem is aggravated by the virtual absence of veterinary services in many places. As a result of the lack of veterinary services, camp extension workers are asked to take over duties from veterinary assistants, a role for which they are not trained.

Draft animal power in Luapula

The low number of cattle in the province poses the major problem for (potential) animal draft power farmers. Most complain about non-availability of cattle, but the major problem is that they cannot compete with the butchers. There are few alternative animal power sources in the province. Many farmers prefer to use oxen rather than cows for draft because:

- they believe that if cows are used their fertility will decrease
- cows are weaker than oxen
- some farmers believe that there are government laws that prohibit the use of cows for draft. It seems that these farmers have misunderstood the law which prohibits slaughter of female cattle.

This has left oxen as the only source of draft power.

Knowledge of handling oxen and use of implements is limited at many farms. Ox-drawn weeding is rare even where ridgers or cultivators are available. This results in high demand for labour even when working with animals.

Supply of implements and spare parts and lack of capable blacksmiths is a problem in many areas.

Farmers also complain about lack of credit for oxen and implements. The situation is worsened by the lack of communication between farmers and lending institutions.

As cattle-keeping is regarded generally as a male issue, access to and ownership of draft animals is difficult for women. Much needs to be done on needs assessment and raising awareness before animal draft power can alleviate the labour burden on women significantly.

Agriculture in Luapula Province in general faces problems with marketing of produce.

Draft animals in relation to crop production

One of the major bottlenecks in crop farming in Luapula Province (and Zambia as a whole) is the shortage of human labour for farming operations. Ox owners as well as non-ox owners depend to a large extent on hired labour for plowing operations, hand weeding, harvesting etc (especially where bigger fields are being cultivated). Limitations on available labour and the low population density on the plateau mean that much land is left uncultivated: less than 2% of arable land at the disposal of the smallholder farmer is being utilised. This indicates the crucial role farm power and mechanisation can play in enhancing the expansion of agricultural production and improved food security at both household and national levels.

Potential solutions

Many areas of the province are populated sparsely and expansion of agriculture is possible. The problems of maize marketing require a shift to other crops like groundnuts, and possibly rice, as alternatives. The shortage of steers for draft animals can be alleviated by introducing the use of cows and donkeys. While cows can do the same work as oxen, with the same implements, donkeys would be particularly useful in transport for those having limited access to cattle, like poor households and women.

There is high demand for hiring of transport, particularly between field and farm in densely populated areas. The efficiency of use of oxen can be increased by training in the use of secondary tilllage implements and for transport.

The province has realised the importance of Lupula in efforts to increase the food production of the country as a whole. As a result, donkey traction has been promoted in three pilot areas with support from the International Fund for Agricultural Development (IFAD). Apart from facilitating acquisition of donkeys, IFAD has addressed a number of issues with regard to donkey culture. Once such issue is ensuring that the target farmers have the necessary knowledge and skills in the management of donkeys and ways of employing them profitably for draft purposes. It is hoped that this will address the labour constraints in agricultural production, especially for women, who have a high workload in both domestic and agricultural tasks.

Index

www.ingramcontent.com/pod-product-compliance
Ingram Content Group UK Ltd.
Pitfield, Milton Keynes, MK11 3LW, UK
UKHW052100290526

12796UKWH00010B/586